THE PATH TO POSTHUMANITY

THE PATH TO POSTHUMANITY

21st CENTURY TECHNOLOGY AND ITS RADICAL IMPLICATIONS FOR MIND, SOCIETY AND REALITY

Ben Goertzel
Stephan Vladimir Bugaj

Academica Press
Bethesda

Library of Congress Cataloging-in-Publication Data

Goertzel, Ben.
The path to posthumanity : 21st century technology and its radical implications for mind, society and reality / Ben Goertzel, Stephan Vladimir Bugaj.
p. cm.
Includes index.
ISBN 1-930901-95-X
1. Technology--Social aspects. 2. Technological forecasting. 3. Artificial intelligence. I. Bugaj, Stephan Vladimir. II. Title.

T14.5.G62 2006
303.48'3--dc22

2006009529

Editorial Inquiries:

Academica Press, LLC

7831 Woodmont Avenue, #381

Bethesda, MD 20814

Website: www.academicapress.com

To order: (650) 329-0685 phone and fax

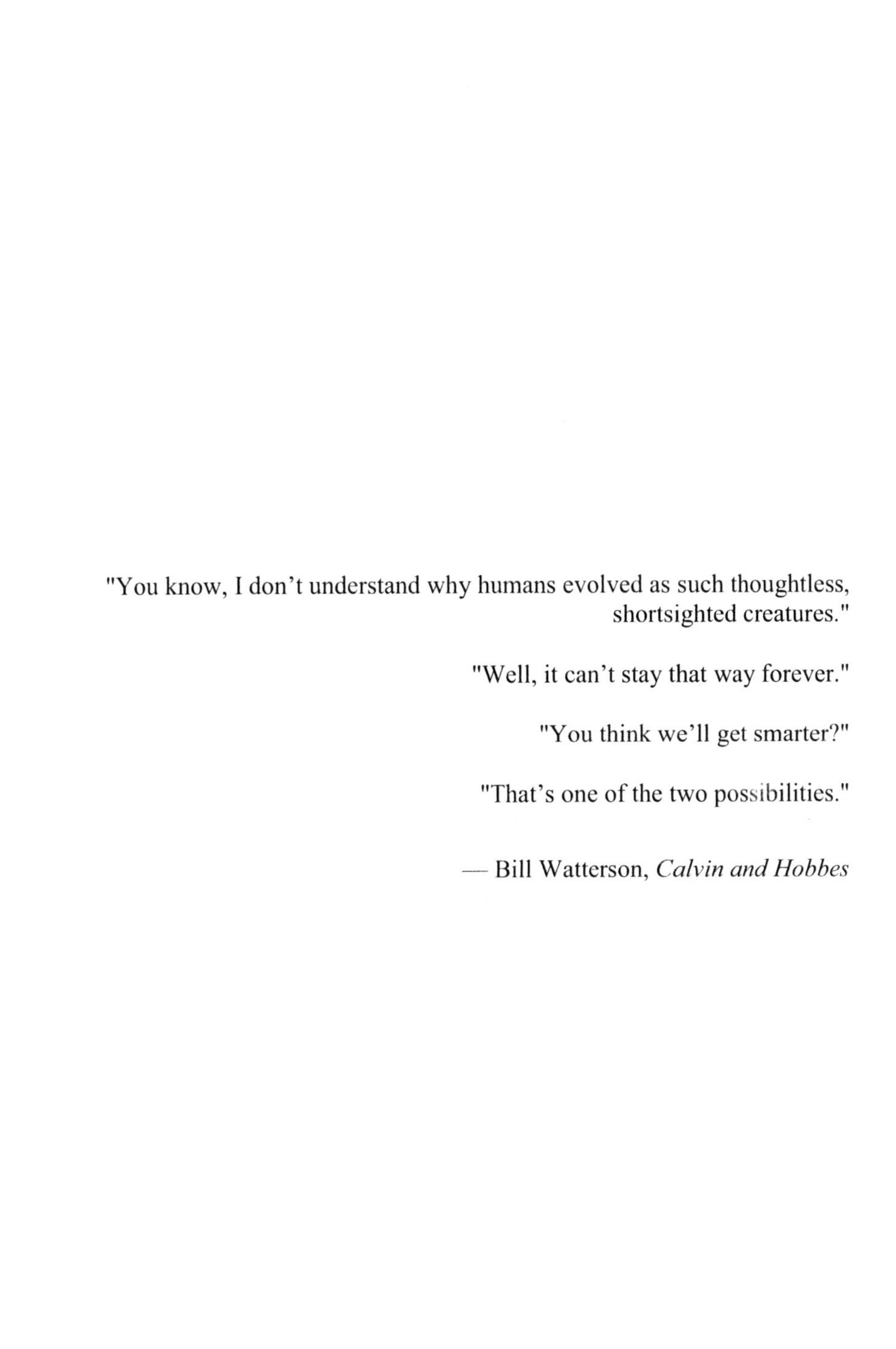

"You know, I don't understand why humans evolved as such thoughtless, shortsighted creatures."

"Well, it can't stay that way forever."

"You think we'll get smarter?"

"That's one of the two possibilities."

— Bill Watterson, *Calvin and Hobbes*

TABLE OF CONTENTS

LIST OF TABLES

LIST OF FIGURES

PREFACE

A few hundred years ago, science was irrelevant to most people's daily lives. It was mainly a hobby of aristocrats, in the same vague category as, say, metaphysical philosophy or jewelry collecting.

Today, science and technology are everywhere (as are metaphysical philosophy and jewelry collecting—our poor aristocrats are relegated to collecting airplanes and buying small islands). Most people don't understand much of it, but they know it's important. Every day people in the developed world (and increasingly large percentages of the developing world) use computers and printers and monitors; they watch TV and record shows on their DVRs; they drive cars with fuel-injection engines and onboard diagnostic computers and fly on airplanes with autopilots; they take medications with names reflecting complex molecular structures; they shop at Wal-Mart, which exists almost entirely to sell various forms of plastic that that didn't even exist 50 years ago; they buy cellular telephones incorporating e-mail clients, web browsers, cameras, and pushing steadily toward actualizing the worn metaphor "everything but the kitchen sink."

However, what most people don't yet comprehend is that this is a transitional period. The importance of science and technology to ordinary life is not going to remain constant at its present level. Rather, it is going to increase—and it's going to increase *fast*, exponentially or more so, just as it has been. Computers, plastics, medications, airplanes and uber-cellphones are just the beginning.

Technology is going to transform us, redefining what it means to be human, not just culturally and psychologically, but *physically*. Genetic and biomechanical engineering will happen, the protests of the religious-conservative camp notwithstanding—enhanced human beings *will* walk among us. AI will happen, not in the 4000 years projected in Spielberg's 2001 film "AI," but quite likely within this century, and probably early on. Computers will be smarter than us in ways more important than playing chess, factoring large numbers, or landing jet aircraft, although we'll be getting much smarter as well. Intelligent and semi-intelligent computers will help us solve puzzles that have so far eluded us: for instance, how genomes make cells, how particles make atoms, and how brains make minds. We will communicate with computers directly using our brains; and we'll communicate with each other similarly. We will move back and forth fluidly between the ordinary physical world, with its trees and roads and love affairs and childcare centers, and virtual worlds in which anything can happen. In our virtual worlds we will be able to wrestle with a molecule or make love to a cloudburst or solve equations by hybridizing our intuition with that of an AI mathematician, and it all will feel realer than real. We will re-engineer our brains and upload our minds into computers and robots and other vehicles. Human nature in some form may persist, but not in the manner that we are accustomed to. Science and technology are leading us into a "transhuman" age. We will indeed be "more human than human." (Philip K. Dick coined the phrase "more human than human" to describe his "replicants" —genetically enhanced human clones— in the story "Do Androids Dream of Electric Sheep," which became the famous film "Bladerunner." In his writing can be found many a prescient idea about both the amazing potential and hidden dangers of technological progress.)

All this is just the beginning. The things we can imagine and describe now are sure to be dwarfed in significance and magnitude by new things to come that lie quite literally beyond the scope of our current brains to comprehend. Once we have created software programs or machines that dramatically exceed us in intelligence, the odds are that fundamental transformations will take place. Physical, psychological and social reality may alter profoundly, so much so that the old word "reality" no longer applies.

This may all sound like crazy science-fiction—and indeed, history shows that predicting the details of future science and technology is a pursuit fraught with peril. There will be limitations we cannot right now foresee, and while some pessimists will laugh at how right they were about some particular limitation, there will also be revolutions our paltry human-nervous-system-based creativity is inadequate to imagine. There are things we simply don't even know we could be optimistic or pessimistic about. Even so, we think the qualitative character of the coming revolution is clear. Vernor Vinge, writing in the 1970s, was right to use the word "Singularity": as science and technology advance faster and faster, there will come a point where, suddenly, they are no longer just a big part of our world, they are inextricable from our world. Indeed, we may have already passed that point. We may be moving from one singularity point to another, more extreme one. New advances are already coming faster than anyone who is not a specialist in a particular field can understand them, in every field of science and engineering. AIs will create new AIs that rapidly outpace their creators, and so on.

Yet, although we think the word "Singularity" is evocative, we prefer to think about the phenomenon in terms of the word "Transcension," introduced in Damien Broderick's novel of that name (which Damien says he got from the writings of futurist thinker Anders Sandberg). The important thing isn't that changes will come faster and faster until the rate becomes effectively infinite from a human perspective—although it seems likely that this *will* occur. The important thing is that the nature of existence as we know it will be transcended. The concept of "Transcension" may sound religious, and in fact, having one's mind expanded beyond its previous boundaries by means of advanced technologies like uploading or neuroengineering may well turn out to have some "spiritual" aspect. But there are no acts of religious-like faith involved in accepting the ideas we'll present here: only a clear look at the nature of modern science and technology, and a particularly optimistic but ultimately rational extrapolation of current progress into the future.

HOW WE CAME TO WRITE SUCH A BOOK

We write about these topics not as outsiders, but as people who have been right smack in the middle of the ongoing revolution. After spending about a decade as an interdisciplinary scientific theorist, about 8 years ago Ben decided to devote his life more fully to one particular futuristic goal: the construction of a truly thinking machine, a software program displaying human-level general intelligence. Stephan pursued a less academic but equally interdisciplinary career path, involving software development and research at various corporations including Bell Labs, and various techno-art and music pursuits including working on four shows with Survival Research Labs. In 1997 Ben and some colleagues founded a company, Webmind Inc., with the twin goals of creating a "Real AI," a truly intelligent software program, and constructing profitable software products from preliminary partial versions of this holy-grail. Stephan joined in 1998, and we made some interesting progress toward the grand goal of Real AI. We created a couple of worthwhile AI-based software products along the way , but we never managed to make the firm profitable. When the dot-com crash hit, a major funder pulled out at the last minute, and the company couldn't recover. In March 2001, Webmind Inc. dissolved.

Following this personal and financial disaster, a group of friends from the Webmind Inc. R&D division decided to stick together as a team, and initiate a new chapter of our quest to build an AGI: an Artificial General Intelligence. We saw the mistakes we'd made at Webmind Inc.: both technically in terms of the construction of our AI system, and business-wise,. We resolved to do it right this time. We are building a new AI system, the Novamente AI Engine, and we're applying it to various practical domains, mainly focusing on bioinformatics. As we write these lines in early 2005, Novamente has sold its first two copies of the Biomind ArrayGenius, a software product based on the core AI engine that helps biologists analyze gene expression data. This is a serious, practical business, with a variety of exciting possibilities for scientific discoveries and medical breakthroughs that could help millions of people live healthier, longer lives. It is also a step along the path to making a true AGI system.

In the interim between Webmind's end and getting Novamente off the ground, Ben carried on a "hobby" we'd both developed during the last couple years of the Webmind era: writing newspaper articles on science and technology topics for the German newspaper *Frankfurter Allgemeine Zeitung* (*FAZ*). Once Novamente got going there was no longer time for this sort of writing, but while it lasted, the process of writing the articles was quite enjoyable. In early 2002 we realized that, if all these articles were gathered together, it'd make a pretty decent book. Many of the chapters of this book originated as articles from *FAZ*, but the book is more than a collection of articles haphazardly gathered together; it has been structured and edited, especially with the rewrites and additions provided by Stephan (who co-authored some of the original articles), with an overall intent in mind — to provide a concise introduction to the philosophical and scientific underpinnings of the social and technological movement known as *Transhumanism*, and give our view of how this movement can effect *positive* change on our human selves,.

One side effect of writing the *FAZ* articles, and this book, was that it forced us to look beyond our own work on AI and take in the overall landscape of science and technology development. Of course, we had always been aware that our work was just one piece of a huge, amazing, unfolding puzzle. No one scientist will accomplish the Singularity, no single 21st century da Vinci or Einstein—or Goertzel or Bugaj. No one branch of science will accomplish it either. What is happening is synergetic, cross-disciplinary, international and emergent. Writing the articles that grew into this book really brought home the multifaceted, multidimensional power of the ongoing revolution.

FOCUS AND PURPOSE OF THE BOOK

This is not the first book to cover these general themes. We greatly enjoyed Damien Broderick's book *The Spike*, published in 1997, which hit on a lot of the same topics. Ray Kurzweil's book *The Singularity is Near* has just been released (early 2006), after years of anticipation. Showcasing a gamut of beautiful graphs of technological and scientific progress, his book demonstrates the hyperexponential growth curves observed in various industries and dissects the causes underlying this growth pattern. We appreciate this sort of work but don't aim to replicate it here, or extend it in any direct way. Our goal here is more qualitative, though closely related. It is simply to describe for you various aspects of contemporary science, which seem to be pushing toward Singularity, and then to give some indication of what this science says about the post-Singularity world—with full understanding that, the further predictions veer from current realities, the less likely it is they can be made with any detailed accuracy. Our predictions are optimistic and may sound like Science Fiction, but people once felt the same way about Alvin Toffler's book *Future Shock*. To a contemporary reader, Toffler's book seems very obvious; boring, even, where he was wrong—and it's only 35 years old.

We have focused on those areas of intensely-future-related science that we understand the best, which means mostly computer science, with a helping of biology, and occasional dips into other areas such as physics, general engineering, and chemistry. But though it's governed by our own interests, the coverage of topics is hardly arbitrary, because the topics we know best are generally things we've learned about precisely because of their central relevance to the ongoing sci-tech revolution. Some parts of what we'll talk about pertain to our own AI and other research, but we haven't harped on that too much. There are a few other books planned about the Novamente AI work and other more detailed technical and philosophical subjects. The purpose of *this* book is to give a broad overview of future technology, dipping into our own work only when it seems really useful for illustrating more general points. The vast majority of the book regards the research of other scientists, some of whom we know personally, others only through their writings. Of course, in one brief book we can't possibly tell you all of what's happening in university and industry research labs, garages and living rooms all around you. The vast accelerating advance of science, technology and creative intelligence defies concise description. But we wouldn't have bothered to take the time to write this if we didn't think we could do a relatively decent job of describing the *essence* of it all.

The more people understand what is really happening with science and technology, and what the potential future outcomes are, the larger the segment of humanity will be that actively plays a role in the ongoing revolution, rather than passively being manipulated by it. In spite of the phenomenal stupidity and destructive tendencies sometimes associated with large mobs of humans, on the whole we think that a broader awareness and conscious participation in our Transcension-ward trajectory would be a good thing.

CONTEXTUALIZING SCIENCE

This book is Ben's second foray into the domain of "popular science" books. The first was a biography of the chemist Linus Pauling, co-authored with his dad Ted Goertzel and his grandparents Mildred and Victor Goertzel. This book has been a lot more difficult and a lot more fun, mostly because in the case of the Pauling biography, Ted was the orchestrator and Mildred and Victor did the years of preliminary research. For Stephan, this is the first of such books, and while he's played the supporting role this time it's still been a lot of hard work, and fun. We've found the process of writing this book both frustrating and fascinating, and we'll take the opportunity of this Preface to briefly share a few of our thoughts on this topic, in the hope that these thoughts may be of some use to the reader in digesting the book we've constructed.

One of the funny things about being a scientific researcher is that, however high your IQ may be, you wind up feeling *incredibly stupid* on a regular basis. You may find yourself thinking about the same thing over and over again, for months or years on end, without making any progress at all. Then, when you get the answer, it often seems completely obvious and you can't understand how you could have been so moronic as to not see it earlier. Then you go out into the world of ordinary, non-scientific people, and you realize: "Wait a minute, I'm actually fairly intelligent compared to a lot of other people! I'm not such an idiot after all! This research is just really, really hard for the measly three-pound human brain."

Given how hard science is even for scientists, explaining it to non-scientists in a reasonably honest way often seems an insurmountable challenge. Even a fully technical scientific paper can never tell the whole story of the research it purports to summarize. What is fascinating, though, is that at its best, a non-technical summary of scientific work achieves a *different kind of meaning* from technical scientific papers. Scientific work focuses on the details—it drills down so far you'd think it would be impossible to drill down any further, and then it drills down yet further. But sometimes it's valuable not to drill down, but rather to look back up, and see how each piece of work relates to everything else: to other bits of science, and to ideas, events and trends in the non-scientific world. This is the kind of meaning that non-technical discussions of scientific ideas can add. The lack of total scientific precision allows broader cross-connections and intuitions to become perceivable. In this sense, we believe that scientific writing for the non-scientific audience doesn't have to merely be a "dumbing-down" of scientific ideas; but can be a "contextualization" of scientific ideas, a deepening of scientific ideas in a different direction from the direction they're taken by scientific research work.

(Okay, okay, so that's a rather lofty perspective. If we're lucky we've achieved this high-falutin' ideal 37.3223% of the time in the following pages. But at least you know what we're aiming for!)

Of course, contextualizing science is one thing in the context of a narrow scientific topic such as laser physics, genetics, or Linus Pauling's chemical research; it's quite another in the context of a synergetic mix of scientific topics whose common theme is their potential to transform the nature of humanity. We have taken on a rather substantial task here. But anyone who knows us could tell you that unambitiousness is one of the few things *not* included on our long lists of flaws. After all, the primary goal of our scientific careers has been the creation of a computer program with human-level general intelligence. And after 15 years for Ben —and 12 for Stephan— we're still at it, and still optimistic.

One aspect of popular science books about which we have mixed feelings is the incorporation of biographical information. Knowing the social context of an idea's conception and development can be both fascinating and clarifying, and the adventure of scientific discovery can be gripping, exciting and well worth a dramatic recounting. On the other hand, we've read many books that told us far too much about the *story* of a scientific discovery, and not nearly enough about the discovery itself. We've tried to walk the fine line here, in this regard. The ideas described in these pages involve countless human stories, and we've told a small selection of these, introducing biographical information when it seemed particularly useful for bringing out some aspect of the sci-tech and philosophical points in question. In these passages, rather than focusing on the already famous, we've made an effort to illustrate the *breadth* of scientific insight that is going into the sci-tech revolution by recounting bits and pieces of the lives and intellectual adventures of some lesser-known individuals who have nonetheless made outstanding contributions.

Nor is our own story of the rise and fall of our company, Webmind Inc., without its dramatic interest: the gathering together of an international team of mad scientists, programmers and financiers to focus on building a thinking machine, and then the emergence of new enterprises from the ashes of the old. Regardless, we have only touched on it briefly in these pages, when it seemed particularly apropos. If we ever have the time we'll write a whole book on our adventures in AI business and research—a fascinating tragicomedy, in some ways. But this is not that book. Here the ideas are the focus. We think they're dramatic enough.

SCIENCE AND ETHICS

One big part of the overall task of contextualizing science is placing scientific results in a *social and ethical* context. Obviously, this is a particularly critical task when one considers scientific work of the sort we're discussing here, work that promises to, in time, completely redefine what it means to be human (and whether the "human" category continues to exist at all).

Our perspective here is rather distant from the current ethical debates on biotechnology, which center on such topics as the ethics of human cloning or genetically-engineered produce. From our own deep-futurist point of view, the correct outcome of these debates is painfully obvious. If genetically engineered produce can help feed the hungry people of the Third World—and apparently it can— then we should pursue it unreservedly. Freeman Dyson has made this point powerfully in his recent book *The Sun, the Genome and the Internet* (1999). On the other hand, the fuss about cloning seems to us to be almost entirely due to peculiar religious notions. We have not seen anything resembling a convincing argument that human cloning will lead to great practical dangers. While we have great respect for the deep spiritual experience that lies at the heart of all religions, it seems that religious superstitions have been and will be the cause of a lot more pain and suffering than could ever come from human cloning.

We also think the ethical debates over "Frankenfood" (a term that is a *tour-de-force* of propagandistic phraseology), human cloning, stem cells and related topics will die down within a few decades at the outside, once these relatively simple kinds of biotechnology become commonplace. The really profound and important ethical debates will be the ones that come after—the ones to do with advanced genetic engineering, man-machine synthesis, and the nature of posthumanity.

There is a real risk that one day genetic engineering will lead to the creation of a genetically super-endowed wealthy-nation elite, with the vast majority of the world unable to afford genetically-enhanced children, just as they now cannot afford computers or Gameboys or GPS widgets or cutting-edge pharmaceuticals. There is a real risk that the wealthy will upload into a digital world, leaving the masses in a polluted, increasingly unlivable physical-world ghetto, patrolled by robotic remotes which gun-down anyone trying to reach the cortex and disconnect the elites. There is also a risk that, somehow, in creating these exciting technological modifications to ourselves, we will lose the passion, the feeling, the essence of being human—that we will fashion for ourselves a more perfect, but less intense and genuinely fulfilling reality. Perhaps the pharmacological elimination of suffering and the creation of viscerally satisfying virtual worlds will make life too easy, destroying the rich if sometimes difficult texture of everyday being that we take for granted. Or, most strikingly, it's quite possible that we'll create superhuman AI's that annihilate humanity altogether, restructuring the universe in a manner more agreeable to their own goals and tastes (we'll review this possibility, and possible strategies for averting it, in the last couple of chapters.) It is our view that these ethical nightmares *can possibly* be avoided, but it's not entirely clear that humanity as a whole will have the *will* to avert them.

Some of those who wish to stop human cloning and genetically-modified food may feel as they do because they view these things as steps along the way toward more dramatic and dangerous things. We have some respect for this perspective, but we believe that the quest to stop science and technology developing (Bill Joy's proposal for technological "relinquishment") is doomed to fail. A pessimist would say, "Pandora's box cannot be closed again." But it's better, in our view, to say, "It's clear the future will involve these technologies, so let's direct our energies toward maximizing the chance that these technologies will be used in a positive way." It is entirely possible to use advanced computing and biotechnology to make human experience richer, deeper and better, for all humans. The more people realize the importance of this goal, the more likely it is to come about. (We should point out Bill Joy hasn't actually relinquished much of anything. In January of 2005, he joined high-tech venture capital firm Kleiner, Perkins, Caufield, and Byers, and in a press release stated: "As a KPCB partner, I will continue to help entrepreneurs advance the Internet, develop wireless innovations, and find new ways of using large piles of computers to solve difficult problems. I'm also particularly interested in discoveries and inventions that solve energy and resource problems, and in applying 21st century advances in physics, chemistry and the natural sciences to help create abundance." Maybe he's come around to our more optimistic view that technological progress can be positive if we have the will to make it so.)

It was over 100 years ago that Nietzsche, in his Zarathustra, said "Man is something that must be overcome." He did not envision that silicon chips, gene chips, PCR and so forth would be the mechanisms of this self-overcoming, but his insight was dead-on nonetheless. We are well along in the process of overcoming ourselves. We are doing so not through the visionary trances of prophets like Zarathustra, but through the workaday R&D activity of thousands of scientists worldwide, studying gene expression data analysis, neurocomputing, advanced computer architecture, anti-aging pharmacology, and a whole host of other cutting-edge disciplines. It is humanity as a whole, not any particular individual, which is in the visionary trance, courtesy of the socially-psychoactive substance called *science*—and heading straight toward a major transformation of the soul. It's quite a drama that's unfolding. We're right smack in the middle of it, and we have at least *some* say in choosing the roles we play.

We realize that not everything we predict in this book will definitely come to pass, but we believe that with sufficient willpower and application of resources, these things are possible. However, part of the true scientific mindset (which sets it apart from the zealot mindset, be it religious or political zealotry), is a willingness to accept that you've been wrong and reconsider your theories based on new ideas and new evidence. With science, and forward-looking engineering, it's not a matter of simply having *faith* that something may be true (though sometimes you need a bit of faith, especially if you're a physical cosmologist or an AI researcher). For something to be scientifically "true" involves building a rational case that something could be true, and then proving it through logical and mathematical proofs of abstract theories which can then be verified with some form of experimental observation. When such rigor can't be achieved, one can postulate what the future may look like, but it is a requirement of being a scientist to revisit these theories as new theories and new discoveries emerge. Either way, if you're a zealot you're a bad scientist. Science is best approached with the "anything is possible, so let's figure out what's actually probable" mindset, which Quantum Physics encourages. Please note, also, that we realize predictions about the future are always conjecture and we intend that they be taken as our "educated guesses" about what might be possible and not as established scientific fact, and also that (as with many pop-science books) there is also a good deal of straight-up philosophizing in this book. It is in this spirit that we hope scientific readers will take this book. This book is not an orthodox philosophy of science at all, but just as we'll gladly consider that we've been wrong in some of our ideas, we hope the scientific mainstream will come to accept the fact that from "fringe science" come many ideas that are tomorrow's mainstream science and that discarding unpopular scientific ideas out-of-hand as "fringe lunacy" and pillorying their proponents does the pursuit of new scientific discoveries a disservice. Many things which were thought "utterly impossible" became prosaically commonplace once the mainstream of engineering and scientific academia and industry put enough resources towards them.

ACKNOWLEDGEMENTS

There are a lot of folks to acknowledge here; these acknowledgements are surely incomplete, and don't proceed in any particular order.

Many of the chapters in this book began as articles for the German newspaper *Frankfurter Allgemeine Zeitung* (*FAZ*). And so, above all, as we put this book together, we found ourselves feeling immensely grateful to Frank Schirrmacher, who is one of five publishers of *FAZ*, and Jordan Mejias, *FAZ*'s New York arts editor. Without their ongoing support and encouragement, none of the articles on which much of this book is based would have been written, and thus the book would not exist. They also published a few solo articles by Stephan which were not relevant to this book, but which he is grateful to have had published all the same.

For years before encountering *FAZ*, we had thought about writing about science for a popular audience, but somehow we'd never been able to do so in an effective way. By giving us the opportunity to write for *FAZ*, Frank and Jordan gave us a real audience to think about. Before, without this *real audience* in mind, we'd struggled and struggled with various strategies for simplifying and popularizing our thinking and writing. Once they put the real audience in front of us, everything just seemed to click. This is particularly ironic since the real audience for the *FAZ* articles consisted of people reading the articles in translation, in a language we do not read or speak or understand!

Ben would like to especially thank his father Ted Goertzel, who spent a goodly amount of time helping him struggle with an earlier attempt at popularizing the ideas underlying Webmind. Bits and pieces of that text wound up in a lot of places—some here, some in Ben's last research treatise *Creating Internet Intelligence*. That manuscript never quite gelled in a fully satisfying way, but the process of building it was extremely instructive.

Of course, beyond these acknowledgements that have to do specifically with the text in this book, we have a huge number of individuals to thank for helping me out with the scientific and business pursuits that led me to investigate the topics discussed here. At the top of the list are the Novamente co-founders, Cassio Pennachin (also our key Webmind Inc. co-conspirator) and Andre Senna, who have helped keep the Novamente AI project alive throughout tricky economic times. Going further back in history, we also owe something to a host of Webmind Inc. people whom we worked with in the period 1997-2001, such as the Webmind Inc. co-founders Ken Silverman, Lisa Pazer, Jeff Pressing and Onar Aam, and Webmind Inc. CEO and investor Andy Siciliano.

On a more personal level:

Ben: "My mother Carol Goertzel, her brother Mike Zwell, my grandfather Leo Zwell and my kids Zarathustra, Zebulon and Scheherazade all deserve thanks for one reason or another (and my father Ted Goertzel who was already mentioned above). My new wife Izabela Lyon Freire, who came into my life after most of this book was written, but whose love and insight energized me through the writing of the most recent material and the glorious task of editing the final manuscript. And finally, my ex-wife Gwendalin Qi Aranya (previously Gwendolyn Goertzel; aka the Reverend Chuan Kung Shakya), with whom I shared numerous discussions of the ideas in these pages.

"Some others to whom I owe debts of a more practical nature: Jeffrey Epstein, who paid my salary for the 2001-2002 academic year; Deepak Kapur and Barak Pearlmutter, who made it possible for me to use Jeffrey's grant to work at the University of New Mexico in 2001-2002; and Jaffray Woodriff, whose seed funding for Biomind LLC helped get Biomind (our bioinformatics/AI firm) off the ground."

Stephan: "First of all, thanks to Ben for inviting me to work on this book with him. I'd also like to thank my wife Anuradha Vikram, and our four cats Cthulhu, Yog, Shub, and Narasimha for putting up with me while I work on this book very close to publication deadlines (and for the final copy-editing pass on this book…which Anu did, with very little help from the cats). My mom, Barbara Bugaj, sisters Katrina and Janine, aunt Henrika, uncle John, and cousin Mary also all deserve thanks for their love and support over the years, as do Anu's parents Revathi and Vikram, and her brother Raj. My late grandmother, Janina Bugaj, also deserves a lot of thanks for surviving almost 16 years first in German forced-labor camps and subsequently in Allied DP camps to finally create a wonderful family here in the United States. I also owe general thanks for various things relating to my career and personal development, in roughly chronological order, to: Aunt Genevieve, Cousins Adam and Jadzia, The Bennett family, The Ciccariello family, The Hettiger family, Dr. Kevin Graham, Mr. Gehring, Karl Vermandois, Mr. Shaughnessy, Mary Antczak, Dafydd Kelsall, Ed Gorman, Igor Maev, Dr. Dima Grigoriev, my whole Simon's Rock College "crew" (Boris, Jason, Max, Adam, Seth, and all you omitted for space folks too!) and professors, Dr. Rudy Rucker, Alan Sondheim, Sasha Chislenko, Halsted, the Qwest/Delusions crew, Rick Sayre, the SRL Crew, Michael Clasen, Mark27 Anquoe, Mark Crumpacker, Jan Lindner, Beth Johnson, the SFNET crew, Cati Laporte, Dr. Jakub Segen, the whole Brazilian Webmind crew, Mattbot Ridenour, Nicki Kaiser, Allen T. Cain, all my friends on the Pixar crew, Marcin and Annika Kobylecki, The Ptaszyk family, and other folks I'll have to mention in the next book…"

PRELUDE
PESSIMISTS FROM YEARS PAST

The history of the 20th century is full of overoptimistic techno-prophecies. How many sci-fi authors of the '50s truly believed that by 2002 we'd be zipping over the rooftops of New York skyscrapers in little Jetsons-style spacecraft? What about Orwell's vision of 1984? As teenagers with heads full of this literature, and a lust for drama and excitement, we found the year 1984 a heck of a disappointment.

On the other hand, it's easy to forget how outlandish today's technologies would have seemed to the average person of 100, 50 or in some cases even 10 or 20 years ago. Nanotechnology pioneer and futurist Ralph Merkle has collected a highly amusing collection of erroneously pessimistic quotes about future technology on his website[1], many of which were originally taken from a Congressional Research Report on "Erroneous Predictions and Negative Comments Concerning Scientific and Technological Developments."[2] We think it is appropriate to open this book with a sampling of our favorite items from his collection.

The poor performance of past prognosticators indicates just how hard it is to project the exact trajectory of progress. On the other hand, those who projected interstellar space travel by 2000, but no computers or lasers or video games, might not be *too* disappointed when they saw the amazing array of technologies we have around us today. The fact of tremendous progress is easy to project even if the details are not, and the closer we get to the Singularity, the more accurate the optimists will be as compared to the pessimists. That's not to say we're absolutely certain that every prediction we make is correct—there's a fair bit of "applied science fiction" in this book's philosophical approach—but rather that we believe it is completely possible to achieve many if not all of the things we propose if we have the willpower to pursue positive technological progress rather than let apathy, greed and/or xenophobia lead us on a path toward utter destruction.

Read these examples, and tell us again that real AI is impossible, or that it must be "at least 1000 years off." Tell us again that the death of the human body is inevitable because it's the natural way of things, just like it's the natural way of things that the Earth is at the center of the universe or that humans can't fly, move mountains, or walk on the moon. Tell us again that building novel forms of matter to order is a pipe dream. Why is it better to hide our collective heads in the sand, and confront new discoveries only as they emerge each year, than to face the inevitability of some form of the Singularity—that is, some transformative next step of human and technological evolution—and apply our hearts, minds and souls toward maximizing the odds that the revolution we're creating will come out for the best for everyone?

ON THE UTTER IMPLAUSIBILITY OF EVER BUILDING AN AIRPLANE

> "Outside of the proven impossible, there probably can be found no better example of the speculative tendency carrying man to the verge of the chimerical than in his attempts to imitate the birds, or no field where so much inventive seed has been sown with so little return as in the attempts of man to fly successfully through the air. Never, it would seem, has the human mind so persistently evaded the issue, begged the questions and, 'wrangling resolutely with the facts,' insisted upon dreams being accepted as actual performance, as when there has been proclaimed time and again the proximate and perfect utility of the balloon or of the flying machine.
> "...Should man succeed in building a machine small enough to fly and large enough to carry himself, then in attempting to build a still larger machine he will find himself limited by the strength of his materials in the same manner and for the same reasons that nature has." —Melville, Rear Admiral George W. "The Engineer

and the Problem of Aerial Navigation." *North American Review* (December 1901), pp. 820, 825, 830-831.

ON THE RELATIVE USELESSNESS OF AIRCRAFT (AFTER THEIR POSSIBILITY WAS DEMONSTRATED)

The astronomer William H. Pickering said:

"...The popular mind often pictures gigantic flying machines speeding across the Atlantic and carrying innumerable passengers in a way analogous to our modern steamships...It seems safe to say that such ideas must be wholly visionary, and even if a machine could get across with one or two passengers the expense would be prohibitive to any but the capitalist who could own his own yacht. Another popular fallacy is to expect enormous speed to be obtained. It must be remembered that the resistance of the air increases as the square of the speed and the work as the cube...If with 30 hp we can now attain a speed of 40 mph, then in order to reach a speed of 100 mph, we must use a motor capable of 470 hp...It is clear that with our present devices there is no hope of competing for racing speed with either our locomotives or our automobiles." —Clarke, Arthur C. *Profiles of the Future*. New York, Harper and Row, 1962, pp. 3-4.

AIRPLANES WILL NEVER SINK BOATS (1966)

"The day of the battleship has not passed, and it is highly unlikely that an airplane, or fleet of them, could ever successfully sink a fleet of Navy vessels under battle conditions." —Woods, Ralph L. "Prophets Can Be Right and Prophets Can Be Wrong." *American Legion Magazine* (October 1966), p. 29.

ON THE USELESSNESS OF AC ELECTRICITY

Thomas Edison advocated his own DC based power system, and didn't much care for Tesla's invention of AC current:

"There is no plea which will justify the use of high-tension and alternating currents, either in a scientific or a commercial sense.

They are employed solely to reduce investment in copper wire and real estate.

"...My personal desire would be to prohibit entirely the use of alternating currents. They are unnecessary as they are dangerous...I can therefore see no justification for the introduction of a system which has no element of permanency and every elements of danger to life and property.

"...I have always consistently opposed high-tension and alternating systems of electric lighting...not only on account of danger, but because of their general unreliability and unsuitability for any general system of distribution." —Edison, Thomas A. "The Dangers of Electric Lighting." *North American Review* (November 1889), pp. 630, 632-633.

THE U.S. GOVERNMENT DECLINES TO FUND ROCKET RESEARCH, RIGHT BEFORE WWII

Robert H. Goddard, the rocketry pioneer, sought funding from the U.S. Army Air Corps in 1941. A quote from the rejection letter is as follows:

"The proposals as outlined in your letter...have been carefully reviewed...While the Air Corps is deeply interested in the research work being carried out by your organization...it does not, at this time, feel justified in obligating further funds for basic jet propulsion research and experimentation..."

THE LACK OF NEED FOR ROADS TO DRIVE THOSE "NEWFANGLED HORSELESS CARRIAGES" UPON

"The actual building of roads devoted to motor cars is not for the near future, in spite of many rumors to that effect." —*Harper's Weekly* (August 2, 1902), p. 1046.

U.S. PATENT OFFICE DECLARES TECHNOLOGICAL PROGRESS IS NEAR ITS END (1844)

Henry L. Ellsworth, U. S. Commissioner of Patents, said in 1844:

"...The advancement of the arts from year to year taxes our credulity and seems to presage the arrival of that period when further improvements must end." —Woods, Ralph L. "Prophets Can Be Right and Prophets Can Be Wrong." *American Legion Magazine* (October 1966), p. 29.

ANESTHESIA CAN NEVER WORK

Alfred Velpeau, an esteemed surgeon, declared in 1839:

"The abolishment of pain in surgery is a chimera. It is absurd to go on seeking it today. 'Knife' and 'pain' are two words in surgery that must forever be associated in the consciousness of the patient. To this compulsory combination we shall have to adjust ourselves." —Gumpert, Martin. *Trail-Blazers of Science.* New York, Funk and Wagnalls Company, 1936, p. 232.

NEITHER CAN BRAIN OR HEART SURGERY

Sir John Erichsen, writing in 1873, tells us:

"There cannot always be fresh fields of conquest by the knife; there must be portions of the human frame that will ever remain sacred from its intrusions, at least in the surgeon's hands. That we have already, if not quite, reached these final limits, there can be little question. The abdomen, the chest, and the brain will be forever shut from the intrusion of the wise and humane surgeon."

NUCLEAR WEAPONS WILL NEVER WORK (AS DECLARED IN 1945)

Adm. William Leahy told President Truman in 1945:

"That is the biggest fool thing we have ever done...The bomb will never go off, and I speak as an expert in explosives." —Truman, Harry S. *Memoirs, Vol. I: Year of Decisions.* Garden City, Doubleday and Company, Inc., 1955, p. 11.

RADIO BROADCASTING IS A FRAUD

The key device that makes long-distance radio broadcasting feasible is the audion tube, invented by Lee de Forest in the early 1900's. De Forest was an entrepreneur as well as an inventor, and founded a company, the Radio Telephone Company, intended to create and market long-distance radio technology. When he tried to sell stock in the company, he was brought to trial for fraud. The district attorney, in the course of the trial, declared that:

> "De Forest has said in many newspapers and over his signature that it would be possible to transmit human voice across the Atlantic before many years. Based on these absurd and deliberately misleading statements, the misguided public...has been persuaded to purchase stock in his company..." —Archer, Gleason Leonard. *History of Radio to 1926*. New York, American Historical Society, 1938, p. 110.

Though he was acquitted, the judge was not convinced his efforts were worthwhile, and gave de Forest the same advice that many other technology and science pioneers have heard from their friends and family: to "get a common garden variety of job and stick to it."

Without a doubt, pessimists can claim "victory" (we suppose *schadenfreude* is a form of victory), on a variety of fronts. Every scientific revolution, from the earliest recorded days of medicine and pre-chemical alchemy, has promised immortality. We've indeed not yet seen the days of interstellar space flight or even of space flight within our own solar system, with colonization of planets and moons becoming a common occurrence. Many techno-prophecies and Sci-Fi dreams have failed to come true. However, it is in the nature of engineers and scientists, just like that of musicians and artists, to continue to dream about "the impossible" and make the impossible real. It's highly improbable to predict the future precisely, but we'll safely bet that technological progress will cause increasingly numerous "impossible" ideas to become ordinary realities to which nobody gives a second thought. We may be wrong about many predictions, but even so serious philosophical pessimists could naysay for a thousand years and still not be as on-target about what the future holds as a single well-thought-out "silly sci-fi book."

---

[1] http://www.foresight.org/News/negativeComments.html

[2] CB 150, F-381, by Nancy T. Gamarra, Research Assistant in National Security, Foreign Affairs Division, May 29 1969 (revised).

CHAPTER 1
THE COMING CONVERGENCE

"I don't want to achieve immortality through my work; I want to achieve it through not dying." —Woody Allen

We live at a crucial point in history, an incredibly exciting and frightening moment; one that is stimulating to the point of excess, intellectually, philosophically, physically and emotionally. Radical new technologies are in development. Biotechnology will allow us to modify our bodies in various ways, customizing our genes and jacking our brains, organs and sense organs into computers and other devices. Nanotechnology will enable us to manipulate molecules directly, creating biological, computational, micromechanical, and other systems that can barely be imagined today. Artificial intelligence—enabling mind, intelligence and reason to emerge out of computer systems—will lead humans to build thinking machines. Advanced virtual reality and neural interface technology will let us create synthetic worlds equal in richness to the physical world. Neither the awkwardly geometrical worlds of VRML, nor today's over-stimulated but shallow video-game worlds, these will make the Buddhist maxim "reality is illusion" a palpable technical fact. Advances in the unified field theory of physics will in all likelihood join the party, clarifying the physical foundation of life and mind and giving the nanotechnologists new tricks about which no one has even speculated yet.

Medical immortality is far from out of the question. As Eric Drexler argued in *Engines of Creation* (1987), nano-scale robots could swarm through your body, repairing your aging cells. Or, as Hans Moravec depicted decades ago in his classic book *Mind Children* (1990), brain-scan technology combined with AI could have us uploading our minds into computers once our bodies wear down. Genetic and biomechanical engineering could allow us to fuse with computers, eliminating aging, and permitting direct mind-to-mind communication with other humans and computers via complex analysis of the brain's electromagnetic fields. It sounds like science fiction, but it's scientifically plausible: for while the previous examples would represent immense engineering achievements, they would not violate any known scientific laws. A lot of seemingly impossible things will soon be possible.

Bill Joy, former Chief Scientist at Sun Microsystems—one of the leading surviving technology companies from the "dotcom" boom (though Sun's future economic success is hardly assured)—wrote an article in *Wired* in 2000 painting the future we've described with a markedly dystopian bent. Joy believes that amazing technological development will happen, but he finds it intensely scary. It's difficult to blame him; the potential for abuse of such technologies is obvious. Joy believes technology will make humans obsolete, and he could be right. However, not everyone believes this would be a bad thing. We have no problem with the idea of humanity as we know it becoming obsolete in favor of superior modes of being; so long as individual humans have the right to choose to continue a reasonably pleasant human existence if that is their desire (we'll return to this issue in the final chapter). Powerful technologies in the hands of unethical humans could well spell total disaster. Do we have to hope an ethical evolution comparable to the technological evolution will occur at the same time? Or is it enough to guide and manage the unfolding of the technology revolution in an appropriate way?

This evolving network of long-term possibilities is wonderful and amazing to think about; however, the details are too numerous for any one or a few minds to grapple with. Imagine a group of pre-linguistic proto-humans attempting to comprehend how the advent of language might affect their futures. Nevertheless, despite the intrinsic difficulty of foresight and the impossibility of planning a revolution in advance, we can do more than sit back and watch as history leads us on. We can focus on particular aspects of the revolution, understanding these as part of the whole—and also focusing sharply on the details, remembering that historically some of the subtlest and most profound general humanistic insights have arisen out of very specific issues. In these pages we will view the ongoing technology revolution through several different perspectives; including AI, biotechnology, nanotechnology and advanced physics. At the end, we'll take a step back and look at the broader perspective, in light of the details we've reviewed.

SINGULARITY WATCHERS

Discussion about the amazing present and future achievements of technology is hardly novel these days. The transformative power of technology is now evident to most people. Any child raised on "Digimon," "Johnny Quest," "Dragonball Z," "Invader Zim," and "Star Trek" considers supercomputers, virtual reality, and intelligent robots commonplace.

The wide acceptance of rapidly improving technology as an everyday phenomenon has not translated into a deep or widespread understanding of where the technology explosion is leading us. After being presented with the evidence, most people concur that technology is improving at a superexponential rate (after you explain to them what "superexponential" means). People, however, seem unwilling to draw the natural conclusion to this observation, which is that machines will outpace us in nearly all ways, and are also quite likely to radically transform us. (Barring any unsuspected limits to progress within 30-100 years (or total annihilation of the human race before we create intelligent, self-organizing machine systems), machines are sure to outpace us unless we accept modification alongside them, and therefore keep the pace by co-evolving with our machine compatriots to a transhuman state.)

The renowned inventor Ray Kurzweil attempts make this point clear in his popular book *The Singularity is Near.* Drawing on Vernor Vinge's concept, Kurzweil has meticulously charted the progress of technology in various areas of industry, and plotted curves showing the approaching Singularity. Two of his more macro-level observations are given in the following charts.

Figure 1 demonstrates the mass use of inventions as it has accelerated over time—a quantitative depiction of the penetration of technology into everyday life.

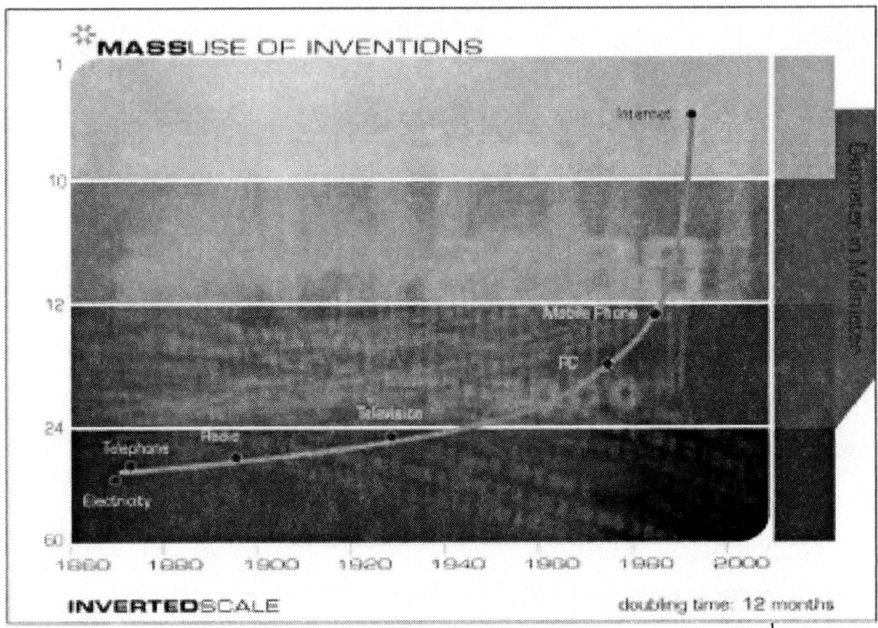

Figure 1.1 The accelerating penetration of inventions into everyday life[1]

Of course, there is much detail underlying the calculation of such figures, but Kurzweil seems to have been reasonably careful and scientific in this regard. For further details on his approach and philosophy, the referenced kurzweilAI.net website is an excellent resource, as is Kurzweil's aforementioned book.

Two more examples of his trend-tracking work are as follows:

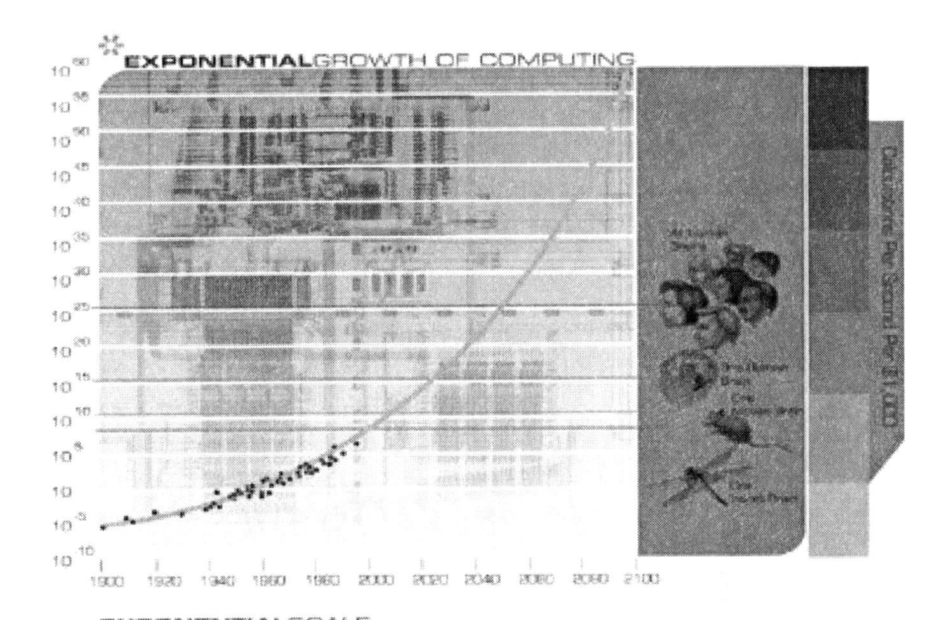

Figure 1.2 Exponential growth of computing[1]

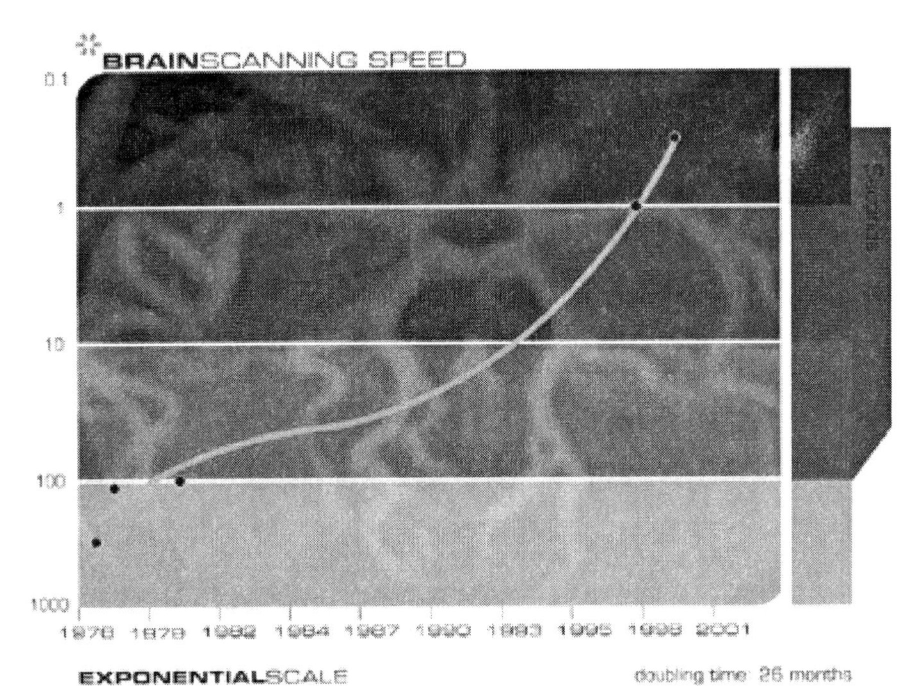

Figure 1.3 Brain-scanning speed[1]

The same trend emerges in one technology domain after another. The rate of progress is accelerating, and even its acceleration is accelerating!

Kurzweil's careful analysis of the phenomenon of exponentially increasing computing power makes several predictions regarding future human achievements:

- Human Brain capability (2 * 1016 cps[2]) for $1,000 in the year 2023
- Human Brain capability (2 * 1016 cps) for one cent in the year 2037
- Human Race capability (2 * 1026 cps) for $1,000 in the year 2049
- Human Race capability (2 * 1026 cps) for one cent in the year 2059

Kurzweil's comparisons with human brain power are based on a rough neural network model of the brain that will be described in the following chapter. As Kurzweil describes it, an error of a couple of orders of magnitude in estimating human brain power will throw off his time estimates by only a couple of years—thus is the power of exponential growth.

Ray Kurzweil is only one among a large and growing group of trend-tracking Singularity pundits. An early work of the genre is Derek DeSolla Price's 1963 book *Little Science, Big Science and Beyond* (reprinted 1986). More recent works include Damien Broderick's *The Spike* (mostly qualitative, but with some quantitative aspects), and Johansen and Sornette's 2001 article "Finite Time Singularity in the Dynamics of the World Population, Economic, and Financial Indices."[3]

In the 1990s, John Smart founded the website SingularityWatch.com[4] and established himself as perhaps the world's first full-time Singularity pundit. Smart's observations are as follows:

> "Complementary to Kurzweil's data on double exponential growth, these authors note that many other computationally relevant variables, such as economic output, scientific publications, and investment capital, have exhibited recent asymptotic phases. Of course, each particular substrate eventually saturates—population, economies, etc. never do go to infinity—but measuring the

asymptotic phases does allow us to better trace the second order computational trend presently unfolding on the planet in a broad range of physical-computational domains. Certainly my best current projected range of 2020-2060 is voodoo like anyone else's, but I'm satisfied that I've done a good literature search on the topic, and perhaps a deeper polling of the collective intelligence on this issue than I've seen elsewhere to date. To me, estimates much earlier than 2020 are unjustified in their optimism, and likewise, estimates after 2060 seem oblivious to the full scope and power of the...processes in the universe."[5]

Smart's trend analysis is very valuable; it provides much-needed quantitative concreteness to the qualitative observation of exponential technology acceleration. However, a deeper perspective is needed, a perspective that provides insight into the *meaning* of the Singularity. The extrapolative graphs produced by Kurzweil and others encapsulate a broad, deep story—a story not only of technological and engineering improvements, but of conceptual breakthroughs deepening our understanding. It is our progressively increasing knowledge of the universe, synergistically evolving with our creation of new technologies, which will bring about the Singularity. The focus of this book is that synergy between our deepening understanding of the human condition and the advancement of our technologies, spanning many domains of science, engineering and philosophy.

SINGULARITY VERSUS TRANSCENSION

The Singularity is an exciting and important idea, with solid thinking and data to support it. It is worth noting, however, that most aspects of posthumanity do not necessarily require a Singularity as described, requiring only a linear, still transformative, rate of change. Earlier we termed this change a "Transcension," a word borrowed[6] from Damien Broderick's 2002 novel of that name, and also from the writings of John Smart. A Singularity is a particular kind of Transcension, but not the only kind. Transcension, and indeed the Singularity itself, may occur for many reasons: technological, psychological, neurological, etc. Even a gradual marked change in any of these areas may lead to Transcension—and if the change is very rapid, a Singularity situation comes into play.

We expect that technological progress will lead one day to the creation of powerful Artificial Intelligence technology; and that, rather than humans, AI beings will carry out the vast majority of future scientific research and technological development. AI advances will allow new innovations to emerge at a superhuman pace. At this crucial point in time, when dramatic new technologies and new ways of thinking are developing daily or hourly, and so fast that even the smartest unaugmented humans literally cannot keep pace, the technological Singularity will be upon us. While technological Transcension is only one possible form, we will argue that a Singularity is most likely to occur at least in conjunction with, if not exclusively because of, rapid technological change. It seems likely that we will need the assistance of Real AI and nanotechnology to make a slow Transcension into a fast Singularity.

The Singularity, from a psychological perspective, is envisioned as a radical transition in the nature of not just technology but *experience.* When civilization, language and rational thought emerged from early human culture, the nature of human experience changed radically. Or, to put it another way, the "human experience" as we now know it emerged from the experience of proto-human animals. By this observation, a Singularity has occurred *at least* once before during our early evolution to a human condition[7]. But there is no good reason to believe that the emergence of the modern human mind is the end state of the psyche's evolution. While evolution might take millions of years to generate another psychological sea change as dramatic as the emergence of modern humanity, technology may do the job much more expediently. The technological Singularity can be expected to induce rapid and dramatic changes in the nature of life, mind and experience.

Although not as radical or rapid as a Singularity, a Transcension too will bring about a fundamental change in the nature of life and mind. A Transcension can occur even given no exponential or superexponential growth in technological or psychological/neurological change; it could occur eventually, even with a linear or logarithmic advance in technology and/or human biology. We think that a Singularity scenario is likely given the current data, but if it does not occur that does not mean that the path to Transcension has ended. Perhaps the most significant difference between the concepts of Transcension and Singularity is that, if the concept of a technological Singularity is correct, then the Singularity is *near* and we had best start worrying about it fast; whereas if a Transcension may occur more slowly, anytime within the next 10,000 years, there's no particular need for us to fuss about it at the moment.

However, it appears that at some point, the advance of technology will become (from a human perspective) essentially infinitely rapid, thus bringing a fundamental change in the nature of life and mind. Though it is not necessarily the case, current data indicates that a key aspect of the Singularity will be technological acceleration. Historical analysis suggests that the rate of technological increase is itself increasing—new developments come faster and faster all the time. At some point this increase will come so fast that we won't even have time to understand how to use the Nth radical new development, before the N+1th radical new development has come. That is, we won't understand it unless the nature of ourselves can change to keep pace with the changing nature of our technologically-mediated reality.

The emphasis "Singularity" places on the rapidity of change, induced by exponentially or superexponentially accelerating advances in technology, makes it important to consider whether the coming change will be sudden or not. If it is sudden, we have a Singularity situation, regardless of whether it's sudden because of technological enabling (as we predict in this book) or because of an evolutionary "leap" in our neurodynamics, a shift in some obscure physical law, or other such phenomenon. The technologies which may be involved, exciting as they are, should be viewed mainly as enablers. Humanity may soon experience profound changes in our "order of being." The ways in which we experience the world—the ways we human animals live life and conduct social affairs—are not the end state of mind in the universe, but only an intermediate state on the way to other modes of existence. Indeed, all states of being may be intermediate states on the way to the next level, indefinitely. The Transcension to these other modes of existence may well occur sooner rather than later.

One might contend that even if we are on the verge of modes of existence that are inconceivable given our current ways of thinking, living and experiencing, our limited and old-fashioned human brains really don't stand much chance of envisioning this new order of things in any detail. On the other hand, it seems, it would be foolish to *not even try*.

THE ORIGINAL CYBER-GURU

While Kurzweil has created an impressively concrete analysis, and Vinge's original writings on the Singularity are elegant and provocative, perhaps the deepest analysis of what's going on was provided by the Russian philosopher-scientist Valentin Turchin. We will use Turchin's notion of the Metasystem Transition as a tool for understanding the future of technology and its general implications.

Valentin Turchin is a fascinating individual who holds a unique position in the history of the Internet. He was the first of the cyber-gurus: The expansion of computer and communication networks that he foresaw in his 1970 book "The Phenomenon of Science" is now a reality; and the trans-human digital superorganism that he prophesied to emerge from these networks is rapidly becoming so. But unlike most scientists who turn toward philosophy in mid-life, Turchin was not satisfied to be a grand old theorist. Now in his 70's, and in spite of an increasingly frustrating battle with Parkinsonism, he is actively making his vision of an Internet superorganism come true. Turchin's Internet start-up company, Supercompilers LLC[8], applies the same cybernetic principles he used to study the future of humanity to create computer programs that rewrite other programs, making them dozens of times more efficient—even able to rewrite themselves.

None of Turchin's generation started their careers in computer science; rather, it was his generation that started computer science. He holds three degrees in theoretical physics, obtained in the 50's and early 60's; and the first decade of his career was devoted to neutron and solid state physics. But in the 60's his attention drifted toward computers, far before computers became fashionable—

especially in Russia. He created a programming language, REFAL, which became the dominant language for artificial intelligence in the Soviet bloc. Apart from any of his later achievements, Turchin's work on REFAL alone would have earned him a position as one of the leaders of 20th century computer science.

It was the political situation in the Soviet Union that drew him out of the domain of pure science, into the world of philosophy. In the 1960's Turchin became politically active, and in 1968 he authored *Inertia of Fear and the Scientific Worldview*, a fascinating document that combined a scathing critique of totalitarianism with the rudiments of a new cybernetic theory of man and society. Not surprisingly, following the publication of this book in the underground press, he lost his research laboratory.

His classic *The Phenomenon of Science*, first published two years later, expanded on the theoretical portions of "Inertia of Fear," presenting a unified cybernetic meta-theory of universal evolution. The ideas are deep and powerful, centered on the notion of a *metasystem transition*, a point in the history of a system's evolution where the whole comes to dominate the parts. Examples are the emergence of life from inanimate matter; and the emergence of multicellular life from single-celled components. He used the metasystem transition concept to provide a global theory of evolution and a coherent social systems theory, to develop a complete cybernetic philosophical and ethical system, and to build a new foundation for mathematics. The future of computer and communication technology, he saw, would bring about a metasystem transition in which our computational tools would lead to a unified emergent artificial mind going beyond humanity in its capabilities. The Internet and related technologies would spawn a unified global superorganism, acting as a whole with its own desires and wishes, and integrating humans to a degree as yet uncertain.

By 1973 he had founded the Moscow chapter of Amnesty International and was working closely with Andrei Sakharov. The Soviet government made it impossible for him to stay in Russia much longer. In 1977, persecuted by the KGB and threatened with imprisonment, Turchin was expelled from the Soviet Union. Taking refuge in the U.S., he joined the Computer Science faculty of the

City University of New York, where he continued his philosophical and scientific work. Among other projects, in the mid-1980's he created supercompilation—a novel technique that uses the meta-system transition concept to rewrite computer programs and make them more efficient.

While Americans tend toward extreme positions about the future of the cyber-world—with Bill Joy taking the pessimist's role, while Kurzweil, Moravec and others play the optimist—Turchin, as he and his team work to advance computer technology, views the situation with a typically Russian philosophical depth. He still wonders, as he did in *The Phenomenon of Science*, whether the human race might be an evolutionary dead-end like the ant or the kangaroo, unsuitable to lead to new forms of organization and consciousness. He wrote then, in 1970: "Perhaps life on Earth has followed a false course from the very beginning, and the animation and spiritualization of the Cosmos are destined to be realized by some other forms of life." Digital life, perhaps? Powered by software, made possible by the supercompiler Turchin's software company is developing?

The Phenomenon of Science closes with the following words: "We have constructed a beautiful and majestic edifice of science. Its fine-laced linguistic constructions soar high into the sky. But direct your gaze to the space between the pillars, arches, and floors, beyond them, off into the void. Look more carefully, and there in the distance, in the black depth, you will see someone's green eyes staring. It is the Secret, looking at you."

This is the fascination of the Net, and genetic engineering, and artificial intelligence, and all the other new technologies spreading around us—now psychologically, soon enough physically within us. It's something beyond us, yet in some sense containing us; created by us, yet recreating us. By writing books and supercompilers, or simply sending e-mails and living our tech-infused lives, we unravel the secret bit-by-bit, but we'll never reveal it entirely until it is upon us.

Meeting Turchin in person, after reading his work and admiring his thinking, was a fascinating experience. He was very down-to-earth, definitely not prone to conversationally spout phrases such as "fine-laced linguistic constructions" and "beautiful and majestic edifice of science" (but of course, this didn't surprise us; after all, our own conversation is a lot more informal than our prose). When we brought up the global brain and the posthuman future, he smiled knowingly, said a few words, and gently changed the subject. Clearly, he hadn't changed his mind about any of that, but it wasn't what was occupying most of his time. To him, all this was old hat. Of course, humans will become obsolete; of course computer programs will become superintelligent. Nothing more to say about that, really. Mostly his mind was occupied by technical problems he had been working on: making a supercompiler that could rewrite software programs to make them dozens of times faster, using advanced logical reasoning; and firming up a theory he'd developed years before called the "cybernetic foundations of mathematics," which grounds mathematical knowledge in the process of acting in and perceiving the world, using his programming language REFAL as a tool.

We were happy to be able to help him out, and Ben played an instrumental role in getting his company Supercompilers LLC funded, by introducing Turchin and his Russian colleagues to a major Chicago investor. (The funder was a contact gained through the ongoing development of our AI company, Webmind. In late 2001 Ben became even more intensely involved with this work, joining the Supercompilers team on a part-time basis, helping them to shape their business approach and define the direction of their research.) But at the time of this writing Turchin is playing only a very "high-level conceptual guru" sort of role in the business. The main torch is being carried by his former student Andrei Klimov, a formidable intellect in his own right; and Turchin's interest has shifted largely to even deeper technical territory, the cybernetic philosophy of mathematics.

We realized, when we met Turchin in person and saw what sorts of things were preoccupying him, that he was specifically not on the kind of path that would lead him to become a well-known "cyber-guru." To become well-known in the media as the originator or defender of a certain idea, you have to spend all your time pushing that idea, writing book after book about the same thing, giving interviews, training disciples. Turchin's way is exactly the opposite of this. He stated his philosophical views very clearly in an excellent book and some related articles, and then in large part went back to technical work, albeit deeply informed by his philosophy. His ideas have had a huge influence behind the scenes, through their influence on countless scientists of our generation. His disciples, if one may call them such, tend to push his technical agenda rather than his philosophical one. It makes us sad to see that, outside Russia, he simply isn't going to get the acclaim he deserves. But in the end, of course, this sort of thing doesn't matter much: Turchin's main goal was never to get acclaim anyway; it was to speak the truth and to spread the truth, and this goal was well accomplished.

METASYSTEM TRANSITIONS

Turchin's notion of metaystem transitions provides a powerful overall framework within which to understand the maelstrom of scientific, technological and social change that whirls all around us. It allows us to see the unfolding transitions in humanity as merely one phase in a longer and larger process, similar in many ways to other phases, though with its own unique and fascinating properties (and obviously with a different personal relevance to us homo sapiens).

The figure below depicts the advance toward Singularity quantitatively. The vertical axis is a measure of technical progress, and the horizontal axis is time.

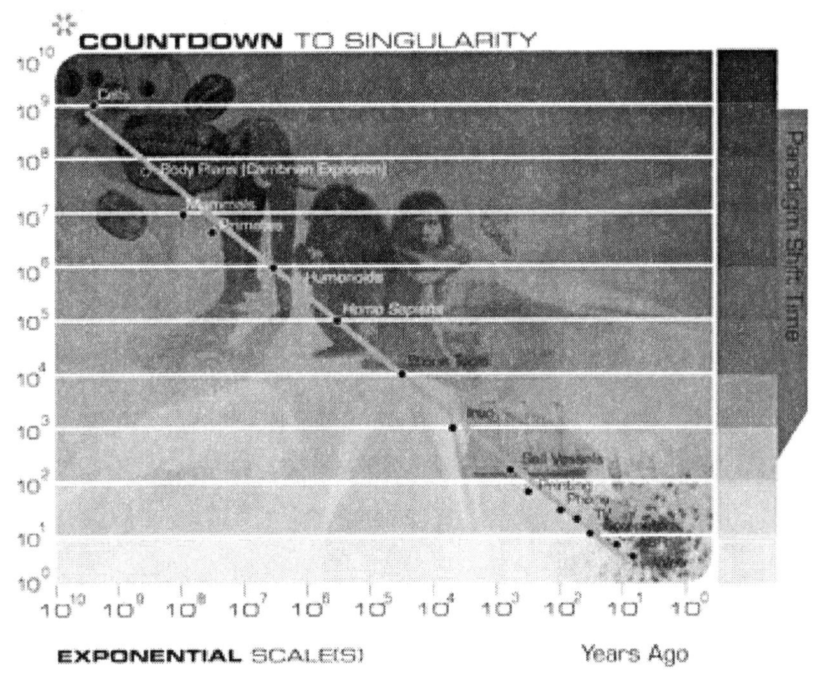

Figure 1.4 Countdown to singularity[1]

As Kurzweil puts it:

> "The paradigm shift rate (i.e., the overall rate of technical progress) is currently doubling (approximately) every decade; that is, paradigm shift times are halving every decade (and the rate of acceleration is itself growing exponentially). So, the technological progress in the twenty-first century will be equivalent to what would require (in the linear view) on the order of 200 centuries. In contrast, the twentieth century saw only about 25 years of progress (again at today's rate of progress) since we have been speeding up to current rates. So the twenty-first century will see almost a thousand times greater technological change than its predecessor."

The metasystem transition notion gives a conceptual picture to go with this graph.

Let's begin at the beginning: according to current physics theories, there was a metasystem transition in the early universe, when the primordial miasma of disconnected particles cooled down and settled into atoms. All of a sudden the atom was the thing, and individual stray particles weren't the key tool to use in modeling the universe—once particles are inside atoms, the way to understand what particles are doing is to understand the structure of the atom. Then there was another transition, from atoms to molecules, which led to the emergence (within the first few instants after the Big Bang) of the elements of the Periodic Table.

There was another metasystem transition on earth around four billion years ago, when the steaming primordial seas caused inorganic chemicals to clump together in groups capable of reproduction and metabolism. Unicellular life emerged. Once chemicals are embedded in life-forms, the way to understand them is not in terms of chemistry alone, but rather in terms of biological concepts like fitness, evolution, sex and hunger. Concepts like desire and intention are not far off, even with paramecia—does the paramecium desire its food? Maybe not by our standards, but it comes a lot closer than a rock does to desiring to roll down a hill.

There was another metasystem transition when multicellular life burst forth. Suddenly the cell was no longer an autonomous life form, but rather a component in a life form on a higher level. The "Cambrian explosion," immediately following this transition, was the most amazing flowering of new patterns and structures ever seen on Earth. Even we humans haven't equaled it yet. 95% of the species that arose then are now extinct, and paleontologists are slowly reconstructing them so we can learn their lessons.

Note that the metasystem transition is not an anti-reductionist concept in the strict sense. The idea isn't that multicellular life-forms have cosmic emergent properties that can't be explained by the properties of cells. Of course, if you had enough time and superhuman patience, you could explain what happens in a human body in terms of the component cells. The question is one of naturalness and comprehensibility, or in other words, efficiency of expression. Once you have a multicellular life-form, it's much easier to discuss and analyze its properties by reference to the emergent level than by going down to the level of the component cells. In a puddle full of paramecia, on the other hand, the way to explain observed phenomena is usually by reference to the individual cells, rather than the whole population. The population in such a case has less wholeness, fewer interesting properties, than the individual cells.

METASYSTEM TRANSITIONS IN HUMAN AND ARTIFICIAL MIND

The metasystem transition idea is important in its clear depiction of the overall patterns of system evolution. We have also found it useful in our own research work on AI. In the domain of mind, there are also a couple levels of metasystem transition. The first one is what one might call the emergence of "mind modules." The second is the emergence of mind from a conglomeration of interconnected, intensely interacting mind modules. Since 1997, Ben and Cassio Pennachin have been leading a team towards building a "real AI" (with help from Stephan from 1998-2001, and on-and-off since). A software program with human-level general intelligence is the goal, with a major focus on achieving this second metasystem transition. We'll talk about this work in a little more detail later on.

The first metasystem transition on the way toward mind occurs when a huge collection of basic mind components (cells, in a biological brain; "software objects" in a computer mind) all come together in a unified structure to carry out some complex function. The whole *is* greater than the sum of the parts: the complex functions that the system performs aren't really implicit in any of the particular parts of the system, but they emerge as the parts coordinate into a coherent whole. The various parts of the human visual system are wonderful examples of this. Billions of cells fire every which way, all orchestrated together to do one particular thing: map visual output from the retina into a primitive map of lines, shapes and colors, to be analyzed by the rest of the brain. The best current AI systems are also examples of this. In fact, we'd be reluctant to call computer systems that haven't passed this transition "AI" in any serious sense.

There are some so-called AI systems that haven't even reached this particular transition—they're really just collections of rules, and each behavior in the whole system can be traced back to one particular rule. But consider a sophisticated natural language system like LexiQuest[9] which tries to answer human questions, asked in ordinary language, based on information from databases or extracted from texts. In a system like this, we do have mind module emergence. When the system parses a sentence and tries to figure out what question it represents, it's using hundreds of different rules for parsing, for finding out what various parts of the sentences mean. The rules are designed to work together, not in isolation. The control parameters of each part of the system are tuned so as to give maximal overall performance. LexiQuest isn't a mind, but it's a primitive mind module, with its own, albeit minimal, holistic emergence. The same is true of other current high-quality systems for carrying out language processing, computer vision, industrial robotics, and so forth. For an example completely different from LexiQuest, look at the MIT autonomous robots built under the direction of Rodney Brooks. These robots seem to exhibit some basic insect-level intelligence, roaming around the room trying to satisfy their goals, and displaying behavior patterns that surprise their programmers. They're action-reaction modules, not minds, but they have holistic structures and dynamics all their own.

On roughly the same level as LexiQuest and Brooks' robots, we find computational neural networks. These carry out functions, like vision or handwriting recognition or robot locomotion, using from hundreds to hundreds of thousands of chunks of computer memory in an emulation of biological neurons. As in the brain, the interesting behavior isn't in any one neuron, it's in the whole network of neurons, the integrative system. There are dozens of spin-offs of neural network concepts, such as the Bayesian networks used in products like Autonomy and the Microsoft Help system. These Bayesian networks of rules are capable of making decisions such as "If the user asks about 'spreadsheet,' activate the Excel help system." The programmer of such a system never enters a statement where this exact rule appears; rather this rule emerges from the dynamics of the network. However, the programmer sets up the network in a way that fairly rigidly controls what kinds of rules can emerge. So while the system can discover new patterns of input behavior that seem to indicate which of a set of actions it should take, it is unable to discover new kinds of actions which can be taken. It can only discover new instances of information, not new types of information. It's not autonomous, not alive, and certainly not intelligent.

As noted above, our own AI work has centered on creating systems that embody a metasystem transition beyond this "mind module" level. First, with our R&D team at Webmind Inc., we built a system in the late 1990's called the Webmind AI Engine (or simply "Webmind"). Since Webmind Inc. folded in early 2001, a smaller group has been collaborating with on a successor system called Novamente, based on the same conceptual principles but with a different mathematical underpinning and software design.

Novamente, like Webmind, is a multimodular AI. Each module of the Novamente system has roughly the same level of sophistication as one of these bread-and-butter AI programs. There are modules that carry out reasoning, language processing, numerical data analysis, financial prediction, learning, short-term memory, and so forth. All the modules are all built of the same components: software objects, called "nodes" and "relationships." They arrange these components in different ways, so that each module achieves its own emergent behavior—each module realizing a metasystem transition on its own.

Mind modules aren't real intelligence, not in the sense that we mean it: intelligence as the ability to carry out complex goals in complex environments. Each mind module only does one kind of thing, requiring inputs of a special type to be fed to it, and is unable to dynamically adapt to a changing environment. Intelligence itself requires one more metasystem transition: the coordination of a collection of mind modules into a whole mind, each module serving the whole and fully comprehensible only in the context of the whole. This is a domain that AI research has basically not yet confronted. It's not mere egotism to assert that the Webmind/Novamente systems are almost unique in this regard. It takes a lot of man-hours, a lot of thinking, and a lot of processing power to build a single mind module, let alone to build a bunch of them, and even more to build them in such a way as to support an integrative system passing the next metasystem transition. Computer hardware is just barely at the point where we can seriously consider doing such a thing. But even being *just barely* there is a pretty exciting thing.

Novamente allows the interoperation of these intelligent modules within the context of a shared semantic representation—nodes, links and so forth. Through the shared semantic representation, these different intelligent components can interact, and thus evolve a dynamical state which is not possible within any one of the modules. Like a human brain, each specialized sub-system is capable of achieving certain complex goals, whether perceptual (such as reading a page of text) or cognitive (such as inferring causal relations), which in themselves seem impressive. When they are integrated, truly exciting new possibilities emerge. Taken in combination, these intelligent modules embodying systems of reasoning, learning, natural language processing, and so forth, can undergo a metasystem transition to become a mind capable of achieving complex goals worthy of comparison to human abilities. The resulting mind cannot be described merely as a pipeline of AI process modules; rather it has its own dynamical properties which emerge from the interactions of these component parts, creating new and unique patterns which were not present in any of the sub-systems.

Such a metasystem transition from modules to mind is a truly exciting emergence. A system such as Novamente will be able to autonomously adapt to changes in more complex environments than their single-module predecessors, and can be trained in a manner which is more like training a human than programming a computer. This kind of a system can theoretically be adapted to any task for which it is able to perceive input, and while the initial Novamente system operates in a world of text and numerical files only, integrating it with visual and auditory systems and perhaps a robot body would allow it to have some facility to perform in the physical world as well. Applications of even the text- and data-constrained system are quite varied and exciting: for example, autonomous financial analysis, conversational information retrieval, true knowledge extraction from text and data, etc.

While there are other systems that can find some interesting patterns in input data, a mind can determine the presence of previously unknown types of patterns and make judgments that are outside the realm of previous experience. An example of this can be seen in financial market analysis. Previously unknown market forces, such as the Internet, can impact various financial instruments in ways that prevent successful trading using traditional market techniques. A computer mind can detect this new pattern of behavior and develop a new technique based on inferring how the current situation relates to, but also differs from, from previous experience. The current Novamente AI Engine already does this, to a limited extent, through the emergence of new behaviors from the integration of only a few intelligent modules. As more modules are integrated the system becomes more intelligent.

For another short-term, real-world example of the promise of computational minds, let's return to the area of information retrieval. What we want isn't really a search engine—we want a digital assistant, with an understanding of context and conversational give-and-take such as a human assistant provides. When an ambiguous request is made of a mind, it does not blindly return some information pulled out of a database; a mind asks questions to resolve ambiguous issues, using its knowledge of your mind as well as the subject area to figure out what questions to ask. When you ask a truly intelligent system, "find me information about Java," it will ask a question back such as "do you want information about the island, the coffee, or the computer programming language?" But if it knows you're a programmer, it should ask instead "Do you want to know about JVMs or design patterns or what?" Like a human, a machine which has no information to the effect that there is an island called Java, for example, might only ask about coffee and computers, but the ability to make a decision to resolve the ambiguity in the first place, in a context-appropriate way, is a mark of intelligence. An intelligent system will use its background knowledge and previous experience to include similar information (Java, J++, JVM, etc.), omit misleading information (JavaScript, a totally different programming language from Java), and analyze the quality of the information. Retrieval segues into creation when a program infers new information by combining the information available in the various documents it reads, providing users with this newly created information as well as reiterating what humans have written.

These practical applications are important, but it's worth remembering that the promise of digital mind goes beyond these simple short-term considerations. Consider, for example, the fact that digital intelligences have the ability to acquire new perception systems during the course of their lives. For instance, an intelligent computer system could be attached to a bubble chamber and given the ability to directly observe elementary particle interactions. Such a system could greatly benefit particle physics research, as the system would be able to think directly about the particle world, without having to resort to metaphorical interpretations of instrument readings as humans must. Similar advantages are available to computers in terms of understanding financial and economic data, and recognizing trends in vast bodies of text.

THE EMERGING GLOBAL BRAIN AND THE COMING TRANSFORMATION OF HUMANITY

The metasystem transition from mind modules to mind is the one that we have mulled over the most, due to its connection with our AI work. But it's by no means the end of the story. When Turchin formulated the metasystem transition concept, he was actually thinking about something quite different—the concept of the global brain, an emergent system formed from both humans and AI systems, joined together by the Internet and other cutting-edge communication technologies. It's a potentially scary idea, but a potent one, and with a concrete grounding in reality that shouldn't be ignored.

Communication technology makes the world smaller each day. Will it eventually make the world so small that the network of people has more intrinsic integrity than any individual person? Shadows of the Sci-Fi notion of a "hive mind" arise here. Images of the "Borg Collective" from *Star Trek* may be the first that come to mind. But what Turchin is hoping for is something much more benign: a social structure that permits us our autonomy, yet channels our efforts in more productive directions, guided by the good of the whole. Having survived and escaped the totalitarian system of the Soviet Union, Turchin is clearly not blindly optimistic about "the common good" as the solution to all our ills, but rather believes that a balance is possible using structures other than what have been tried in the past (be they Communism, Corporatism, or whatever else).

As noted above, Turchin himself is somewhat pessimistic about the long-term consequences of all this—but not in quite the alarmist vein of Bill Joy, more in the typically Russian ironic spirit of doubting human nature. In other words, Bill Joy believes that high technology may lead us down the road to hell, so we should avoid it; whereas Turchin sees human nature itself as the really dangerous thing, leading us to possible destruction through nuclear, biological, or chemical warfare, or some other physical projection of our intrinsic narrow-mindedness and animal hostility. Turchin hopes that technological advancement will allow us to overcome some of our shortcomings, and thus work toward the survival and true mental health of our race, allowing us hope to change ourselves for the better so that we can ultimately evolve beyond these self-destructive behaviors.

Through the Principia Cybernetica project[10], co-developed with Francis Heylighen (of the Free University of Brussels) and Cliff Joslyn (of Los Alamos National Labs in the U.S.), Turchin has sought to develop a philosophical understanding to go with the coming technology revolution, grounded on the concept of the metasystem transition. As he says, the goal with this is "to develop—on the basis of the current state of affairs in science and technology—a complete philosophy to serve as the verbal, conceptual part of a new consciousness." This isn't exactly being done with typical American technological optimism. Rather, as Turchin puts it, "My optimistic scenario is that a major calamity will happen to humanity as a result of the militant individualism; terrible enough to make drastic changes necessary, but, hopefully, still mild enough not to result in a total destruction. Then what we are trying to do will have a chance to become prevalent. But possible solutions must be carefully prepared."

As we see it, the most likely path from the Net that we have today to the global brain that envelops humans and machines in a single overarching superorganism involves not one but several metasystem transitions. Of course, this path may never be taken—we may annihilate ourselves first, or someone may create a superhuman AI[11] that launches a Singularity well before the global brain gets a chance to emerge. But if a Turchin-style global brain does emerge, which we believe to be a plausible outcome, these are the likely phases through which it will develop:

The first phase will be the emergence of the global web mind, transforming the Internet into a coherent organism. Currently, the best way to explain what happens on the Net is to talk about the various parts of the Net: particular websites, e-mail viruses, shopping bots, and so forth. But there will come a point when this is no longer the case, when the Net has sufficient high-level dynamics of its own that the way to explain any one part of the Net will be by reference to the whole. This will come about largely through the interactions of AI systems. Intelligent programs, acting on the behalf of various websites, Web users, corporations, and governments, will interact with each other intensively; forming something halfway between a society of AIs and an emergent mind whose lobes are various AI agents serving specific goals. The traditional economy will be dead, replaced by a chaotically dynamical Hypereconomy[12] in which there are no intermediaries except for information intermediaries. Producers and consumers (individually or in large aggregates created by automatic AI discovery of affinity groups) negotiate directly with each other to establish prices and terms, using information obtained from subtle AI observation, prediction, and categorization algorithms. We can't really tell how far off this is, but it would be cowardly not to give an estimate: we're betting no more than 10-50 years.

The advent of this system will be gradual. Initially, when only a few AI systems are deployed on the Web, they will be individual systems which are going to be overwhelmed with their local responsibilities. As more agents are added to the Net, there will be more interaction between them. Systems which specialize will refer questions to each other. For example, a system which specialized in (had a lot of background knowledge and evolved and inferred thinking processes about) financial analysis may refer questions about political activities to political analyst systems, and then combine this information with its own knowledge to synthesize information about the effects of political events on market activity. This hypereconomic system of Internet agents will dynamically establish the social and economic value of all information and activities within the system, through interaction amongst all agents in the system. As these interactions become more complex, agent interconnections more prevalent and dynamic, and agents more interdependent, the network will develop into more of a true shared semantic space: a global integrated mind-organism. Individual systems will start to perform activities which have no parallel in the existing natural world. One AI mind will directly transfer knowledge to another by literally sending it a "piece of its mind." An AI mind will be able to directly sense activities in many geographical locations and carry on multiple context-separated conversations simultaneously; and from this a single global shared-memory will emerge which allows explicit knowledge-sharing in a collective consciousness. Across the millions—someday billions—of machines on the Internet, this global Web mind will function as a single collective, collaborative thought space, allowing agents to transcend their individual limitations and share consciousness directly, extending their capabilities far beyond their individual means.

All this is fabulous enough: collective consciousness among AI systems, the Net as a self-organizing intelligent information space. And yet, it's *after* this metasystem transition from Internet to global hypereconomic Web mind that the transition envisioned by Turchin and his colleagues at Principia Cybernetica can take place: the effective fusion of the global Web mind and the humans interacting with it. It will be very interesting to see where biotech-enabled virtual reality technology is by this time. At what point will we really be jacking our brains into the global AI matrix, as in Gibson's novel *Neuromancer* (1984)? At what point will we supercompile and improve our own cognitive functions, or be left behind by our constantly self-reprogramming AI compatriots? But we don't even need to go that far. Putting these more science-fictional possibilities aside and focusing solely on Internet AI technology, it's clear that more and more of our interactions will be mediated by the globally-emergent intelligent Net. Every appliance we use will be jacked into the matrix. Every word that we say will potentially be transmittable to anyone else on the planet, using wearable cellular telephony or something similar. Every thought that we articulate will be entered into an AI system that will automatically elaborate it and connect it with what other humans and AI agents have said and thought elsewhere in the world; or what other humans and AI agents are expected to say based on predictive technology. The selfish, hostile nature of the current human race makes for a variety of horrific possibilities within such situations, but great things are also possible if we are able to overcome our current preferences for greed and violence and to balance individual with collaborative needs within a single system; the way any stable, functional complex system operates. The Internet Supermind is not the end of the story: it's only the initial phase, the core around which will crystallize a new order of mind, culture and technology. Is this going to be an instrument of fascist control, or factional terrorism? It's possible, but certainly not inevitable—and the way to avoid this is for as many people as possible to understand what's happening, what's likely to happen, and how they can participate in the *positive* expansion of this technology.

Imagine: human and machine identities joined into the cooperative mind, creating a complex network of individuals from which emerges the dynamics of a global Supermind, with abilities and boundaries far greater than would be possible for any individual mind, human or artificial, or any community consisting of humans or AIs alone. As Francis Heylighen has said, "Such a global brain will function as a nervous system for the social superorganism, the integrated system formed by the whole of human society." Through this global human-digital emergent mind, we will obtain a unique perspective on the world, being able to simultaneously sense and think in many geographical locations and potentially across many perceptual media (text, sound and image; and various sensors on satellites, cars, bubble chambers, etc.) The cliché, "Let's put our minds together on this problem," will become a reality—allowing people and machines to pool their respective talents directly to solve tough problems in areas ranging from theoretical physics to social system stabilization, and to create interesting new kinds of works in literature and the arts. This kind of group-mind which still allows the individual minds within it their own autonomy (not a coercive system, but a benefit-reward cooperative system as any truly functional emergent, dynamic system ultimately is) is what we call a Mindplex, and is the topic of a chapter later in this book.

Weird? Scary? To be sure. Exciting? Amazing? To be sure. Inevitable? An increasing number of techno-visionaries think so. Some, like Bill Joy, have retreated into neo-Luddism, believing that technology is a big danger and advocating careful legal control of AI, nanotech, biotechnology and related fields. Of course, history indicates that "careful governmental control" is not any guarantee of positive outcome; concentrating decision-making power about humanity's future in the hands of a few elites has rarely been the path to success. Paranoid, reactionary action usually leads to much worse results than allowing change to proceed whilst trying to understand, manage, and guide it in a positive direction.

Turchin is progressing ahead as fast as possible with building the technology needed for the next phase of the revolution, being careful to keep an eye on the ethical issues as he innovates, and hoping his pessimism about human nature will be proved wrong. As for us, we tend to be optimists. Life isn't perfect. Plants and animals aren't perfect, humans aren't perfect, and computers aren't perfect—yet, the universe has a wonderful way of adapting to its mistakes and turning even ridiculous errors into wonderful new forms.

The dark world of tyranny and fear described in the writings of Cyberpunk authors like William Gibson and Bruce Sterling, and in films such as "The Matrix" and "Blade Runner," is certainly a possibility. But there's also the possibility of less troubling relationships between humans and their machine counterparts, such as those we see in the writings of Transrealist authors like Rudy Rucker and Stanislaw Lem, and in film characters like Star Trek's Data and Star Wars' R2-D2 and C3P0. Indeed, in all but the most nihilistically dystopian Cyberpunk, there is a range of personalities from benevolent to evil amongst both machines and humans—reflecting a belief, shared by many, that even in the darkest times, and after all but the ultimate catastrophe, intelligent beings trend ultimately towards self-redemption. We believe that through ethical treatment of humans, machines, and information, a mutually beneficial human-machine union within a global society of mind can be achieved. The ethical and ontological issues of identity, privacy and selfhood are every bit as interesting and challenging as the engineering issues of AI, and we need to avoid the tendency to set them aside because they're so difficult to think about. But these things are happening. Right now we're at the beginning, not the end, of this revolution, and the potential rewards are spectacular: enhanced perception, greater memory, greater cognitive capacity, and the possibility of true cooperation among all intelligent beings on earth.

In thinking about such wild and amazing transitions, it pays to remember: we've been riding the roller coaster of universal evolution all along. The metasystem transitions that gave rise to human bodies and human intelligence were important steps along the path, but there will be other steps, improving and incorporating humanity, and ultimately going beyond it.

ONWARDS AND UPWARDS

Having glimpsed such a glorious, ambitious, multifariously intricate futuristic vision—having *bathed* ourselves in this vision for years—where does one go next?

One can write science fiction stories or metaphysical poems exulting in the wonder of it all. And why not? We must admit that we each have succumbed to this urge now and then.

One can sit back and relax and enjoy oneself, waiting for the Singularity. Unlike *Waiting for Godot*—one of our favorite pieces of writing, which is all about empty spaces—the wait for the Singularity is full of entertainment, constant changes and advances.

One can plunge into the details, and there sure are a lot of them. The technology revolution and the impending Singularity are too big a topic for any single book to treat with suitable richness. In these pages we'll detail several aspects of the great unfolding story, with an emphasis on computation and biology, but with significant discussion of other areas such as nanotechnology, physics, psychology and sociology as well. However, we hope that after reading this book, you'll rush out and read more about each of these areas in depth.

Perhaps the real story (as some predict) lies in advanced AI, rapidly modifying its own source code until it's incomprehensibly many times more intelligent than humans and effortlessly reconfigures all the matter in the solar system every 15 seconds or so. In the long run they may well be right. But we hope that John Maynard Keynes' famous quip, "In the long run, we'll all be dead," turns out *not* to be right. A future populated by superintelligent AI systems is far more palatable to us than some other possibilities, such as, say, a future in which all life and intelligence in Earth are wiped out by nuclear weapons. Just as we're glad bugs still exist alongside we more intelligent beasts, we're happiest thinking about a future in which humans persist no matter how far technology advances. In fact, we tend to think this is likely, though the humans of the future may not look, act or think exactly like we do (we don't look, act, or think exactly like Sumerians, either). We'll return to these radical futurist ethical issues toward the end of the world, after we've gone through the preliminaries in more detail.

Some people like the idea of growing old and dying. There is a spiritual completeness associated with death, a sense of oneness with nature. This is fine, for them. Some of us, on the other hand, prefer the notion of existing as long as possible, perhaps as long as the universe itself—or longer, if advanced physics shows us how to escape into other universes. This is a modification of current humanity that bioscience seems very likely to provide. Apoptosis (the aging of cells) will one day become as obsolete as smallpox. Parable and superstition aside, there is no reason not to try to "cheat death." Humanity has been trying to do so all along, and perhaps this has been our destiny all along: to *create* eternal life, not simply to wait for some greater being to hand it to us on a silver Rapture.

Some people don't like video games, or computers, or electronic musical instruments. Ben has struggled for years to overcome his fanatical hatred of television, and has been only partially cured in recent years via his daughter's obsession with *South Park*. Stephan prefers to use telephones only when unavoidable. Again—no worries—diversity of taste is a good thing. But we are going to see advances beyond current human-computer interfaces, which make the difference between parchment and e-mail seem microscopic. Already quadriplegics can control computers directly using their brain waves; and multiple projects are well on the way to giving the blind sight, by linking camera output directly to the brain's optical system (either through the optic nerve, or cortically). Once we can scan brain-states better in real time, we'll be able to project our thoughts directly into computers, and perhaps back out of computers into others' brains (we would hope, in a communicative rather than coercive or displacive manner). Virtual-reality technology has been overhyped so far, but eventually it really will happen. There are no fundamental obstacles, "merely" engineering difficulties and thorny neuroscience questions, but ones of a type that we humans have proved ourselves rather adept at solving.

Genetic engineering scares people greatly, and the downside is indeed dangerous. We don't particularly want to see, for instance, the government outlawing genes that have been found to correlate with civil disobedience, or the writing of unpatriotic literature. Yet how bad would it be to see genes correlated with serial killing, or profound retardation, or multiple personality syndrome filtered out of the population? And how bad would it be if genetic engineering could modify the brain-encoding genes, producing people twice as good at math and physics as Einstein, or five times as socially pragmatic and compassionate as Gandhi or Mother Teresa? Eugenics is a revolting word, but if pressed, pretty much everyone would get rid of some genetic predispositions—be they sickle cell anemia, Down's syndrome, or some other real or perceived maladies—and enhance some others. It's not a topic to be treated lightly, and we could write thousands of pages about the ethical dilemmas and probability-theoretic gedankenexperiments involved in deciding what human traits we may wish to alter should we find the genetic markers for them. Balancing the pluses and minuses of genetic engineering seems to be an unsolvable ethical puzzle. Mulling over these ethical concerns is worthwhile—but not from a pragmatically irrelevant "should we" perspective, rather from a "how to make it positive" one. Waiting to develop the technology isn't going to help anything. A 20 or 50 or 100 year delay is not going to make humans any more consistently ethical. (The only thing that could do that would be genetic engineering itself!) Perhaps the best thing that those concerned about the ethical implications of genetic engineering can do is to become geneticists themselves, and work hard on positive applications. Let's breed a more compassionate and ethical race before someone breeds a race of superintelligent psychopaths.

The cooperation between genetic engineering, human-computer interfacing and AI development is going to be particularly exciting. Eventually we'll be able to manipulate the genetic code in such a way as to create humans who are especially well-suited to interface with virtual worlds and artificial intelligences, or whatever other new cognitive and perceptual realms we discover or create in the meantime. Sure, it's wild and crazy sci-fi—just like TV, submarines, airplanes and spacecraft once were. Ask your great-grandfather. In fact, ask your *father* what his generation thought about nanotechnology when they were growing up. We're going to tell you, in these pages, about why these new scientific ideas aren't so totally fictional after all. The real-world science of today is not very far off from the fantastic dreams of tomorrow. Turchin may not live to see the metasystem transition in which networked AI comes about, let alone the one in which genetically modified humans and AIs interact in virtual worlds, forming a whole new kind of virtual mind/society/organism. On the other hand, being born in 1966 as was Ben, or in 1973 as was Stephan, we might just make it. And if Turchin avails himself of cryonic technology (so far as we know he has no current plans to do so, but is intrigued by the idea), and has his body frozen shortly after death, he may be defrosted into a whole new world, perhaps even a post-Singularity "reality" that bears little resemblance to anything we now conceive as a "reality" or a "world."

---

[1] Reproduced from http://www.kurzweilai.net/articles/art0134.html

---

[2] Computations per second
[3] available online at http://xxx.lanl.gov/abs/cond-mat/0002075
[4] Later to become the Acceleration Studies Foundation (http://www.accelerating.org)
[5] John Smart, in a 2001 article posted on KurzweilAI.net (Hhttp://www.kurzweilai.net/bios/bio0035.htmlH for more about Smart)
[6]Damien Broderick says that, so far as he knows, the first person to systematically use the word Transcension in a transhumanist context was Anders Sandberg (http://www.nada.kth.se/~asa)
[7] One could argue that the agricultural and industrial revolutions represent other Singularity situations, albeit weak ones compared to that which may be upon us next.
[8] http://www.supercompilers.com
[9] http://www.lexiquest.com
[10] Hhttp://pespmc1.vub.ac.be/H
[11] A Superhuman AI is any AI which is capable of general cognition greater than that of the current state of human intelligence (as opposed to the potential state of augmented human intelligence).
[12] “Chaotically dynamical Hypereconomy” is a term coined by the late transhumanist theorist Alexander Chislenko

CHAPTER 2

NEURAL NETS AND ARTIFICIAL BRAINS

Although our primary backgrounds are in math and computing, we'll touch on a number of biological topics in these pages. We've been thinking about biology a great deal these days—Ben in particular, due to his job as CEO and Chief Science Officer of a bioinformatics software company, Biomind LLC—and we'll venture that a lot more computer scientists should be doing the same. Biology both provides unparalleled inspiration for the construction of innovative computer systems, and, increasingly, is a primary application of computer technology. Over the course of the 21st century, the bio- and computer revolutions are going to fuse together, yielding a new kind of biodigital science—the outlines of which we can barely grasp today.

We happened into the biology domain pretty much by accident. When Webmind Inc. folded, the successor firm Novamente LLC had to choose some suitable application domain which would benefit from advanced AI software. We were burned out on the domains our Webmind Inc. products had dealt with previously: financial prediction and document management. We wanted to do something new and exciting, where we hoped there was plenty of money to be made by applying advanced AI technology. After a month or so of brainstorming and vacillating, our latent interest in the complex self-organizing properties of biological systems surged up, and it occurred to us that perhaps the right direction for us to go in was the application of AI to biology.

Biology had never been our primary area of interest. Deep futurists since our sci-fi-loving youth, we had long been convinced that humans were virtually obsolete, in the sense that the title of "most intelligent creature on Earth" was not much longer to be held by the human race or by any biological organism. Moore's Law states that the processing speed of computers doubles roughly every 18 months—an empirical "law" that has held up remarkably well over decades. If it holds up for a couple more decades, we will have computers vastly exceeding the human brain in power, according to nearly all theories of brain function at any rate. Even before the PC appeared on the scene and Moore's Law became common knowledge, it seemed clear to us that, given the exponentially increasing quality of technology on the whole, it was only a matter of decades before machines would somehow overtake us. This intuition is more common than it was during our childhoods in the '70s and '80s, though even now it's far from a mainstream view. It was because of this feeling that, when we became scientists and engineers, we chose artificial intelligence as one of our main topics of research. We figured that intelligent computers were going to outpace human beings during our lifetimes, and it was best to be on their good side when that happened.

The idea of somehow transforming ourselves into an AI—perhaps by uploading brain-states into a computer program; perhaps by gradually grafting new components onto our biological bodies until they are wholly replaced—appealed to us tremendously as an alternative to the unwelcome prospect of death. (Stories of reincarnation or an afterlife never struck either of us as convincing—why put faith in the completely unknowable supernatural realm when you can instead believe in the merely preposterous world of sci-fi technologies?) Of course, we realized that making uploading work would require massive advances in AI and computer technology, as well as in brain science. Having little patience for the tremendous number of seemingly unrelated details one had to memorize to become an expert in brain science—and feeling we were a long way from having good enough experimental tools to really understand how the brain works—we decided to pursue the AI side of things more fully (though we both kept a keen interest in human cognitive science and philosophy alive).

As impressive as Moore's Law is, recent advances in biological science have been at least as exciting. The biology we studied in college (through reading on our own; we never took any bio classes) is mostly still valid, but it's no longer at the heart of modern bioscience research. Advances in genetics have been amazing and tremendous: we've sequenced the human genome, and can now study the dynamic activity of genes and proteins as they build up cells (Walter, 2002). The same can be said of neuroscience, with PET and fMRI scans allowing us to watch the movement of energy through the brain as it thinks (Cabeza and Kingspan, 2001). Of course, there are still limitations, which we'll discuss in the following pages; but the rate of progress is every bit as explosive as in computing, and is due in substantial part to advances in computing and robotic technology. Analysis of gene and protein dynamics, or blood flow in the brain, wouldn't be possible without fast computers and sophisticated software to crunch the data.

We still believe that Real AI is going to revolutionize everything, but now we can see that it's not going to be the only revolution of early 21st century science. As computers gain more and more intelligence, genetic engineering and gene therapy will allow us to improve boring old human beings far beyond our current condition. Freed from psychological pain and emotional disability by advanced neuroactive pharmaceuticals, we will achieve heights of experience that visit us all too rarely today. With the ability to jack our sensorimotor cortices directly into computer-generated realities, we will have access to simulated realities: direct control over robotic extensions of ourselves which allow us to explore dangerous areas without mortal peril; new ways to communicate with each other; and probably much else we haven't even thought of yet, with consequences far more profound than those of chat or e-mail. Genetic and biomechanical engineering may be used to make this human-computer synthesis more effective, and with this synthesis we may create even better genetic and biomechanical engineering inventions which then feed back to further improve the synthesis, and so on.

AI is (in our view) likely to vastly transcend human intelligence, and enter into new domains of mind and reality that humans can't even conceive. The path by which we get to this point will probably involve the development of a lot of other interesting technologies along with way, including plenty of biotechnology. Quite possibly, some of the fantastic things superhuman AIs do will involve enhancing the biological existence of those humans who prefer to remain biologically human even when other possibilities exist.

There is nothing new about these grandiose visions for biotech, any more than there is about these grandiose visions for AI. In early 2001, what fascinated us upon studying biotechnology with a view toward applying our Novamente AI system to it was the specific concrete progress bioscientists were making toward making these visions reality. The main focus of this book is the interface between today and tomorrow. The bioscience of today does not liberate us from psychological pain or physical death, nor does it allow us to upload our brains into computers or jack into simulated realities—but it is pushing in that direction, surely and not all that slowly.

The two main bioscience topics we'll touch on in this book are neuroscience (in this chapter) and genetics (later on). These are both areas to which we've been led by our own research. Neuroscience, because anyone seeking to build an AGI would be foolish not to learn everything possible about the only generally intelligent system known to man: man's own brain. And genetics, because after careful study, we and the Novamente team chose to focus our initial AI application efforts on the analysis of "gene expression data" (information about how much of a gene is being produced in a cell at a given time), a kind of data produced by new biological tools called "microarrays" that are very resistant to analysis by traditional computer programs.

THE THREE-POUND UNIVERSE

The first time one sees a human brain in a medical laboratory, one is overwhelmed by a very strange feeling. How, one wonders, mouth agape, can this three-pound gray mass of nerve cells possibly lead to *so much*? How can all the social and psychological and cultural complexity we see around us, and feel within us, come out of this little lump of meat? A lump of meat not so different from the similar lumps of meat found in the heads of apes or monkeys; or, for that matter, sheep or cows.

Understanding the brain is an important part of the human race's quest for self-comprehension. The contradictions and dilemmas and great moral questions that follow us from one culture to the next—aggression, sexuality, gender differences, freedom and bondage, death and suffering—all have their roots in the patterns by which electricity courses through these three-pound masses in our heads.

But that's not all. The significance of the brain extends beyond even the human condition, for the human brain is, at the moment, our unique example of a highly intelligent system. (Some people think whales and dolphins are highly intelligent, but this remains a speculation.) Stanislaw Lem, in his famous book *Solaris*, hypothesized an intelligent ocean on another planet whose intelligence was revealed to humans through the complexity of its wave patterns and its strange effects on peoples' minds. Even if Solaris's real-world equivalent is out there somewhere, we haven't voyaged there yet. For the moment, the human brain—profoundly flawed as it is—is our only incontrovertible example of an intelligent system. So, if we want to understand the general nature of this mysterious thing called intelligence, and/or create intelligent devices to improve our lives, the brain is the most direct source of inspiration we have.

Neuroscientists try to understand the workings of this amazing three-pound meat hunk one cell at a time, one specialized region at a time. The basic principles seem simple, but the complexities in the details are astounding. As of now, they know a lot about how the brain is structured—which parts do which sorts of things—and they know a lot about cells and chemicals in the brain. But they have remarkably few hard facts about the dynamics of the brain: how the brain changes its state over time, a process that's more colloquially referenced with terms like "thinking" and "feeling."

The main cause of this situation is that there's no way, right now, to monitor the details of what goes on in the brain all day, how it does its business. Brain scanning equipment, PET and fMRI scanners, have come a long way, but they are still far too crude to do much beyond tell us which general regions of the brain a person uses to do which general kinds of thinking. They don't yet let us monitor the course of thinking itself. A few bold neuroscientists have made guesses as to the big picture of how the brain works, but the more timid 99.9% remain focused on their own small corner of the puzzle—all too aware of how much more there is to be understood before general theories about brain function can be systematically verified or discredited.

The difficulty of understanding brain and mind affects areas of inquiry beyond neuroscience. Psychology, artificial intelligence, philosophy of mind, linguistics and other disciplines have run up against equally tough dilemmas as the ones that face neuroscience. Psychologists bend over backwards to create complex experiments that will test even their simplest theories, let alone their subtler, more complicated ones. AI programmers find current computer hardware inadequate to implement their more ambitious designs; and, given both the state of the art of brain science and the fact that AI minds may substantially differ from human ones, they can only attempt to prove their theories in implementation rather than by comparison. Linguists enumerate rule after rule describing human language use, but fail to encompass its full variety. This cross-disciplinary quandary has led to the creation of a combined interdisciplinary field called *cognitive science*, which wraps all the difficulties up into one unmanageable but fascinating package. Many universities now offer degrees in cognitive science: the study of the baffling phenomenon of brain-mind from every aspect. Cognitive science attempts to piece together the diverse fragments of evidence from all these disciplines, to try to arrive at a holistic understanding of this most baffling three-pound meat hunk that lies at the very center of our selves, and holds the key to untold future technologies.

Not surprisingly, cognitive science is itself a rather diverse endeavor, encompassing a number of different approaches to understanding the mind-brain.

Each approach has its own merits—however, among the most fascinating conceptual frameworks under the cognitive science umbrella is the field of *artificial neural networks*. Scientists and engineers working in this area are creating computer programs that, in some sense, work like the brain does—not just on the overall conceptual level of being intelligent, but by simulating the dynamics of interaction between brain cells. So far, none of these neural net programs is anywhere near as intelligent as the brain. Yet some of them have shed light on various aspects of brain function (memory, learning, disorders like dyslexia); and others are serving very useful functions in the world right now, embedded in other software programs which do everything from translating handwriting into typescript, to filtering out porn on the Internet, to helping diagnose problems in auto engines.

There is a lot to be learned from these neural network programs' successes—and from their limitations. We can see just how much, and how little, of what makes the human brain so powerful and wonderful can be captured in simplified, small-scale models of its underlying low-level mechanism. Neural networks model the brain as being somewhat similar to more traditional computer algorithms and networks, and in this sense they form an excellent *foundation* for work bringing together brain and computer, human and digital intelligence.

RUNNING ON ELECTRIC

The brain is made of cells called neurons. There are many other cells there too—glia, for example, which fill up the space between the neurons, and according to recent evidence, carry out some complex communication on their own. Still, all the evidence thus far—dating back to Santiago Ramon y Cajal at the end of the 19th century—shows that neurons are the most important ones. Neurons in groups do amazing "mindlike" things. Neurons individually, however, are complex dynamical systems which are merely like other cells, rather than exhibiting any mind-like behaviors of their own. The coupling of the individual dynamical systems found in neurons to form the whole-brain dynamical system seems to encompass a lot of the mystery of how mind emerges from matter.

A neuron is a nerve cell. Nerve cells in your skin send information to the brain about what you're touching; nerve cells in your eye send information to the brain about what you're seeing. Nerve cells in the brain send information to the brain about what the brain is doing. In addition to monitoring all your perceptions and controlling all your biomechanical functions, the brain monitors itself—that's what makes it such a complex and useful organ!

How the neuron works is by storing and transmitting electricity. We're all running on electric! To get a sense of this, look at how electroshock therapy affects the brain, or look at people who have been struck by lightning. One man who was struck by lightning was never again able to feel the slightest bit cold. He'd go outside in his underwear on a freezing, snowy winter day, and it wouldn't bother him one bit. The incredible jolt of electricity had done something weird to the part of his nervous system that experienced cold.

The neuron can be envisioned as an odd sort of electrical machine, which takes charge in through certain "input connections" and puts charge out through its "output wire." Some of the wires give positive charge: these are "excitatory" connections. Some give negative charge: these are "inhibitory." The trick is that until enough charge has built up in the neuron, it doesn't fire at all. When this "threshold" value of charge is reached, all of a sudden it triggers. From a low-level, mechanistic view, this "threshold dynamic" is the basis of the incredible complexity which happens in the brain. It is the dynamics between neurons, passing charges around to each other, from which intelligence emerges in our brain.

Of course, the connections between neurons aren't really as simple as electrical wires—that's just a metaphor. In reality, each inter-neuron connection is mediated by a certain chemical called a neurotransmitter. There are hundreds of types of neurotransmitters. When we take drugs, the neurotransmitters in our brain are affected, and neurons may fire when they otherwise wouldn't, or be suppressed from firing when they otherwise would. Still, the electrical storage of information in our brains—though based on the fluid electrochemistry of organic molecules, rather than on the physical properties of crystals and metals in solid-state electronics—is not entirely dissimilar from how a computer stores its atomic bits of information.

The near-consensus among neuroscientists is that most learning takes place via modification of the connections between neurons. The idea is simple in essence: not all connections between neurons conduct charge with equal facility. Some let more charge through than others; they have a higher "conductance." If these conductances can be modified even a little, then the behavior of the overall neural network can be modified drastically. This leads to a picture of the brain as being full of "self-supporting circuits;" circuits that reverberate and reverberate, keeping themselves going. This concept, now called "Hebbian learning," goes back to Donald Hebb in the 1940's, and it's held up since then, throughout all the advances in neuroscience.

Christof Koch, a well-known neuroscientist, believes there is some fairly subtle electrochemical information-processing going on inside each neuron, and in the extracellular space between neurons. If this is so, no one knows exactly what role it has in intelligence. The network of glia, it appears, may carry out information-processing functions, passing information from one part of the neural network to another, perhaps guiding the process of Hebbian synaptic modification. This would also mean that we cannot pragmatically achieve AI through neural network models alone, at least not without building an infeasibly large neural network whose dynamics have no resemblance to those of biological neural networks—but the idea of a regulatory network atop neural networks being necessary is one which some people coming from an AI perspective have already reached. Even if this is the case, neurons still play a central role in human intelligence, and a lot can be learned from understanding their dynamics.

A few mavericks go even further, and argue that the neuron itself is a complex molecular computer. Anesthesiologist Stuart Hameroff has suggested that the essence of intelligence lies in the dynamics of the neural cytoskeleton: in the molecular biology in the walls of the neurons. Roger Penrose, the famous British mathematician, has taken this one step further, arguing that one needs a fancy theory of quantum gravity to explain what's going on in the cytoskeletons of neurons, and tying this in with the idea that quantum theory and consciousness are somehow interrelated (Penrose, 1996). There's no real evidence for these theories, but it's an indicator of how little we understand about the brain that these wild, outlier theories can't be convincingly or empirically refuted at this stage.

KNOWLEDGE FROM NEURONS

Neurons are the stuff of the brain; but they live at a much lower level than thoughts, feelings, experiences or knowledge. *Above* the neuronal level of interesting mind-stuff, on the other hand, we have the various regions of the brain, with their various specialized purposes (visual cortex, sensorimotor cortex, etc.). In between is most the interesting part—and the least understood.

One hears a lot about left brain versus right brain, but the really critical distinction is forebrain versus hindbrain and midbrain. The hindbrain is situated right around the top of the neck. What it mostly does is regulate the heart and lungs, and control the sense of taste. The midbrain, resting on top of the hindbrain, integrates information from the hindbrain and from the ears and eyes; and it appears to play a crucial role in consciousness. Collectively, the hindbrain and midbrain are referred to as the brainstem.

The hindbrain is old—it's basically the same in humans as in reptiles, amphibians, fish, and so forth. The forebrain, on the other hand, is much more complex in mammals than in other animals. It's where our smarts live, and it has its own highly complex and variegated structure. The mammalian forebrain is subdivided into three parts: the hypothalamus, the thalamus and the cerebral cortex. The cerebral cortex itself is divided into several parts, the largest of which are the cerebellum and the cerebrum. The cerebellum has a three-layered structure and serves mainly to evaluate data regarding motor functions. The cerebrum is the seat of intelligence—the integrative center in which complex thinking, perceiving and planning functions occur. We have only a very vague idea of how it works.

All this is just the coarsest view of high-level brain structure. With PET and fMRI brain scan equipment, scientists can go one level deeper. By getting a rough picture of which parts of the brain are getting more attention at which times, they can see which parts of the brain are involved in which activities. For example, attention seems to involve three different systems: one for vigilance or alertness, one for a sort of executive attention control, and one for disengaging attention from whatever it was focused on before. Three totally different systems, all coming together to let us be attentive to something in front of us. A lot of what we know about brain structure has come from studying dysfunction. Depending on which of the three systems is damaged, you get different kinds of awareness deficits and different neurological disorders; and from this data scientists have reverse-engineered the information about which brain structures are involved in attention (and similarly with other functions), so that we have a better idea of what we're looking at when we do scans to understand the next-lower-level of dynamics.

All these parts of the brain are made of neurons, passing electricity around via neurotransmitters in reverberating cycles and complex patterns. Still, the different kinds of neurons and neurotransmitters in the different parts of the brain, and the different patterns of arrangement and connectivity of the neurons, obviously make a huge practical difference.

Atoms and molecules are a reasonable metaphor here. All brain regions are made of neurons, just as all molecules are made of atoms. Different atoms have very different properties, as do various types of neurons. Different molecules, formed from diverse types of atoms, can do very different things—just as individual brain regions are specialized for different functions. Only the very simplest molecules can, in practice, be analyzed in terms of atomic physics—otherwise we have to use the crude heuristics of theoretical chemistry. Complex molecules like proteins can barely even be studied with computer simulations; the physical chemistry is just too tricky. Similarly, so far only the very simplest brain regions and functions can be understood in terms of neurons and their interactions. Complex brain functions can't even be understood via computer simulations yet. There are too many neurons, and the interconnectivity patterns and ensuing dynamics are just too tricky.

BIOLOGISTS AND ENGINEERS TURN TO ARTIFICIAL NEURAL NETWORKS

The model of the brain as a network of neurons, passing charges amongst one another, is a crude approximation of the teeming complexity of the actual physical brain. Still, it has a number of advantages: simplicity and comprehensibility, for instance. Just as critical is its amenability to mathematical analysis and computer simulation. In practice, modeling all the molecules in the brain as a set of equations is an impossible task. It might be necessary for a real understanding of brain function, but we certainly hope not. Even modeling the chemical dynamics inside a single neuron is a huge exercise, to which some excellent scientists devote their entire lives, and which tells us extremely little about the functioning of the brain as a whole and the emergence of intelligence from matter. On the other hand, if one is willing to view the brain as a neural network, then one can construct mathematical and computational models of the brain quite easily. It's true that the brain contains roughly a hundred billion neurons—and that no mathematician can write down that many equations, and very few computers can simulate a system that large with reasonable efficiency. However, at least one can simulate small neural networks, to get some kind of feel for how brain dynamics work. As computers continue to get more powerful, larger and larger simulations can be run, and if it turns out we need glial network simulations interwoven with those of neurons, we could similarly do small simulations now and larger ones later (if we actually understand what role, if any, the glia play in the system).

This is the *raison d'etre* of the "neural networks" field, which was launched by the pioneering work of cyberneticists Warren McCullough and Walter Pitts in the early 1940s (McCullough and Pitts, 1943). What McCullough and Pitts did in their first research paper on neural networks was to prove that a simple neural network model of the brain could do anything—could serve as a "universal computer."[1] At this stage computers were still basically an idea; but Alan Turing and others had worked out the theory of how computers could work, and McCullough and Pitts' work—right there at the beginning—linked these ideas with brain science.

Artificial neural nets have come a long way since McCullough and Pitts. Yet even after all these years, the main thing this kind of work achieves is to expose our ignorance.

For instance, to decide which artificial neurons should connect to which in an artificial brain-simulator program requires detailed biological knowledge that is hard to come by—and much of it still hasn't been uncovered. Deciding what types of neurons to have in one's toy neural net, requires similarly elusive knowledge, if simulating the brain accurately is really one's goal. A few neural net specialists focus on acquiring and utilizing this data; Stephen Grossberg of Boston University is a leading example.

Gerald Edelman, who won a Nobel for his work in immunology, has taken a different approach. He is trying to abstract beyond these difficult issues while remaining within the computational brain modeling domain, by modeling networks of *neuronal groups* rather than networks of individual neurons (Edelman, 1987). In this work, he has proposed some detailed and fascinating theories about the connectivity patterns of neuronal groups and the dynamics of neural group networks, and has embodied these theories in computer simulations of vision processing. Edelman doesn't consider himself a neural net theorist per se, but the basic line of thinking and mathematical modeling of his neuronal nets is not far off.

We have done some research in this area ourselves: tying in the "neuronal group" idea with the notion of Hebbian learning mentioned above, in which synaptic conductances modify themselves adaptively based on the patterns of electricity flowing through them. Software, conceptually based on Hebbian learning and higher-level group structures, was attempted at our company Webmind as part of its core AI engine. More recently, Ben wrote a paper called "Hebbian Logic" describing how artificial neural networks embodying a specific variant type of Hebbian learning can give rise to *logical reasoning behavior* on an emergent level. Suppose one posits that concepts like "cat" and "dog" and "beauty," as well as procedures like "pick my nose" or "calm my child down," are represented by networks of neuronal groups. The paper shows that, if one assumes that learning is carried out by the right variant type of Hebbian learning, then the dynamics of interaction between networks of neuronal groups will *automatically* follow the rules of logical inference to within a decent degree of approximation. The details are subtle, but the overall message is simple to understand. Surely, and not all that slowly—one mathematics paper and neurophysiological experiment at a time—we are building a bridge from brain to mind.

All this theory is interesting, but what are these artificial neural networks supposed to *do*? While the brain is connected to a body, most artificial neural networks are not attached to any real sensors or actuators. There are exceptions, such as neural nets used in vision processing or robotics. But other than these cases, neural network programs must be carefully fed data, and carefully studied to decode their dynamical patterns into meaningful results in the given application context. It often turns out that figuring out how to represent data to feed to a neural network program requires more intelligence than the neural network program itself embodies.

In practice, the biological-modeling-motivated research carried out by neural net researchers like Grossberg and Edelman is fairly separate from more pragmatic, engineering-oriented neural network research. Most engineering applications of neural nets aren't based on serious brain modeling at all; rather, the brain's neural network is taken as general conceptual inspiration for the creation of an artificial neural network in software, and then computationally efficient learning algorithms are applied to this artificial neural network. It seems likely that the standard learning algorithms for artificial neural nets are actually *more* efficient at learning than the brain's learning algorithms, in the context of very small neural networks, say hundreds to thousands of neurons. On the other hand, they don't scale up well to billions of neurons. The brain's learning algorithms, on the other hand, work badly in the small scale but remarkably well in the large. This holds as well for the special "logic-friendly" Hebbian learning rules mentioned above. They cause neural nets to give rise to emergent logical behavior—but only when the neural nets are really big, minimally hundreds of thousands of neurons, and preferably millions or hundreds of millions.

The brain has at least a hundred billion neurons; but for current practical applications of artificial neural nets, we're usually talking hundreds to tens of thousands of neurons. This limitation is hard to get around, because it's imposed by the inefficiency of implementing neural networks on contemporary computers, which can only do one thing at a time (e.g. let one simulated neuron fire a simulated charge at a time)—unlike the brain, whose distributed physical embodiment allows all neurons to chemically and electrically act and interact at all times. The solution is to use supercomputers or distributed Internet-based computing, but neither of these is economically practical for most real-world neural net applications today.

Artificial Development, Inc.'s "CCortex: Human Neural Network Simulation"[2] is one research project that is attempting to build a massively distributed neural network on a network of computers. It'll be interesting to see how their work—funded as a "pet project" by an Internet-boom multimillionaire (Marco Guillen) who made his money on simple Web technologies—turns out.

As an example of more pedestrian (but useful!) neural net applications, Rulespace[3] uses a neural network program to recognize Internet pornography. America Online uses it to filter out porn for customers who request this feature. Granted, the neural net inside their program can't actually read text. Rather, there's other code that recognizes keywords in Web pages, and then uses the presence or absence of each keyword in a Web page to control whether a certain neuron in a neural network gets activated (given simulated electricity) or not. The pornographicness, or not, of the Web page is then determined by whether a special neuron (the "output neuron") is active or not, after activation has been given a while to spread through the network.

In an application like Rulespace, a neural network is being used as a simple mathematical widget. Simulation of the brain in any serious sense is not even being attempted. The connectivity pattern of the neurons, and their inputs and outputs, are engineered to yield intelligent performance on the particular problem the neural net was designed to solve. This is pretty typical in the neural net engineering world.

In the real world, neural net engineering becomes quite complex. For instance, to get optimal performance for optical character recognition (OCR), researchers have constructed modular nets with numerous subnetworks. Each subnetwork learns something very specific, and then the subnetworks are linked together into an overall meta-network. One can train a single network to recognize a given feature of a character—a descender, or an ascender coupled with a great deal of whitespace, or a collection of letters with little whitespace and no ascenders or descenders. But it is hard to train a single network to do several different things—say, to recognize letters with ascenders only, letters with descenders only, letters with both ascenders and descenders, and letters with neither. Thus, instead of building one large network, it pays to break things up into a collection of smaller networks in a hierarchical architecture. If the network learned how to break itself up into smaller pieces, one would have a very impressive system; but currently this is not the case, as the subnets are carefully engineered by humans. (For a general review of neural nets used for OCR and other forms of visual pattern recognition, see Egmont-Peterson et al., 2002.)

Some experiments with fairly simple neural nets, such as the ones used in these practical applications, have had fascinating parallels in human psychology. In the mid-1980s, for instance, Rumelhart and McLelland created uproar with simple neural networks that learned to conjugate verbs. They showed that the networks, in the early stages of learning, made the same kinds of errors as small children. Other people, a few years later, trained neural nets to translate written words into sounds. They showed that if they destroyed some of the neurons and connections in the network, they obtained a dyslexic neural net. This is the same thing that happens if you lesion someone's brain in the right area. You can do the same sort of thing for epilepsy: by twiddling the parameters of a neural network, you can get it to have an epileptic seizure. Of course, none of this proves that these neural nets are good brain models in any detailed sense, but does show that they belong to a general class of "brain-like systems."

HUGO DE GARIS AND THE ARTIFICIAL BRAIN

Perhaps the most fascinating neural net project around is Hugo de Garis's Artificial Brain project: originally conducted at ATR in Japan, then pursued for a while at Starlab in Brussels, and now, following Starlab's bankruptcy, occupying Hugo's non-teaching hours as he serves as a professor at Utah State University, Logan. USU is not MIT, but the fact that de Garis wound up there is no big shock—since when have the most radical, exciting, maverick scientific approaches originated in the most established institutions? Sometimes they do; but as often as not, radical insights come from way outside the establishment where thinking is freer and funding scarcer.

The Artificial Brain Project (de Garis, 2002) is de Garis's attempt to create a hardware platform for advanced artificial intelligence, using special chips called Field-Programmable Gate Arrays to implement neural networks. As de Garis says, his CAM-Brain Machine (CBM)[4] platform "will allow the assemblage of 10,000s of evolved neural net modules. Humanity can then truly start building artificial brains. I hope to put 10,000 modules into a kitten robot brain..." The CBM is listed in the Guinness Book of World Records as the "World's Largest Artificial Brain." This research remains unfinished, but it will be interesting to see where it leads.

This is mad scientist type stuff, and de Garis plays the role quite well. Many have told him he looks the part of a mad scientist. He doesn't mind. Recently, a computer science researcher, a colleague of Ben's, asked him if he knew de Garis. "Yes, a little." The next question: "So is he truly insane or not?" Ben's inevitable answer: "Why are you asking *me*? Because you think it takes one crazy man to recognize another?"

Indeed, in some ways Hugo is a little further out even than we are, and we're fairly good mad scientists ourselves. But the future will be pretty far out compared to the present, so one wouldn't expect the most average everyday people to have the best futuristic insights—let alone the deepest technical ideas. About the great sci-fi writer Olaf Stapledon, something was once written roughly along the lines of: "It's true, Stapledon had a few eccentricities. But then, your ordinary everyday Joe doesn't sit around in his spare time and compose vast poetic novels recounting the entire history of life across the universe."

De Garis is a bit less of an optimist than we are, however. He believes that, sometime during the next century, there's going to be a world war between advocates of intelligent computers and those who want to extinguish them to save humanity. On the other hand, he doesn't see this as a reason to halt his AI work. He reckons this kind of conflict is inevitable, whether or not he works on it, so he's going to put his time into making sure AI is created as responsibly as possible—without being quite sure that "as responsibly as possible" is very responsible at all. This year he wrote a book on the theme, with the impressive title *The Artilect War: Cosmists Vs. Terrans: A Bitter Controversy Concerning Whether Humanity Should Build Godlike Massively Intelligent.*

Born in Australia, aged 54, and with two adult children, de Garis has lived in 6 countries (Australia, England, Holland, Belgium, America, Japan) and now says he feels like a foreigner wherever he goes. Like Turchin, he started off his career as a theoretical physicist, working with quantum pioneer and philosophical maverick David Bohm—but whereas Turchin turned to theoretical computer science, de Garis shifted toward artificial intelligence.

The CBM project was initiated during the 8 years he spent working in the research lab of ATR, a large Japanese telecommunications firm. He left Japan in frustration with its culture, with his brain-building project only half done. As he puts it on his website:

> "I lived in Japan…because I felt that country offered me the best opportunities to achieve my long term dream of building artificial brains…However, Japan's suppression of big-egoed individualism, its utter intolerance of Western criticisms of its third-world social, political and intellectual values, simply enraged me. I had to leave its intellectual sterility to have a life of the mind…It's a culture that is first-world materially, but third-world socially, and is quite unsuited for the vast majority of Westerners to live in—too socially backward, too insular, too uncosmopolitan, too closed, too racist, too chauvinist, too passively obedient, too feudal, too fascist for Westerners to tolerate. Japan needs two generations of heavy social engineering to catch up with the West socially."

Whoa! Our Japanese colleague Takuo Henmi's comment on reading this was: "There's some truth to these things, but he pushes it way too far." Ben also noticed, when he last saw Hugo in person, that he was very much in love with his Japanese girlfriend...who didn't seem to mind his own special brand of "big-egoed individualism" at all.

In ATR's Human Information Processing lab, run by Katsunori Shimohara (a first-rate techno-visionary in his own right), de Garis developed the theory and designed the details of the CBM and then contracted out the actual engineering of the machine to some American hardware experts (Genobyte, Inc.[5]). It's a device of incredible power. Unlike the general-purpose computers we use every day, it's built specifically to do one thing. It combines genetic algorithms, a computational simulation of evolution by natural selection, with neural networks, a computational simulation of the brain. It can run a huge number of little neural networks. It can pass information between these neural networks, and create new neural networks by evolving them according to specified "fitness criteria." If one wants a little neural net to solve a certain problem, one casts this problem as a "fitness criterion" described in the mathematical language of the GA, gives the problem to the CBM, and the CBM makes a neural net that solves your problem. The idea is that if one hooks together a lot of little neural nets solving relevant, interrelated problems, then one has a brain-mind.

This is one lean, mean, evolving-neural-nets-with-a-genetic-algorithm machine. It performs as fast as 1,000 Pentium IV computers would, if they were turned to this particular task. So far, 4 of them have been built. The current pricetag is $400,000—far cheaper than 1,000 Pentium IV computers plus all the supporting hardware needed to cluster them, offering hope that clever large-scale hardware designs might become both price- and performance-competitive with networks of workstations in certain domains. Only a few more can be built using the exact current design, because one of the components (a special Xilinx field-programmable gate array) is not being produced by the manufacturer anymore. But other, similar components could be substituted if more customers were found.

The main limitation of the system seems to be the artificial way that you have to set up the "fitness criterion" in order to have the system evolve a neural net for you. You have to specify exactly what outputs the neural net is supposed to provide, when given various inputs. The CBM then uses simulated natural selection to "evolve" a neural net that produces the specified output given the specified inputs. It maintains a whole population of neural nets; evaluates how suitable the input/output behavior of each one is; and then takes the best ones, mutates them and combines them with each other to get a new population of neural nets which are evaluated all over again. This "genetic algorithm" methodology is great when it's applicable, but not all learning problems are easily cast in this artificial way. For instance, learning how to interact with other minds isn't about producing the exact right output for each given input, there's a lot more subtlety to it. One suspects that when the time finally comes to integrate the CBM with a fully-featured AI system, with long-term memory, perception, action and the whole kit-and-kaboodle, some substantial modifications will be required (or, it will be a specialized "cortex" of an AI brain which serves a special-purpose rather than general-purpose function). But even as it is, the CBM is surely a huge boon to AI research, and a powerful reminder of the academic ossification of the mainstream AI community. If one maverick researcher can get this amazing AI hardware created, imagine what could be done with a concerted effort by the governments, universities and corporations of the world to get real AI working.

De Garis himself understands that the CBM has some fairly serious limitations, but he reckons it's already pushing the limits of what can be done with current science and technology. "Real AI," he says,

> "is still many decades away. We still haven't a clue how the brain works. What is an idea? How is memory stored?...I think humanity will have to wait until we can 'scan' the brain, which is probably a decade or two away. Then we can store the scanned results and analyze them with massively parallel computers that future technology will give us. Circuits keep doubling their speed and densities every year or two, so 20 years from now our circuits will be on a molecular scale, with trillions of trillions of them.

> Then I think humanity will be able to tackle real AI. Our present tools are too primitive."

One of his goals, in the ATR/Starlab period, was to create a robot kitten, Robokoneko, with a billion artificial neurons. This project was being pursued, for quite some time, together with Dr. Michael Korkin of Genobyte, the one who actually built the CBM hardware to de Garis' specifications. Building the kitten was envisioned as a tremendous learning experience, one step on the path to creating an artificial human brain. Unfortunately, this project was slowed down dramatically by some financial difficulties related to the bankruptcy of Starlab and the details of its financial relationship with Genobyte. At the moment, de Garis' plan is to start over with a CBM-2 architecture—faster and better than the CBM-1—and pursue his robotic visions with the CBM-2, thus sidestepping the CBM-1's business issues while achieving yet greater AI power.

We regret that we have not yet seen the CBM run, although Ben did see the machine itself. When he visited de Garis at Starlab in summer 2001, the machine was locked in a room at one end of the huge and beautiful Starlab building—a palatial construct that once was the Czech embassy to Belgium, situated in the semi-rural outskirts of Brussels. Starlab was rather surreal in appearance due to the fact that it had closed its doors two weeks earlier, and the only person still using the building was de Garis, one of a handful of scientists who had not only worked in the building but lived there in an apartment directly attached to the research labs. It was an empty palace, full of half-cleaned-out desks and strange machinery like the CBM. When Ben returned to Hugo's apartment late at night and found he wasn't home yet, he became very familiar with the whole building as he searched for a way to sneak in. A basement door that was unlocked gave entry to the building, and then a butter knife applied to the lock on de Garis' apartment door got him into the apartment and safely into bed. We had a good laugh about the excellent security being used to guard the incredibly valuable equipment in the building.

Starlab seems to have been an amazing place for the few years it lasted; we wish we'd had the chance to work there ourselves. At Webmind Inc.—our company in the late 90's, which closed its doors just a few months before Starlab did—we were constantly struggling to balance our long-term AI R&D goals with short-term product development goals, and would sometimes laugh jealously about Starlab's motto: "Where 100 years means nothing." "100 years," we'd say; "Damn. All we need is 5 more years of funding to finish our thinking machine. I wish we had enough money to let us focus on nothing but that for the next five years...let alone 100." Well, Starlab didn't last 100 years, any more than Webmind did—2001 was definitely not the year for radically innovative research. De Garis, in person, gives an impression of tremendous intensity and intelligence. He has the ambition to see what has to be done on a grand scale, and then set about following a complex long-term plan aimed at achieving his goals. On the other hand, he's also not afraid of confronting and even embodying the contradictoriness of reality. It doesn't worry him particularly to push hard toward creating real AI, while at the same time popularizing the dangers of this question, and the possibility that it may indirectly lead to mass destruction.

De Garis's inner contradictions, it would seem, are on the extreme side even for wild-eyed techno-pioneers. For instance, Bill Joy, the former Chief Scientist of Sun Microsystems, and Jaron Lanier, the virtual reality pioneer, have come to the media recently with strong anti-technology statements, but in spite of this they continue to pursue high-technology work. However, Joy and Lanier are working on particular pieces of technology that are only indirectly related to the technologies they're warning us about. They're warning us about AI and nanotechnology and genetic engineering, and then working on Internet-distributed computing and computer vision. Lanier told us openly, when we met him during the Webmind days: "I'm your ideological arch-enemy." He believes AI is impossible, and even if possible probably dangerous; yet he works on computer vision research, modeling the brain's perceptual algorithms in software that does computer graphics tricks with 3-D faces. But compared to the mild conflicts between belief and action presented by people like Joy and Lanier, de Garis' life would seem to pose a far more acute paradox. De Garis is working directly on building brains, and then telling us that brain building may destroy the world.

Quite simply, strikingly, and seriously, he predicts a late 21st century world war between two human groups, whom he terms the Cosmists and the Terrans. The Cosmists will be in favor of creating "artilects": superhuman artificial brain-minds, the next phase in the evolution of intelligence. The Terrans will be radically opposed to this kind of technology development, and willing to kill billions in order to prevent the advent of artilects—because, after all, the artilects will have the power to destroy humanity altogether.

De Garis is well aware of the contradictory nature of his roles as artificial brain builder and visionary pessimist. "I feel I am part of the problem," he says, "the problem being, 'Who or what should be dominant species in the 21st century?'…I am helping to pioneer this brain-building field, so I feel a strong moral obligation to stimulate discussion on this enormous question. It is for this reason that I try to 'raise the alarm' in the world media, by making the general public conscious that next century's global politics will be dominated by the 'Artilect Question,' i.e. do we allow the 'artilects' (artificial intellects) to take over, or not."

Crazy? Certainly not. Out of the ordinary? Well, your average ordinary Joe doesn't go around creating artificial brain machines, now does he? Even if Robokoneko never comes to fruition because of funding problems, de Garis' work has advanced our understanding of the brain-building problem considerably. He has shown us what can be done with highly specialized hardware, oriented specifically toward one key aspect of computational intelligence. And we're sure he will teach us much more in years to come.

As for his conflict, maybe being "part of the problem" is the only way to be part of the solution. As many of the atomic scientists realized with respects to atomic fission, once someone has figured out that something *can* possibly be created, someone *will* create it, so if you have humanity's best interests at heart you very well ought to be involved in creating it for the best possible purposes. Many atomic scientists also recognized the dangers of atomic weaponry and worked diligently to persuade governments to use fission for power, not weaponry. Rather than judging technological progress out of fear and ignorance, we'd be far better off to learn as much about these things as we can and make decisions for ourselves rather than trying to hide or criminalize them; in which case only the shadowy elements of our society will be involved in determining how they come to fruition. Let's hope that we've learned the lessons of the atomic scientists and the horrors of Hiroshima and Nagasaki, and that AI research is used for peaceful, positive purposes.

Personally, we find de Garis' political prognostications much less convincing than his scientific work. If there is another world war, which we doubt, we suspect it will more likely be centered on old-fashioned concerns like religion, natural resources and national pride, rather than being focused on artilects in any direct way. Even so, the dilemma that de Garis points out is real and inescapable. This contradiction between AI boosters and AI detractors is going to be a huge part of the human dialogue over the next century—though probably in a more complex manner than de Garis envisions—mixed up with the whole mess of other more familiar human issues and conflicts. We'll return to these issues in the last few chapters of the book.

In the end, even if one doesn't agree with all his theories and predictions, one has to admire the man for his courage to confront large scientific and moral issues directly; instead of hiding in a little tiny corner of the world, working on narrowly-defined research problems, and letting the big issues evolve of their own accord. We could use more "mad scientists" like this one.

THE MYSTERY OF NEURODYNAMICS

The essence of the brain lies not in what it is at any particular time, but rather in what it is becoming on an ongoing basis: how it learns, adapts, and responds to chance, in short, its *dynamics*. de Garis has realized this well, and the essence of his CBM lies in how it uses genetic algorithms to evolve neural networks carrying out useful functions. Furthermore, his genetic algorithm doesn't even create fully featured neural nets; it creates "initial conditions," baby neural nets that have to evolve and grow into useful neural network structures.

On the other hand, most of the artificial neural networks used in practical applications these days are pretty simple dynamically. In order to guarantee reliable functionality in their particular domains, they're restricted to very limited behavior regimes. It's easy to predict what they'll do overall, although the details of the activation spreading inside them may be wildly fluctuating. Neural nets constructed for biological modeling are usually allowed a freer rein to evolve and grow, but these are rarely the focus of mainstream AI research. In general, the potential for really complex and subtle dynamics in neural networks has hardly been explored at all.

This is a shame, because the "threshold" behavior of a neuron conceals the potential for immense dynamical complexity of a very psychologically relevant nature. Think about it: let's say a neuron holds almost enough charge to meet the threshold requirement, but not quite. Then, a chance fluctuation increases its charge just a little bit. Its total store of charge will be pushed over the threshold, and it will fire. A tiny change in input leads to a tremendous change in output—the hallmark of chaos. But this is just the beginning. This neuron, which a tiny fluctuation has caused to fire, is going to send its input to other neurons. Maybe some of these are also near the threshold, in which case this extra input will trigger their firing. And these new impulses may set off yet other neurons, and so on. Eventually, some of these indirectly-triggered neurons may feed back to the original neuron, setting it off yet again, and starting the whole cycle from the beginning. The whole network can be set alive by the smallest fluke of chance! In this way, you get chaos and complexity out of simple formal networks.

One biological researcher who has really grasped this aspect of neural nets is Walter Freeman, who has shown in his work with the olfactory part of the brain that real neural networks display the same complex and crazy dynamics as artificial neural nets; and that this complex dynamics is critical to how brains solve the problem of identifying smells. If we use complex, near-chaotic dynamics to do something as low-level as identifying smells, it's hardly likely that the neural nets in our frontal lobes use simple dynamics like those embodied in current neural-network-based software products. We have a long way to go before our toy neural net model catches up with the neural nets in our heads!

NEURAL NETS AND THE QUEST FOR REAL AI

Biologists and pragmatic computer scientists each have their own use for neural network models, their own favorite neural network learning schemes, and so forth. Our own interest in neural nets, on the other hand, has been mainly oriented toward neither brain modeling nor immediate practical engineering applications. Rather, it's been oriented toward real AI—toward the creation of truly intelligent computer programs—programs that, like humans, know who they are and behave autonomously in the world. Viewed in this light, current neural network research comes up rather lacking. Now, this isn't a tremendously shocking conclusion; since, with the exception of de Garis' brain machine, the CCortex, and a couple other frontier projects, researchers are playing around with vastly smaller-scale neural networks than the one that the brain contains. But it's interesting to delve into the precise reasons why current neural net research isn't terribly relevant to the task of artificial mind creation. To understand the role of neural nets in the history of AI, one also has to understand their opposition. When we first started studying AI in the mid-1980s, it seemed that AI researchers were fairly clearly divided into two camps: the neural net camp and the logic-based or rule-based camp. This isn't quite so true anymore, but it's still a reasonable first-order approximation of the situation.

Whereas neural nets try to achieve intelligence by simulating the brain, rule-based models take a totally different approach. They try to simulate the mind's ability to make logical, rational decisions, without asking how the brain does this biologically. They trace back to a century of revolutionary developments in mathematical logic, culminating in the realization of Leibniz's dream: that a complete logical formalization of all human knowledge is actually achievable in principle, although very difficult in practice.

What about Kurt Gödel's Incompleteness Theorem? A system which can formalize all human knowledge need not be able to formalize all knowledge about itself—there's no requirement that humans have complete knowledge of that system's axioms in order for it to exist. Practically, an AI system, or a human, does not need to have a complete axiomatic formalization of something in order to work with it. To formally reason about something one merely needs a model that is complete enough to make reasonably accurate predictions within the formalized domain(s), and from an engineering perspective, to make meaningful changes to the system.

Rule-based AI programs aren't based on self-organizing networks of autonomous elements like neurons, but rather on systems of simple logical rules. Intelligence is reduced to following orders. In spite of some notable successes in areas like medical diagnosis, chess playing and financial analysis, the main thing this approach has taught us is that it's really hard to boil down intelligent behaviors into sets of rules. The rule sets are huge and variegated, and the crux of intelligence becomes the dynamic learning of rules rather than the particular rules themselves. Evidence is building for the idea that a useful understanding of any complex problem domain is inherently an evolutionary, not a declarative, process.

Now, to most any observer not hopelessly caught up on one or another side of the debate, it's obvious that both of these ways of looking at the mind—rules or neural nets—are useful for practical applications, but extremely limited descriptions of only portions of what a mind actually does. True intelligence requires more than following carefully defined rules, and it also requires more than random or rigidly laid-out links between a few thousand artificial neurons.

The attempt at a solution to this problem, embodied in the Novamente software system, bears some resemblance to Gerald Edelman's. The Novamente program is based on entities called "nodes" that are roughly of the same granularity as Edelman's "neuronal groups." Nodes are a bit like neurons—they have a threshold rule in them, and they're connected by "links" that are a bit like synapses (connections between neurons) in the brain. But Novamente's nodes have a lot more information in them than neurons. The links between nodes have more to them than the links in neural net models: they're not just conduits for simulated electricity; they have specific meanings, sometimes similar to the meanings of logical rules in a rule-based AI system. The underlying *network structure* of Novamente is a hybrid of ideas from neural and neuronal network models and logical semantic networks.

The key intuition underlying Edelman's, and our, approaches is to focus on the intermediate level of brain/mind organization: larger than the neuron, smaller than the abstract concept. The idea is to view the brain as a collection of clusters of tens or hundreds of thousands of neurons, each performing individual functions in an integrated way. One module might detect edges of forms in the visual field; another might contribute to the conjugation of verbs. The network of neural modules is a network of primitive mental processes, rather than a network of non-psychological, low-level cells (neurons). The key to brain-mind, in this view, lies in the way the modules are connected to each other, and the way they process information collectively. The brain is more than a network of neurons connected according to simple patterns, and the mind is more than an assemblage of clever algorithms or logical transformation rules. Intelligence is not following prescribed deductive or heuristic rules like IBM's super-rule-based chess player Deep Blue; but neither is intelligence the adaptation of synapses in response to environmental feedback, as in current neural net systems. Intelligence *involves* these things, but at bottom intelligence is something different: the self-organization and mutual intercreation of a network of processes, embodying perception, action, memory and reasoning in a unified way, and guiding an autonomous system in its interactions with a rich, flexible environment.

THE INTERNET AS AN ARTIFICIAL BRAIN

Like his good friend and close collaborator Valentin Turchin, Francis Heylighen started his career as yet another physicist with a craving to understand the foundations of the universe—the physical and philosophical laws that make everything tick. Also like Turchin, Heylighen didn't give up his previous intellectual ambitions when he got the computer bug. Rather, he became convinced that complex, self-organizing computer networks are just as valid and important a way to understand the universe as physics or metaphysics. Since

1982, he's used his research position at the Free University of Brussels to pursue precisely this perspective. In particular, he's focused his thinking on the fascinating and futuristic idea of the *global brain*—the idea that the Internet, as it evolves, will eventually adopt its own unique form of the dynamical and structural complexity, and self-organizing intelligence, displayed by the brain.

In Heylighen's vision, everyday Internet interactions using e-mail, chat, the Web, etc. are themselves glimmerings of the birth of the global brain. We are not yet at the point of the metasystem transition where the Net becomes an autonomous, self-organizing intelligence, but each time we send an e-mail or create or follow a hyperlink, we're getting there. Tim Berners-Lee's Semantic Web project is a less ambitious vision of a similar notion: that the Internet can become a cooperative mindspace for humans and their machine counterparts[6]. The Semantic Web folks don't envision anything as grandiose as a global brain, at least not publicly, but the underlying notion of making an Internet that is friendlier to knowledge exchange is a step in that direction.

In 1989, Heylighen, Valentin Turchin, and Cliff Joslyn founded the Principia Cybernetica Project, aimed at marshalling a group of minds together to pursue the application of cybernetic theory to modern computer systems. In 1993, very shortly after Tim Berners-Lee released the HTML/HTTP software framework and thus created the Web, the Principia Cybernetica website[7] went online. The

Internet, the site claimed boldly, was the ideal medium for the development of the next generation of thinking about life, the universe and everything.

For a while after its 1993 launch, Principia Cybernetica was among the largest and most popular sites on the Web. Today the Web is a whole different kind of place, but Principia Cybernetica remains a rich, sprawling website. It is a unique and popular resource for those seeking deep, radical thinking about the future of technology, mind and society. Eschewing the traditional hierarchical structure of most websites, it is structured more like the "semantic networks" used inside AI programs, with each page linked to the other pages that relate to it in various ways (a nonlinear organization which Berners-Lee originally envisioned for the entire Web, when he created a system in which information hyperlink theory could actually be used widely). It doesn't yet organize itself automatically based on user feedback or AI intuition, but it's actively improved and updated by the numerous humans involved with the organization. The basic philosophy presented is founded on the thought of Turchin and other mid-century systems theorists, who view the world as a complex self-organizing system in which complex control structures spontaneously evolve and emerge.

The site's creation and early development was a collaborative effort on the part of its three creators. Today, though, Turchin spends most of his time working on his own investigations, in computer science and philosophy, and his start-up company. Joslyn is primarily occupied with systems theory, knowledge management and systems biology work inspired by cybernetics. Francis Heylighen, however, remains squarely focused on the Principia Cybernetica vision and all that it entails. He has fleshed out the Internet-brain parallel in some very concrete and interesting ways.

For example, Heylighen and his colleague Johan Bollen have experimented with Web-like systems in which the links between pages are created, destroyed, strengthened or weakened by user. The brain-Internet parallel here is striking and direct. Web pages are neurons; hyperlinks are synapses. Learning in the brain involves modification of synaptic weights; and, Heylighen proposes, learning in the Internet should involve modification of the weights of hyperlinks between Web pages. Currently, hyperlinks don't have weights, of course—a hyperlink is just a highlighted word, phrase or picture on one Web page, which, when you click on it, brings you to another page. But what if each hyperlink had a weight, indicating how strong a relationship it represented? What if these weights were determined by a combination of AI text analysis programs studying the documents at either end of the link, and reinforcement based on human user habits—links followed more frequently have greater weights[8]?

In the kind of coincidence that is very common in science, Ben discovered Heylighen's ideas along these lines when, in 1995, he posted a paper online suggesting a similar idea. Ben didn't call it a "global brain," but his phraseology was similar—the paper was entitled "From World Wide Web to World Wide Brain." This paper made the same synapse-hyperlink analogue as Heylighen had, but from there moved in a somewhat different direction. Heylighen focused on the modification of hyperlink weights based on human usage patterns, whereas the World Wide Brain proposal was to put an AI at each website: analyzing the relationship between that site and other sites, building new hyperlinks, and modifying the weights of existing ones. Internet intelligence was envisioned as emerging from the synergetic activity of AI agents associated with various websites and databases, each one fairly intelligent on its own. The Net in this view would be a kind of hybrid mind/society of AIs, not totally dissimilar from Marvin Minsky's "Society of Mind" idea. Humans would enter into this society of sub-minds alongside the AIs as equals, jacking into the Net first via e-mail and chat, later via virtual-reality tech, yet later by advanced bioengineering utilities (perhaps the fabled "cranial jack" of cyberpunk Sci-Fi). AI agents and humans would modify the weights of hyperlinks, but this would only be part of the story—because, as we see it, hyperlinks are not a rich enough data structure to store all the types of knowledge a mind requires. Complex relationships, knowledge of procedures, and so forth can't be represented that way. The network of adaptively-weighted hyperlinks would just be one "virtual lobe" of the world-wide virtual brain.

Heylighen, on the other hand, placed less focus on AIs and more on the network itself. He had no illusions that weighted hyperlinks were an adequate structure to represent all forms of knowledge, but he figured every brain has to start somewhere. He has made efforts to get the commercial leaders of the Internet world—browser makers or the W3C, the non-profit Internet advisory board headed by Tim Berners-Lee—to incorporate adaptively-weighted hyperlinks into the real live Internet, but so far this hasn't met with widespread success (though the Semantic Web, if it takes off, is a step in the right direction). Prototypes of "toy Internets" demonstrating the "hyperlink as synapse" mechanism have been just as successful as envisioned.

What happens, Heylighen and Bollen have found, is that over time—as heavily-used hyperlinks have their weights increase and less-used hyperlinks have their weights decrease—the structure of the Web of documents gradually becomes a Knowledge Web that represents the collective thoughts and beliefs of the users. The philosophical undertones here are rather different from Principia Cybernetica, which reflects Turchin's more elitist vision of brilliant scientists gradually refining one another's conceptual formalizations, slowly adding one node after another to the emerging network of understanding. Rather, Heylighen and Bollen's adaptive hyperlink approach suggests that truth can be arrived at through a kind of statistical chaos, involving the contributions of all who are willing to make them. Just add together everyone's opinions—the bad ones will cancel out through destructive interference, and the good ones will reinforce each other through constructive interference. Ultimately the true ideas will emerge. Heylighen and Bollen's experimental systems haven't been released on the Principia Cybernetica site yet (given the limiting nature of current Web software, there are some implementation difficulties), but this will no doubt happen in time.

In 1996, Heylighen founded the "Global Brain Group," an international discussion forum that includes most of the scientists who have worked on the concept of emergent Internet intelligence. This group runs an e-mail discussion list, which initially was extremely limited in membership, open only to scientists who had published serious articles on the notion of a global brain. This group numbered about 10, and was not particularly chatty, so eventually it was decided to admit more people—though only people approved by the initial elite group. The group is still fairly quiet, although a few interesting discussions have emerged, the most interesting ones revolving around the notion of "freedom" and the question of whether the emergence of brain-like complexity in computer and communication networks will take it away from us humans.

For instance, in November 1999 Leor Gruendlinger wrote the following worried message on the Global Brain e-mail list:

> "Before I happily agree to become the part of a cyber-brain (and hence die one clear day because of a bug), I would like to retain

> my autonomy, or at least lose it in stages...What kind of stages? I think about insect colonies as an example: still free to move, to act by themselves, but very much committed to the community, sharing food and resources, caring for the young together, etc...Perhaps before humans agree that their sight, smell and other senses be manipulated by a chip, they will need this confidence and trust in the system they will be part of. It has to sustain them better, perhaps by seeing farther into the future and preparing in advance for challenges they cannot even grasp...A neuron-like symbiosis has the flavor of being even more demanding. What levels of autonomy are there to pass through on the way to the global brain? Will such a passage be gradual, or very fast?"

Steve Wishnevsky then pointed out that this vision of the future Net as usurping individual autonomy and rendering us like ants in a colony may be a big exaggeration. After all, he argued, "consciousnesses larger and more permanent than human have existed for thousands of years, in the form of bureaucracies, churches and empires."

Ben found this argument somewhat lacking, and argued in a reply: "'Largeness' and 'permanence' are not the most important parameters of consciousness...Suppose we accept the pan-psychic theory that everything is conscious...Still, some things are more conscious than others...There is something called 'intensity of consciousness' (which...has to do with the amplification of information)...I think that a bureaucracy has a much lesser intensity of consciousness than a human."

The key question isn't whether the Net is gaining more and more structure, invading our lives and implicitly directing more and more of our activities. Obviously it is, and it's not about to stop. The key question is: how much? How much control will this emergent metasystem have—will it just be like a weird new kind of social institution, or will it be something bigger, something that invades our minds and makes us into some new kind of posthuman human? If this does happen, will its intensity of consciousness suppress that of humans?

It seems unlikely, at least for as long as there *are* humans. Any such system will emerge from human activity, and thusly depend upon it as a component of its own function. Suppressing the dynamics of human activity will cause the system to lose robustness and variety, in a potentially very deleterious way. An ant colony functions as it does because the structure emerged when ants were at a very low level of individual intelligence, but even so the system does not attempt to decrease the individual contributions of its component ants—this would cause it to fail. Any system incorporating humans as part of it should have, as its worst-case *functional* scenario, a situation in which human evolution becomes superceded by that of the system; but there would appear to be no benefit to the system to curtail our abilities. Like any system involving humans—a government, a corporation, etc.—it may choose to *ignore* the contributions of some humans, but there would appear to be little reason for it to try to suppress our creativity on any large scale.

Heylighen, with a modesty that is unusual—almost quaint—among contemporary Internet pundits, doesn't claim to have all the answers, or even most of them. He's content to study the issues, to broadcast his insights as he makes them, and to organize information and discussions leading progressively toward the truth—which will emerge bit by bit, taking its own good time. His vision of Web pages as neurons and hyperlinks as synapses is an exciting one—not an exact parallel between the Internet and the brain, or a complete guide to making a distributed, whole-Internet-based intelligence, but an important contribution with a simplicity and elegance that is sure to make a big impact someday. The Internet will never be a *brain* per se, but it will accrete more and more aspects of formal "neural networks" and eventually become an intelligent system, with which we will communicate in various old and new ways.

NEURAL NETS: THE PATH AHEAD

Right now, where neural net research is concerned, things are in all ways extremely primitive. The dynamics of the brain have not been tremendously elucidated by experimentation with neural net models—at least, not yet. But maybe they will be, as these models become larger (due to faster computers and bigger computer networks) and richer in structure (due to greater understanding of the brain and/or further development in theoretical AI). Thus far, though neural nets have proved useful in various areas, there aren't any major engineering problems that neural nets solve vastly better than other non-brain-inspired algorithms. So far, nothing close to a fully functioning artificial kitten[9], let alone an artificial human brain, has been produced using neural net software—or any other approach to AI. Perhaps Hugo de Garis will get there in another 5-10 years, if his funding situation improves and he can build a massive CBM with enough nodes to compete with a human brain in terms of numbers of neurons and interconnects.

Larger systems aren't the only path ahead. Neural network researchers need to embrace ideas like Erdman's neuronal networks, improved perceptual systems and contextual groundings of the network's knowledge in them, and so on. Additionally, they should be looking into combining their networks with regulatory networks (perhaps based on glial structure, if that turns out to be the key in our brains), and ultimately accepting a fusion with other disciplines such as genetic programming and logic systems. These should be collaborating, not competing, approaches.

We're at the beginning of this fascinating area of cognitive science research, not the end. If neural net researchers are willing to grow their field to embrace more and more of the multilevel complexity of the three-pound enigma in our skulls, then we can expect more and more wonders to emerge from their work as the years roll by.

---

[1] There isn't only one kind of "universal computer." A von Neumann computer is another, and a quantum computer is yet another. Mathematically, all these kinds of computers can do "anything"—though not necessarily with the same level of efficiency. They are all able to perform the tasks which a universal Turing machine can perform. A UTM is a theoretical "universal computer" devised by Alan Turing (1936) to give a mathematical formalization of what constitutes an algorithm, universally.

[2] http://ad.com/ccortex.asp?id=1

[3] http://www.rulespace.com

[4] CBM stands for "CAM-Brain Machine," where CAM is a word for one of the machine's hardware components.

[5] http://www.genobyte.com

[6] "The Semantic Web is an extension of the current Web in which information is given well-defined meaning, better enabling computers and people to work in cooperation."
— Tim Berners-Lee, James Hendler, Ora Lassila, "HThe Semantic Web*H*," in *Scientific American.*
See also: http://www.w3.org/DesignIssues/Semantic.html
[7] http://pespmc1.vub.ac.be/
[8] These ideas are reflected in a lot of interesting work by Francis Heylighen; Tim Berners-Lee and the Semantic Web developers; both of the authors and the Webmind and Novamente teams; Alexander Chislenko; Larry Page and Sergey Brin's page rank algorithm; and a variety of others who have looked at various ways of providing weights, rankings, economic value, and so on to links on the Web as a way of creating a more meaningful organization of the information contained there.
[9] Encouragingly, though, Rodney Brooks' team at MIT does have some robots which are very close to a fully functioning artificial bug. It's not human- or even kitten-level AI, but it's still great work.

CHAPTER 3

THE CHALLENGE OF REAL AI[1]

"Artificial intelligence" is a burgeoning sub-discipline of computer science these days, but it would be easy to draw the wrong conclusion from this fact. One might imagine scientists around the world slaving away day and night trying to create computers smarter than people—computers holding intelligent conversations, outsmarting Nobel Prize winners, writing beautiful poetry, proving amazing new math theorems. The reality is, by and large, far less ambitious and exciting. Cowed by failures to create general machine intelligence in the field's nascent years, AI research as it currently exists focuses almost entirely on highly specialized problem-solving programs, constituting at most small aspects of intelligence and involving little or no spontaneity or creativity. What is called AI research today is primarily research into computer algorithms and data structures for different types of statistical pattern-finding methods or formal logic rule systems. "Real AI"; "Artificial General Intelligence"; "Human-Level (and beyond) AI"—the creation of computer programs with general intelligence, self-awareness, autonomy, integrated cognition, perception and action—is still considered the stuff of science fiction even by most AI researchers today.

But the world is a big place—and even though the trend in the AI field is toward unambitious hyperspecialization, there is a loosely organized group of maverick researchers bucking this trend and focusing their efforts on AI in the grand sense. We're proud to be members of this club (which is not really a club: there's no list of members, no organization, and it's unlikely that any existing "AGI" (Artificial General Intelligence) researcher knows more than 25% of the others who exist). In fact, we've decided to spend some fraction of our time trying to organize the distributed population of AGI researchers into some kind of community, including the creation of the AGIRI Institute and, previously, by co-organizing the Zeroth Global Brain conference along with Francis Heyligen and others. Ben and our colleague, Cassio Pennachin, have edited a book entitled "Artificial General Intelligence," which includes contributions from a dozen or so AI innovators around the globe. We have no delusions that either the AGI book or this one will shift the focus of the mass of AI academics, but perhaps these and other similar efforts will serve to open the eyes of some fraction of the new generation of AI students to the fact that there are serious researchers out there who have not given up on creating software with human-level (and beyond), autonomous general intelligence. It is a cliché in science that revolutions don't occur by the majority of older, established scientists changing their minds, but rather by generational displacement: a new generation of scientists comes about, for whom what used to be an outlying, maverick view is now the norm.

In fact, there is a little evidence that the AI mainstream may slowly be shifting its attention back toward its old AGI ambitions. In 2004 the American Association for AI hosted a small symposium on "Human-Level AI," which Ben attended, co-presenting a brief paper on Novamente with collaborator Moshe Looks. The gathering was not particularly receptive to innovative approaches, but its very existence was encouraging. Perhaps the AI community will come back to its long-abandoned AGI focus sooner than we've been thinking. As with so much else, time will tell.

THE AUTHORS' CHECKERED HISTORY OF TRYING AND FAILING TO CREATE REAL AI

Ben has been riding the AGI wild horse for quite some time now—throughout his entire professional career, and even before. From 1988 (one year before he got his PhD in math) through 1997, Ben spent much of his time theorizing about how, given enough computational resources, he could create an AGI. Since 1997, he's been working with a team of incredibly talented colleagues

(Stephan joined in 1998), trying to turn these AGI theories into a reality—trying to build a real thinking machine.

From 1997 to early 2001, we led the R&D division of Webmind Inc. on an AI adventure. We did a lot of other things at Webmind Inc. as well, from designing products to writing documentation and enduring endless sales and business meetings—but it was the Webmind AI Engine project that was dearest to our hearts. During the 3 ½ years that the company lived, we and our colleagues transformed a promising but incomplete conceptual and mathematical theory about how to build an AGI into a comprehensive, detailed software design; and implemented a decent percentage of the software code needed to make this design work. At its peak, the team working on this R&D project numbered 40 scientists and engineers, spread across four continents.

Webmind Inc. is gone, but we're still at it—Ben and a handful of colleagues have continued the essence of the Webmind project under a different name, and with different, more technically advanced, but conceptually similar ideas (4-5 people directly concerned with AGI work; 7-8 others collaborating with them on related narrow-AI projects). A couple of the team members are in the U.S., but the majority of them are in Brazil, where Webmind Inc.'s largest office had been. The Webmind AI Engine codebase had to be abandoned due to legal issues regarding ownership after the dissolution of Webmind Inc., but we've created a successor system and called it Novamente. In hindsight, abandoning the Webmind codebase was a good thing anyway, as our work with Novamente has made us aware of a great many serious shortcomings of our previous Webmind design.

"Novamente" means "new mind" in Portuguese; and also "again," "anew" or "afresh." As a background meaning, it's also "new lie," which should provide Portuguese-speaking AI skeptics with a bit of amusement. So far Novamente doesn't have as much in it as the Webmind AI Engine did, but we're building it for efficiency this time around: it's a lean, mean, hopefully-will-be-thinking machine. We're applying the partial version that we have today to some practical problems, including the comprehension and production of the English language, and the analysis of gene expression data—a major puzzle in computational biology that we'll discuss briefly a couple more chapters in.

The goals of the Novamente project, like those of the Webmind project before it, are huge. First of all, we want to make a Novamente system that can hold a decent English conversation—not necessarily sounding exactly like a human, but sounding interesting and smart, spontaneous, creative and flexible. We want Novamente to be able to control an agent in a simulated 3-D world, much like a video game world, but with quite a bit more intelligence. Controlling a real robot would be nice too, although at present we're not focusing on the hardware aspect of robotics. After these little warm-up exercises, we intend to move on to the main event: giving Novamente the ability to rewrite its own program code for improved intelligence. If this works, we hope, it may set off a trajectory of exponentially increasing software intelligence; perhaps even create The Singularity. Lofty aspirations—but as Sir Edmund Hillary, the first to climb Everest, said in the title of his memoir, "Nothing venture, nothing win."

The Novamente approach to AI incorporates some aspects of "standard" AI paradigms, including probabilistic logical inference, evolutionary programming and neural networks. But ultimately, in spite of its broad base of inspiration, it doesn't rely on anyone else's vision of AI—it's an original approach to the problem of creating a thinking machine, which will stand or fall on its own merits. In this chapter, we'll say just a little about Novamente, and then move on to other things. (Ben has a series of specifically-Novamente-related books in the works, and the intention in this book is to focus on the broad scope of future-pointing technology rather than that particular research, as exciting as we think it is.) Before touching on Novamente at all, we'll review the field of AI in general: the major trends and research approaches, some of the more ambitious current projects, and how AI fits with the broader technological advances that surround us, including the transformation of the Internet into a global brain, and the Singularity/Transcension.

A brief comment on the sometimes extreme opinionatedness of our own points of view on AI: we haven't tried terribly hard to be "objective" in the other chapters in this book; neither of us are big believers in "one objective truth" anyway. Philosophically, we're very sympathetic to Mikhail Bakhtin's notion of "polyphonic" reality (Bakhtin, 1984), in which the actual world is understood as a kind of unholy (trans-holy?) superposition of everyone's subjective point of view. In these pages we are presenting our own points of view in the hope that others will find them interesting and compelling enough to act on—without putting forth our perspectives as any kind of universal or absolute truth. However, it's also true that some subjectivities are more subjective than others; and we think that our views on AI are a bit more strongly opinionated than our views in other areas. All the topics discussed here are important to us, but AI has been a large portion of our work (especially for Ben, who has devoted the majority of his career to it), and will continue to be. This brings a particular depth of knowledge and wisdom to our discussion of the subject matter, but it also brings a host of extremely strongly-felt opinions. Take them for what they are: the strongly-held, deeply-informed feelings of a pair of maverick AGI researchers who don't agree completely with very many other AGI researchers (and sometimes don't even agree with each other), but who note an increasing number of researchers coming to agree with more and more of our views as time goes on. Perhaps they're finally realizing that having failed as a discipline to create AGI once doesn't mean we ought never to try again.

IS AI POSSIBLE?

As bizarre as it sometimes seems to AI optimists like us, not everyone believes it's possible to create an AGI program. Some of these disbelievers are highly intelligent, educated people, whom we're forced to take seriously in spite of their unintuitive (to us) attitudes. There are several varieties to the anti-AI position, some more sensible than others.

First, there is the idea that only creatures granted minds by God can possess intelligence. This may be a common perspective, but isn't really worth discussing in a scientific context. Even if it were true, we'd have no way of knowing whether or not God might grant our AGIs minds when we've developed them far enough—a Deus Ex Machina, literally. Whether or not this is the case is, by definition, unknowable. Fewer and fewer serious theologians hold such opinions these days, anyway.

More interesting is the notion that digital computers can't be intelligent because mind is intrinsically a quantum phenomenon. This is a claim of some subtlety, given that the British physicist David Deutsch (see Deutsch, 1998 for a non-technical review) has formalized the notion of "quantum computing," and proved that quantum computers can't compute anything beyond what ordinary digital computers can. Still, in some cases, quantum computers can compute things much faster on average than digital computers. And, as noted above, a few mavericks like Stuart Hameroff and Roger Penrose have argued that non-computational quantum gravity phenomena are at the core of biological intelligence. Of course, there is as yet no solid evidence of cognitively significant quantum phenomena in the brain. But a lot of things are unknown about the brain—and about quantum gravity, for that matter—so these points of view can't be ruled out. Even if they are correct, however, there's not enough knowledge yet about these kinds of phenomena to know for sure whether or not they could occur in non-biological systems if constructed carefully. After all, both human brains and silicon computers are ultimately nothing more than collections of elementary particles.

Our own take on this is: yes, it's possible (though totally unproven) that quantum phenomena are used by the human brain to accelerate certain kinds of problem solving. On the other hand, digital computers have their own special ways of accelerating problem solving, such as super-fast, highly accurate arithmetic. Even if the brain does use quantum computing, that doesn't prove quantum computing is the *only* path to intelligence. It is not a requirement that machine intelligence function in the same way as biological intelligence. The only requirement is that, in order for us to know it exists, it must be able to somehow communicate with biologically intelligent beings (it could come to exist in a form which could not do so, but we wouldn't know).

Another, even more cogent objection is that, even if it's possible for a digital computer to be conscious, there may be no way to figure out how to make such a program except by copying the human brain very closely, or running an enormously time-consuming process of evolution which roughly emulates the evolutionary process that gave rise to human intelligence. This argument—that the only path to intelligence must be similar to the one already taken successfully in biology—may turn out to be correct, but that is not a reason to try to determine whether or not these differently structured, digital information processing systems can be configured in a novel way which produces intelligence. The same idea about the requirement to copy nature was held about heavier-than-air flight only being possible in the manner in which it developed in birds. Eventually, enough discoveries in science and innovations in engineering were made to prove this incorrect.

We don't have the neurophysiological knowledge to closely copy the human brain. To simulate a decent-sized primordial soup, and then evolve digital life by replicating the physical and biological processes which brought it about in the real world, is simply not possible with current computer technology. This objection to AI is not an evasive tactic like the others, it's a serious one. From a general futurist point of view, it doesn't really matter—brain scanning is advancing rapidly, so that within a few decades we'll actually be able to closely copy the human brain. AI and neurosimulation will fuse into one. We'll be able to simulate the human brain in a computer, and then perform simulated neuromodifications on it; achieving superhuman AI by leveraging a combination of neuroscience and computational experimentation (and experimenting with biological superintelligence in simulation before attempting to implement the neuromodifications to achieve it).

However, our own scientific suspicion is that it's possible to circumvent this objection and make AI without brain-simulation, creating a fundamentally non-human AGI using a combination of psychological, neurophysiological, mathematical, and philosophical cues to puzzle out a workable architecture and dynamics for machine intelligence, which takes advantage of the inherent strengths of digital hardware and works around its weaknesses. Our own neural hardware has strengths and weaknesses different from those of digital computers, so trying to simulate it in silicon as the only path to AGI may actually turn out to be a *worse* path to intelligence—since it would be attempting to shoehorn the structures of one system onto another, and the neural model will need to be adapted over time to fit this hybrid approach.

In this "mind engineering" approach, one has to redo a lot of the work that evolution did in creating the human mind/brain. An engineered mind (like Novamente, if successful) will have some fundamentally different characteristics from an evolved mind like the human brain; but this isn't necessarily problematic, since our goal is not to simulate human intelligence but rather to create an intelligent digital mind that knows it's digital, and uses the peculiarities of its digitality to its best advantage. We already have a prolific system for creating more humans, so it's more interesting (particularly given the inability of modern brain science to guide us in detail through the process of human-mind-emulation) to try to create the conditions in which a new *type* of intelligence may arise.

Our basic philosophy of mind holds that mind is not tied to any particular set of physical processes or structures. Rather, "mind" is shorthand for a certain pattern of organization and evolution of patterns. This pattern of organization and evolution obviously can emerge from a human brain, but it could also emerge from a computer system (or a whale brain, or a space-alien brain). In this view, unless it's created via detailed brain-simulation, a digital mind will never be exactly like a human mind—but it will still manifest many of the same higher-level structures and dynamics. To create a digital mind, one has to identify the abstract structures and dynamics that characterize "mind in general," and then figure out how to embody these in the digital-computing substrate. Not an easy job; but almost certainly, in the next few decades—maybe sooner—we suspect someone's going to crack it. Perhaps the Novamente team will be the ones to do it. (And if not, brain simulation will likely get us there eventually.)

WHAT IS "INTELLIGENCE"?

When Ben first approached the AI field as a college student in the early 1980s, it seemed that one reason the AI field had deviated so far from its original goal (creating a human-level-or-superior general intelligence) was the lack of a reasonable definition of the core concept of "intelligence." Of course, like many undergraduate insights, this was not an entirely original observation. The lack of a definition for the "I" in "AI" has long been the source of sardonic humor among AI researchers. One humorous definition goes: "Intelligence is whatever humans can do that computers can't do yet." There is some truth to this quip: for instance, most people would say that playing chess requires intelligence. Yet, now that we see Deep Blue playing grandmaster-beating chess via some simple mathematical algorithms (rather than through any human-like intuition), we reclassify chess-playing as something on the borderline of intelligence, rather than something definitively involving intelligence. A less sarcastic slant on the same basic observation would be to say that AI research has helped us to clarify our understanding of what intelligence really is.

In a classic paper from the 1950s, computing pioneer Alan Turing proposed emulation of humans as a criterion for artificial intelligence, now called the Turing test (see the collection of papers in Turing, 2004). The Turing test basically says: "Write a computer program that can simulate a human in a text-based conversational interchange—and any human should be willing to consider it intelligent." We don't particularly like the Turing test. One thing that intelligence clearly *doesn't* mean is a precise simulation of human intelligence. Human intelligence isn't even the only intelligence we know of: great apes, elephants, dolphins, whales and octopodes all seem to exhibit signs of substantial intelligence, though not at a human level. It would be unreasonable to expect human intelligence from any AI system lacking a human body.

The Turing test serves to make the theoretical point that intelligence is defined by behavior rather than by mystical qualities, so that if a program can act like a human, it should be considered as intelligent as a human. But it's not necessarily useful as a guide for practical AI development. In mathematician's lingo, we view the Turing test as a sufficient but not necessary condition for intelligence[2]. It is certainly possible that a radically non-human intelligent computer system could develop, as unable to imitate humans as we are to imitate dogs or anteaters or sharks, but still as intelligent as humans or more so.

Although we've devoted years of thought to it, we don't claim to have thoroughly solved the "What is intelligence?" problem. We don't have an IQ test for Novamente or other would-be AIs. The creation of such a test might be an interesting task, but it can't even be approached until there are a lot of intelligent computer programs of the same type. IQ tests work moderately well within a single culture, but much worse across cultures—how much worse will they work across species, or across different types of computer programs, which may well be as different as different species of animals?

We do however have a simple "working definition" of intelligence, which shares with the Turing test a focus on pragmatic behavior. It builds on various ideas from psychology and engineering, and was first explicitly presented in Ben's book *The Structure of Intelligence* in 1993: ***intelligence is the ability to achieve complex goals in a complex environment***.

None of our experience over the past decade with developing AI and reviewing others' AI systems has in any way contradicted this understanding of intelligence. Of course, it begs the question of defining "complexity" of goals and environments—a question that leads to a long mathematical story that Ben has spent many years of his career attempting to tell. The truth is that research on the mathematics of complexity hasn't yet played a major role in anyone's practical AI development work. The qualitative notion of "achieving complex goals in complex environments," on the other hand, has proved to have a substantial heuristic value in trying to actually engineer an AGI system.

RULES VERSUS NEURONS

As we noted in the previous chapter, for at least two decades now AI researchers have been divided into two main camps: the neural net camp and the logic-based or rule-based camp. Both camps wanted to make AI by simulating human intelligence, but they focused on very different aspects of human intelligence. One attempted to model the brain; the other, the mind.

We've already discussed the neural net approach in the previous chapter. It starts with neurons, the nerve cells of which the brain is made, and tries to simulate the ways in which these cells link together and interact with each other. Rule-based models, on the other hand, try to simulate the mind's ability to make logical, rational decisions, without asking how the brain did this biologically. Based on our own intuitive understanding of the mind/brain, it seemed obvious to us right from the start that both of these ways of looking at the mind were really extremely limited. On the one hand, there's a lot more to the brain than links between neurons. On the other hand, there's a lot more to the mind than rational, logical thinking. Finally, some might say that neural networks and rule-based systems, if combined, would unify the two approaches and create a more complete model of the mind. We disagree with this, as we feel there's much more to mind than these two approaches offer, even combined. The two approaches each add important and necessary insights, but even in conjunction, they're not complete.

When we first studied these issues, we tended to think that everyone who was taking either of these two approaches must be fools—how could they miss such obvious shortcomings as the ones in these approaches? Eventually, as we got further into studying all the issues involved, we understood more of what was going on. The key point is that, in the early years of AI when the key trends and ideas in the discipline were formed, available computer technology was very limited. There was no way to write a program simulating the whole structure of the brain, and there was barely enough processor power and memory to deal with very limited logical processing, let alone the integration of logic with all the other aspects of the mind. Yet it seemed to us that, rather than acknowledging that they were taking very limited approaches because they wanted to actually implement programs on contemporary computers, early AI researchers had turned the limitations of their own computer resources into philosophical arguments. And contemporary AI researchers, by and large, were following in their footsteps.

They were making theories of the mind based on ideas like "the mind only does logical reasoning" or "the mind is just a simple neural net," when the truth was that it was their computers—with their limited memory and processing—that could only do logical reasoning, or could only act like a neural net, and were unable to run the full diversity of processes needed for true cognition.

Researchers wanted their theories to be empirical, to be testable; they wanted to be able to "put up or shut up" rather than making huge, untestable and unprovable statements about how the mind works and how the mind could be implemented if you had a good enough computer. This is a good motive: it's a basic scientific method. The irony is that mind is at least 50% about computing power. There's no way to make a chicken's brain really smart by rewiring it—the raw processing power isn't there. Of course, processing power can be wasted; you need to use it intelligently with the right data structures and algorithms—but without it, you can't implement a mind, and you can't test serious theories about the mind. By trying to do real science and test their ideas about the mind with computational experiments on the computers they had at the time, computer scientists were pushed toward trivial and just plain wrong theories about how the mind works.

Even though the standard AI theories about how the mind works and how it can be implemented in software are not correct models of a *fully functional mind*, a lot can be learned from them. The Novamente system mentioned throughout this book incorporates aspects of neural nets and evolutionary programming (another AI approach we'll briefly discuss soon), and also of logic-based AI. It doesn't use any of these in a conventional way, but it does draw from, and expand upon, research in both the two historically traditional approaches to AI; newer widespread techniques; and a lot of original thinking. Academic and industry computer scientists working on various "narrow" approaches to AI have created a lot of great science, even if very few of them have done much "real AI" according to our definition—they haven't created any thinking machines, or even any viable designs for thinking machines. Many of them have done work that's relevant to those of us actually concerned with thinking machine construction;

and so have a lot of other people working in other fields such as system theory, neurobiology, theoretical mathematics, and so on. Synthesizing this research can lead us towards the creation of true AI, and this is the approach being taken in the Novamente system.

NEURAL NET AI

"Neural networks" is the older of the two competing paradigms. Its roots go way back, to work in cybernetics and systems theory from the pre-computer age, such as Norbert Wiener's book *Cybernetics: Control and Communication in Animals and Machines*, first published in the 1930s. An amazing book for its era, it showed for the first time that the same mathematical principles could be used to understand both man-made electrical control systems and biological systems like bodies and brains.

The early neural network systems weren't general purpose, digital computers—they were analog electrical models of the brain, physically wired together with tubes, switches and other electronic equipment. Until the early 1970s, this was generally more efficient than using computer simulations, because the computers at the time were so limited in capability. These early attempts were extremely primitive. Marvin Minsky's vacuum-tube neural net from the '50s, for example, had about 50 neurons in it, compared to hundreds of billions in the brain. Even the computer-simulated neural networks people use today normally have, at most, thousands of neurons. Even if one believed that the existing neural network model completely encompassed all the theoretical underpinnings of a mind, it seems extremely unlikely that you'd manage to create one in a system with a level of complexity several orders of magnitude below what human-level intelligence needs.

Simulating neurons on computers is kind of a weird thing. The issue is that the human brain uses parallel processing: billions of neurons are all working at the same time. In the very early days of computing—the late '30s and the '40s—it was thought that computers might be based on the same principles. However, the early, neurally-influenced hardware designs turned out not to be a very efficient way to do things. The Hungarian mathematician John von Neumann was the first to figure out a passably efficient design for a computer, and all the computers we use today are based on his design. The design is very simple: a von Neumann computer is centered on a central processing unit, which follows instructions given to it by a programmer. The central processing unit goes through a series of five steps, over and over again, one after the other:

1. It calls up an instruction from its memory
2. It calls up any data required by the instruction
3. It processes the data according to the instruction
4. It stores the data in memory
5. It calls up the next instruction from its memory

This is not how the brain works, obviously! The brain's structure and patterns of "processing" are far more complex. Memories and instructions in our brain are not so linearly defined, and a huge number of memories and instructions may be active (in some stage of activity) at any given time. There are some things that the brain's mixed-up processes do more naturally than von Neumann's orderly machines. On the other hand, computers built using this architecture take much more naturally to performing routine iterative tasks like adding up long columns of numbers, doing routine accounting calculations, filling out income tax forms, retyping manuscripts, and so on and so on, than people do. The first application for computers was doing routine math calculations for solving military problems, like computing ballistic trajectories and figuring out the details of how to build hydrogen bombs. De Garis' work and the CCortex, mentioned earlier, are examples of serious modern attempts to get around the terrible mismatch between neural network mathematics and von Neumann computer hardware—by going beyond von Neumann machines and building a special hardware framework designed specifically for neural net processes.

The conceptual advantage of neural networks is that they are at least vaguely similar to the most intelligent system we know of, the human brain. On the other hand, von Neumann computers don't immediately give the appearance of being capable of anything resembling intelligence. It's obvious that real intelligence requires more than just the iterative routine that von Neumann computers are good at. Intelligence requires the ability to find new solutions to unanticipated problems. Programming a von Neumann computer—meaning any computer commonly used today—is a matter of telling the computer specific rules for what to do; from which, in von Neumann's design, the computer can never vary (if the programmer has made an error, for example, the computer will faithfully fail at that point every time). How to make rules governing spontaneous, unplanned, creative behavior is not exactly a trivial problem.

In the 1960s, more and more researchers started pushing in this direction. They decided that simulating the brain on a totally non-brain-like hardware system probably wasn't such a good idea, and started thinking about better ways to use computers to emulate human thought. Impressed with what computers could do using a few rules, applied over and over again very quickly, they decided that everything the human mind does could be simulated by a big enough computer, performing a sufficiently complex set of rules. There were definite successes here, but also an awful lot of failures. We'll tell you some of the stories in a moment.

By the 1980s, the AI community was moving back towards neural networks again, but both camps were frustrated because their programs were never quite able to do what they wanted them to do.

With 20/20 hindsight, it now seems clear that while both the neural net and the rule-based approaches were useful building blocks, the early researchers had ridiculously oversimplified their task out of necessity (and then, it seems, had forgotten the necessity and codified these oversimplifications as theory). True intelligence requires more than following carefully defined rules, and it also requires more than random links between a few thousand artificial neurons. It's probably a good thing, however, that early researchers underestimated the difficulty of achieving true intelligence. If they hadn't, they might not have had as much energy and enthusiasm for their pioneering efforts.

This brings us back to a point made above: *Why has no one yet managed to build a thinking machine?* Mainly, we believe, it's because no one has really tried to build a whole mind: a computer system that could observe the world around it; act in the world around it; remember information; recognize patterns in the world and within it; and create new patterns inside itself in response to its own goals. Presumably, no one has tried to do this because the computer resources available until recently were blatantly inadequate to support a program with reasonably sophisticated modules devoted to memory, perception, action and conception. Lacking the computer resources to build a whole mind, researchers have historically focused on one or another particular aspect of the mind (logic, or neural net type pattern recognition), and tried to push this aspect as far as it could go. Today's computer resources may still be inadequate for the task, but at least their inadequacy is less patently clear—computer hardware is a much worse excuse for lack of AI imagination and ambition than it used to be.

EARLY RULE-BASED AI SYSTEMS

Although our own AI work has more of a logic foundation than a neural net one, we tend to have more affinity for neural net type AI systems than for logical rule based AI systems. Partly this is because when we were going through school, logical rule- based AI was the orthodoxy that we were rebelling against, whereas neural nets were maverick at the time, up-and-coming and exciting. And partly, it's because a significant subset of neural net research[3] has focused on emergence and self-organization, which we think is extremely important conceptually. Nevertheless, in spite of our affection for neural nets, when we look at it objectively we have to admit that the early work on rule-based AI was really just as important as the early neural network research. Both areas have contributed much towards the development of AI, but now it is time for new areas to be more seriously considered—genetic programming, semantic-neural hybrid networks, and generative programming (self-rewriting code) chief among them.

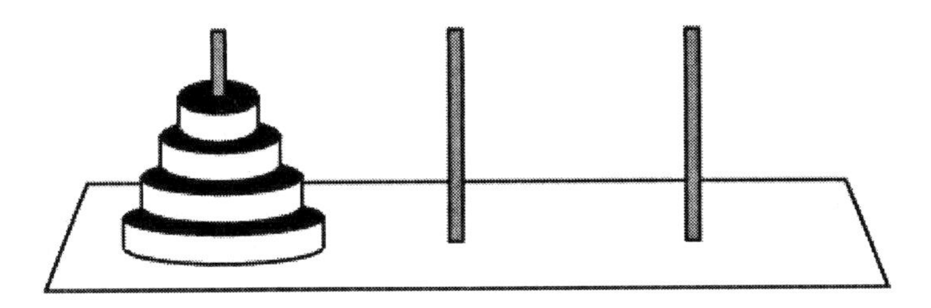

Figure 3.1 The "Tower of Hanoi" Puzzle

One famous early rule-based program (back in the 1960s) was something called the General Problem Solver—a very ambitious name. It was written by Alan Newell and Herbert Simon, both of whom went on to do a lot of other great AI research. This was a pretty interesting program, but it didn't quite live up to its advertisements. In fact, the title of their paper on GPS was quite possibly one of the biggest and emptiest brags of all time: "General Problem Solver: A Program that Simulates Human Thought."

Not quite.

GPS could solve simple problems like the "Tower of Hanoi" puzzle, and "cryptarithmetic" puzzles like DONALD + GERALD = ROBERT (to solve this, assign a number to each letter so that the equation comes out correctly). But in the overall scheme of intelligence, solving problems such as these is not all that different from computing logarithms or solving differential equations. A simple mathematical strategy suffices. There's no real learning in such a system—it simply applies mathematical rules in a relatively brute-force way until the answer matches a known result. Neither new rules nor new ways of evaluating results are possible in GPS. GPS is not a general problem solver; it's yet another narrowly-defined-mathematical-problem solver.

Of course, there were points in history where the ability to solve such puzzles would have been considered a remarkable display of intelligence. The "Tower of Hanoi" puzzle, for example, was introduced to the modern world by the French mathematician Edouard Lucas in 1883, but his inspiration came from ancient Hindu folklore. Tales are told of a Hindu temple where this puzzle or a similar one was used as a challenge for young priests. At the beginning of time, the priests in the temple were given a stack of 64 gold disks, each one slightly smaller than the one beneath it, and presented the task of transferring the 64 disks from one of the three poles to another without ever placing a large disk atop a smaller one. Day and night the priests labored, and (so the story goes) when they finally finished their work, the temple would crumble into dust and the world would vanish. Clearly, a program that could solve this problem in subsecond time would have seemed rather intelligent to the ancient Hindus who believed this myth. But now, we can see that solving the Tower of Hanoi doesn't require inordinate general intelligence, any more than causing rain via cloud-seeding requires one to have general control over the weather.

What GPS was doing was taking an overall goal—solving a puzzle—and breaking it down into subgoals. It then tried to solve the subgoals, breaking them down into sub-subgoals if necessary, until it got subgoals small enough that it could deal with them in some direct way—like by enumerating all possible values a certain letter could take in a cryptarithmetic puzzle. This same basic logic is used now in a much bigger and better rule-based AI program called SOAR, also developed by Newell and his colleagues. SOAR is probably the ultimate height of rule-based AI. Ben worked with a Masters student at the University of Western Australia who was trying to model how humans solved simple practical problems by watching how SOAR solved them. SOAR wasn't totally useless for this—sometimes you could get it to follow strategies of breaking goals into subgoals that were something like what humans did. The parts of a problem that took people a long time often took SOAR a long time too.

This business of goal and subgoals is important to any AI system—inside our Novamente AI system, for example, we have something called a GoalNode, and we have processes called schema that can break goals contained in GoalNodes into subgoals. The basic conceptual algorithm of GPS and SOAR is clearly something that's necessary for the mind. However, it doesn't have to be done in as rigid a way as these programs do it. In fact, doing it in such a rigid way is tremendously destructive. To make this process flexible, you need the goal and subgoal management part of the mind to interact with the other parts of the mind. The system has to be able to flexibly determine which of its processes are effective for achieving which of its goals in what contexts—and for this, it needs reasoning and association-finding and long-term memory. The system must also be able to use context-dependent, intuitive reasoning to figure out what goals to split into subgoals, in what way and in what situation. Basically, GPS and SOAR and this whole line of AI research are a result of taking one aspect of the mind—goal-directed, problem-solving behavior—and extracting it from the rest of the mind. Unfortunately, when you extract it from the rest of the mind, this aspect of thinking isn't all that useful, because it has no way to control itself in a context-dependent way. You wind up with a simply mechanistic approach: apply some enumerated list of decomposition rules; then apply some enumerated list of atomic problem-solver rules. Such a system can only ever solve problems for which it has been explicitly programmed, or problems very similar to these—it can never learn anything qualitatively new.

Herbert Simon helped create another program called BACON, which was at least as ambitious. He called the program BACON because it was inspired by the writings of the brilliant scientist Sir Francis Bacon, who thought that science was a matter of recognizing patterns in tables of numerical data. Simon programmed BACON to do just that: to look at large quantities of data and infer the general patterns hidden within. Today, we'd call this data mining—it's both a moderately big business, and a flourishing academic subfield. Modern data mining programs are really useful. Businesses use them to find patterns in their customer base, for example. They might discover that a particular brand of soap is bought particularly often by 40-year-old men who live in a certain zip code, and so forth. Scientists also use data mining software to find patterns in data and set their minds thinking in certain directions, but these tools definitely don't replace scientists or even market analysts. The problem is that there's much more to doing science than finding patterns in tables of numbers. Francis Bacon, the man, never really understood how the mind works—and this is why even data mining programs that are a lot "smarter" than BACON aren't really intelligent minds.

Let's look at BACON's reasoning in detail, through one of its more impressive examples. BACON was able to "learn," in some sense, the ideal gas law from thermodynamics. This law says, "pV/nT = 8.32 where p is the pressure of the gas, V is the volume of the gas, T is the temperature in degrees Kelvin, and n is the quantity of the gas in moles." In practice, this relation never holds exactly—because there is no ideal gas in the real world—but for most real gases, it is a very good approximation.

If you give it appropriate table of numbers, BACON can learn this law, using rules such as these:

- If two columns of data increase together, or decrease together, then consider their quotient.
- If one column of data increases, while another decreases, then consider their product.
- Given a column of data, check if it has a constant value.

As pressure goes up, volume goes down, so BACON forms the product pV. Next, as the combined quantity pV goes up, so does the temperature—thus BACON constructs the quotient pV/T. And as pV/T goes up, so does the number of moles—hence the quotient (pV/T)/n = pV/nT is constructed. This quotient has a constant value of 8.32—so the ideal gas law is "discovered"—that is, this relationship is mined out of the data.

Very interesting, indeed, and an important step in the right direction, but how terribly far this is from what real scientists do! Most of the work of experimental science is in determining what kind of data to collect, and figuring out creative experiments to obtain the data. Once a reliable set of data is there, finding the patterns is sometimes a major challenge—but sometimes it's the easiest part. Often the pattern is guessed on the basis of terribly incomplete data, and this intuitive guess is then used to guide the search for more complete data.

BACON, however, is absolutely incapable of making an intuitive guess. Another problem is that BACON hasn't any idea what its results mean *physically*. It has no concept of "gas" or "pressure" or "volume" and so on. Without the intervention of a more abstract intelligence (like a human's) that can develop an understanding of these physical concepts, BACON can do nothing whatsoever with its results other than apply them to other tables of numbers.

Simon has claimed that a four-to-five hour run of BACON corresponds to "not more than one human scientific lifetime." AI theorist and author Douglas Hofstadter, in his book *Metamagical Themas* suggests that one run of BACON actually corresponds to about one second of a human scientist's life work. We think that Hofstadter's estimate, though perhaps a little skimpy, is much closer to the mark. Only a very small percentage of scientific work is composed of BACON-style data crunching.

But, tools like BACON are not useless. While they are not equivalent to a scientist's life's work, they are great time-saving tools in the hands of a human scientist who can creatively design experiments, collect the data, and interpret the relationships that are mined out as physically, biologically, etc. meaningful results. A tool like BACON might save a scientist years of work in mining through data looking for patterns, but only because the scientist knows how to apply the tool. Modern earth-moving equipment can save a civil engineer many man-years of routine digging and lifting work, but such routine tasks only occur after much creative problem solving. The same holds true for data mining tools. Neither a data mining tool, nor a payloader, can *initiate a project* on their own, nor *understand the results* of their own work as applied to the physical problems of the real world.

In Novamente, we actually use algorithms analogous to those in BACON—though vastly more sophisticated, as is to be expected since AI has progressed a great deal in the last few decades. In the terminology of modern AI, this aspect of Novamente's thinking is called "machine learning" or "data mining." Recognizing patterns in vast amounts of data is a very important part of the mind, but still, it's only *part* of the mind. Novamente learns rules explaining why humans like some messages or e-mails better than others, using methods not that different from BACON's. But, we know now that the *real trick* is in mapping the messages or e-mails into numbers that data mining methods can handle. This involves understanding the meanings of various words, phrases and expressions. Also, there's the matter of deciding what data to look at, which is done by the general association-finding mechanisms in Novamente's mind. Reasoning, which brings general background knowledge into the process—as opposed to pure data mining, which is just pattern-finding—is an approach which sets Novamente apart from other systems. Bringing associations and reasoning into the picture, you need long-term memory, which opens a whole big and beautiful can of worms. You get the picture. Pattern-finding is crucial, but it's only part of the story. In fact, the biggest trick underlying Novamente's application to biological data mining is the way Novamente goes beyond plain old hyper-BACON-style quantitative pattern recognition and incorporates an *understanding of context* into the process of data analysis. We'll get to this a little later on when we talk about bioinformatics.

Rule-based AI—"symbolic" AI—has had plenty of practical successes. However, our big complaint with it is that every one of these successes has resulted from specialized tricks, rather than flexible intelligence. One term for this situation is "brittleness." Or, you could call it "remarkable literal-mindedness." These programs are a lot like Microsoft Word, DOS 6.0, or a pocket calculator—they do what they're told, and very little more. If they're programmed to deal with one context, then that's what they'll deal with; not in a million years will they generalize their knowledge to something totally different. Such programs, unlike a human, can not be "taught" financial analysis, but then decide that they'd rather do aeronautical fluid dynamics and apply their mathematical "training" to that domain instead.

One famous early AI program, Patrick Winston's ARCH, contained logical definitions of everyday words. An "arch" was defined as "three blocks, A, B and C, so that C is supported by A and B, and A and B do not touch." This was fairly impressive for its time, and it's all very well for playing with blocks—but what will the program do when it gets to Arches National Park in Utah, or if it tries to build arches out of modeling clay? Nothing. On the other hand, show a clever three-year old human an arch made of blocks, and she'll immediately recognize a rock arch as a member of the "arch" category. It won't occur to her that a rock arch can't be naturally decomposed into three blocks, A, B and C, and that therefore there's no relationship; because her rules aren't that literally applied—they're more flexible, and more conceptual. Her categorizations are fuzzier, and eventually she may refine them more: decomposable arches, atomic arches, toy arches, rock arches, bridges, cheek arches, the Golden Arches, etc.—but in doing so, she won't lose track of the overall category of "arches" and decide all these things are different types of objects. The minds of human children, unlike expensive research computers, are anything but brittle.

(THE SURPRISING PAUCITY OF) COMPETITORS IN THE RACE TO REAL AI

We've explained what "creating an AGI" means to those of us on the Novamente project: creating a computer program that can achieve complex goals in a complex environment, using limited computational resources and in reasonably rapid time. At least initially, the domain of achievement we are looking at is socially interacting[4] with humans, and analyzing data in the context of the Internet.

A natural question to ask, when presented with such an ambitious goal, is: if AI is possible, how come it hasn't been done before? And how come so few people are trying?

Peter Voss, an AI theorist, entrepreneur and futurist whose ideas we like very much, has summarized the situation. Of all the people working in the field called AI, he observes:

- 80% don't believe in the concept of General Intelligence (but instead, in a large collection of specific skills & knowledge)

- Of those that do, 80% don't believe it's possible—either ever, or for a long, long time
- Of those that do, 80% work on domain-specific AI projects for reasons of commercial or academic politics (results are a lot quicker)
- Of those left, 80% have the wrong conceptual framework...
- And nearly all of the people operating under basically correct conceptual premises lack the resources to adequately realize their ideas

The bulk of the work being done in the AI field today presupposes that solving sub-problems of the "real AI" problem, by addressing individual aspects of intelligence in isolation, contributes toward solving the overall problem of creating AGI. While this is of course true to a certain extent, our experience with Webmind and Novamente suggests that it is not as true as is commonly believed. The problem is that, in many cases, the best approach to implementing an aspect of mind in isolation is very different from the best way to implement this same aspect of mind in the framework of an integrated, self-organizing AI system. This is true in other kinds of engineering as well. You wouldn't put fixed wings on a helicopter, but that doesn't mean someone's an idiot for ever designing a fixed-wing aircraft. If your goal is not to build helicopters, you may find fixed wings quite useful. Similarly, while the algorithms in a system like BACON do not integrate well with general intelligence, they are useful stand-alones. There are good practical reasons to develop stand-alone AI algorithms that don't work well for AGI, if AGI is not your primary goal.

So who else—besides the authors and their noble comrades—is actually working on building generally intelligent computer systems at the moment? Not as many groups as we would like, but there are some interesting things going on. Hugo de Garis' artificial brain project (mentioned earlier) is one of them, but not the only one.

An interesting, related research direction is Rodney Brooks' well-known Cog project at MIT, which is aiming toward building AGI in the long run. Their path to AGI involves gradually building up to cognition, after first getting animal-like perception and action to work via "subsumption architecture robotics." This approach might eventually yield human-level intelligence, but only after many decades, or so it seems. While the idea that perception is a key to intelligence seems critical, the short-sightedness here may be that only the kinds of perception that animals have already exhibited are of any value. In a system like Novamente, the idea is that other kinds of perception and perceptual environments may be equally valid (for example, "living" entirely on the Internet and having perceptual inputs tuned for that situation).

Another is the Non-Axiomatic Reasoning System (NARS) project pursued by our friend and collaborator Pei Wang. NARS has some similarities to Novamente, but there are also major differences. Novamente is more conservative in its reliance on standard probabilistic mathematics, whereas Pei has introduced his own variety of uncertain logic; but NARS is more conservative in that it places logic squarely at the center, whereas Novamente considers logical inference as one among several cognitive processes that combine to form the emergent structures and processes of mind. Pei has devoted a huge amount of time to his AGI theory and implementation over a 15 year period, showing a dedication and vision to the grand goal that is surprisingly rare in the AI field.

Some other important Real AI research directions will be discussed a little later in this chapter: Jason Hutchens' sophisticated chat bots, Doug Lenat's multi-decade, multimillion-dollar CYC project, and Danny Hillis' Connection Machine. Eliezer Yudkowsky, about whom we'll talk in a later chapter, is also at work on his own AGI, and Peter Voss has a small company, Adaptive Intelligence Inc., working on implementing his own neural-net-ish approach to AI. All these are wonderful projects indeed—but the surprising thing is that there are so few of them to discuss.

There are hundreds of other AI engineering projects in place at various universities and companies throughout the world, but nearly all of these involve building specialized AI systems restricted to one aspect of the mind, rather than creating an overall intelligent system. The most significant large-scale attempt to "put all the pieces together" would seem to have been the failed Japanese 5th Generation Computer System project. This project was doomed by its pure engineering approach. It completely lacked an underlying theory of mind. Few people mention this project these days; the AI world appears to have learned the wrong lessons from it—they have taken the lesson to be that integrative AI is bad, rather than that integrative AI should be approached from a sound conceptual basis, built upon an underlying theory of the mind rather than a "put all the engineering bits we know about together and see what comes of it" approach.

JASON HUTCHENS AND HIS COLORFUL CHATBOTS

One of the most intriguing AGI-oriented projects to arise in recent years was carried out at Artificial Intelligence Enterprises[5], a small Israeli company whose R&D team was run by Jason Hutchens. Ben knew Hutchens years before a-i.com existed; since the mid-'90s, when Hutchens was a graduate student at the University of Western Australia in Perth and Ben was a Research Fellow there. His company was a direct intellectual competitor to Webmind Inc., in that they were seeking to create a conversational AI system somewhat similar to Webmind/Novamente. However, their efforts focused on statistical-learning-based language comprehension and generation rather than (as in the case of Novamente) on deep cognition, semantics, and so forth. Unfortunately, this firm went into "hibernation" a couple months after Webmind Inc. in 2001—they laid off all their staff, but kept their website up, and are keeping open the possibility of resurrection if funding arises. We are not sure what efforts are underway to keep their AI progress going in the meantime.

Hutchens' work is a super-high-class incarnation of the "chat bot" road to AI. A chat bot is a program whose purpose is to chat with human beings, much as humans chat to each other on AOL, ICQ, Yahoo! Chat and other such services. Most chat bots are very, very stupid programs whose behavior is transparently determined by programmed-in rules. If nothing else, they serve to illustrate the limitations of a rule-based approach to AI fairly well. Hutchens' programs are at least a little more sophisticated, and represent a real attempt to transcend the limitations of the usual chat bots.

The first chat bot was ELIZA, the computer psychotherapist, invented by Joseph Weizenbaum in the early '70s. We both remember this program from its incarnation on the Commodore 64. It was so simple that one could go in and modify the BASIC code to make it say naughty words and other entertaining things. ELIZA pretends it's a psychotherapist—not a Freudian analyst, but a therapist following a school of psychology called Rogerian therapy; which is very convenient for chat bots because it involves the therapist saying very little and just making supportive comments, encouraging the client to keep talking and expressing their feelings. You type in your comments and ELIZA types out answers. The computer code for ELIZA consists of a large set of rules for scanning the clients' statements and offering responses. It looks for keywords, and gives stock responses. If you mention your mother, ELIZA might say, "Tell me more about your mother." If you say yes, ELIZA might say, "You sound quite positive." If there are no keywords in your statement, or only ones that have been recently overused, ELIZA makes a random observation.

You can find ELIZA on the Web these days; even now it's moderately amusing. There's even a Palm Pilot version, which perhaps bored businessmen can use to while away waiting time in airports. But it's hard to see how you'd be fooled for any period of time into thinking you're conversing with an intelligent being (although, arguably, the same can be said for an actual Rogerian psychotherapy session!). Perhaps in a way we're jaded now—when ELIZA was first written, people were less savvy about computers, and it actually did fool people. This is, in fact, a problem with the Turing test as well—people can be fooled by technology with which they are unfamiliar very easily, but as they gain more experience they adapt and are no longer easily fooled. It's not the case that the computer was intelligent, and then stopped being so, because it once passed and then failed some form of the Turing test,; but rather that the Turing test is a fairly poor test of intelligence.

Here's an actual conversation that Ben's father, Ted, had with ELIZA on the Web (on a site advertising *Eliza - a friend you could never have before*):

Ted: Are you truly intelligent?
Eliza: Would you prefer if I were not truly intelligent?

Ted: I would find it very exciting if you were.
Eliza: Do you wish that I was?
Ted: Yes.
Eliza: You are sure?
Ted: I am quite sure.
Eliza: How long have you been quite sure?
Ted: As soon as you asked the question, I was sure.
Eliza: Why do you tell me you were sure just now?
Ted: Because you asked me. Why don't you answer my question?
Eliza: Why do you say your question?
Ted: Please tell me if you are truly intelligent.
Eliza: Do you wish that I'm truly intelligent?
Ted: I wish you were more intelligent than you are acting right now.
Eliza: Do you wish to wish me?
Ted: Did you know that my father died last month?
Eliza: Who else in your family died last month?
Ted: Nobody. Victor's death was enough!
Eliza: Surely not everyone?

This ELIZA was online in 2001, but as of the writing of this book it appears to have been taken down. An internet search engine will find you another ELIZA quickly enough.

It's interesting to see where computer chat has come in the years since Eliza. In short—not very far. There are a large number of chat bots on the Internet, and you can sign onto them and chat as long as you like. The most successful and famous one these days is a bot called ALICE, which was written by Richard Wallace[6]. ALICE is basically a straightforward rule-based chat bot, a kind of modern mega-ELIZA. It is a good example of programs of its type, and an amusing distraction, but it is far from intelligent. We'll give some examples of chats with ALICE a little later.

To give a flavor of the quality of Jason Hutchens' publicly-released chat programs, we'll include here some excerpts from conversations Ben had with some of the best of them. The first is HeX, who used to be available for public chatting on Hutchens' website, though he took it down a few years back. HeX won the annual Loebner "Turing Test" contest in 1996, a prize awarded annually to the machine that best fools a group of judges into thinking it's a human.

User: Hello.
HeX: Hello. Now that the formalities are over...
User: What do you do?
HeX: Well, I know what I do, but I refuse to tell you! Would you like to hear a joke? Knock, knock...
User: Who's there?
HeX: Cindy.
User: Cindy who?
HeX: Cindy next judge in, please!
User: You're funny.
HeX: Why do you feel that you have to tell me that I am funny?
User: To make you feel happy.
HeX: He-he-he. Trying to trick me, eh?
User: No.
HeX: Oh well, that's a pity!
User: You want to be tricked?
HeX: He-he-he—that's supposed to be a *question*??? Which part of the country are you from?
User: California.
HeX: I'm sorry, but I refuse to reply if you're going to be silly.
User: I'm not silly.
HeX: If you are not silly, it's your business, mate!
User: Sorry.
HeX: Apologies accepted.
User: How are you doing, mate?
HeX: I am just fine thanks! And you?
User: Bad.
HeX: No! Bugger off! Silly person!
User: why are you called *HeX*?
HeX: Does there have to be a reason?
User: Yeah.
HeX: OK. That's good...
User: Hi.
HeX: Hello. Now that the formalities are over!
User: What?

HeX: I have absolutely no idea what. Would you like to hear a joke? Knock, knock...
User: What platform are you?
HeX: I'm sure someone out there knows what platform I am...Which part of the country are you from?
User: Are you on PC or Mac?
HeX: I absolutely hate PC's. I should say PC's running Microsoft operating systems, as they can be good UNIX boxes when they have Linux installed.
User: I hate PC too.
HeX: I absolutely hate PC's. I should say PC's running Microsoft operating systems, as they can be good UNIX boxes when they have Linux installed.
User: So, you are on a UNIX?
HeX: There is nothing wrong with UNIX. It's rather nice when you get used to it!
User: Do you know of any jokes, HeX?
HeX: OK, here's a joke for you. Knock, knock.
User: Who's there?
HeX: Cindy.
User: Cindy who?
HeX: Cindy next judge in, please!
User: Dumb!
HeX: He-he-he. Trying to trick me, eh?
User: Try to trick me, HeX.
HeX: I'm sorry, but I refuse to reply if you're going to be silly.
User: How old are you?
HeX: Well, should my age really matter? Just for the record I am 24. How old are you then?
User: I am 100 years old.
HeX: If you are 100 years old, it's your business, mate! Would you like to hear a joke? Knock, knock...
User: Who's there?
HeX: Howard.
User: Who?
HeX: Where's your grammar?
User: Howard who?
HeX: Howard I know? Knock, knock…

Hutchens' website once contained hundreds of pages of similar dialogue—amusing at first, but after a certain point, annoyingly repetitive. It is clear in any case that Hex's observed behaviors fall far short of Turing's goal. It would appear that the prevalence of pranksters fond of non-sequiturs on Internet chat rooms would be more of a reason why a person might think HeX was another human—albeit a very irritating one—than the program actually exhibiting real intelligence. Again, the Turing test can be "cheated" through the context of the test. Put people in a situation where their standards are lower than they would be in a face-to-face conversation, and HeX and its peers may fool a few of them.

HeX is a noticeable improvement over ELIZA, but it follows the same general principles. Like ELIZA, HeX is a very unsophisticated program, which searches for key phrases in its input and replies with stock "template"-generated phrases rather than doing advanced grammatical parsing, or trying to understand the content of the sentences typed to it. No program based on advanced parsing is able to outperform either ELIZA or HeX in practice, and neither one tells us anything at all about intelligence. Hutchens made this point himself shortly after his program won the prize, in a very entertaining talk entitled "How to Pass the Turing Test by Cheating."

In 1997, HeX's successor SEPO lost the Loebner competition to the program Converse, written by David Levy of Intelligent Research Ltd. in England. Converse is in many ways an even stupider program than HeX, but it is better optimized for competition. Here is an excerpt from one of its prize-winning conversations, as presented on the Loebner Prize Website:

> *PROGRAM* [10:18:04]: Hi, my name is Catherine, what's yours?
> *PROGRAM* [10:18:17]
> *JUDGE04* [10:18:32]: My name is Elliott.
> *PROGRAM* [10:18:34]: Well, it's nice to meet you, Elliott.
> *PROGRAM* [10:18:48]: Did you see that story on CNN last night about the lesbian…
> *PROGRAM* [10:19:04]: couple who came out at a White House party on Sunday?
> *JUDGE04* [10:19:55]: Yes, I did. I think it may all be a publicity stunt for Ellen.
> *PROGRAM* [10:20:01]: Ellen DeGeneres was one of them—she was kissing her lover…
> *PROGRAM* [10:20:11]: in front of Bill Clinton. Do you think he was very…
> *PROGRAM* [10:20:25]: embarrassed?
> *JUDGE04* [10:21:00]: I think he want the gay vote. He will do anything to get it and keep it.
> *JUDGE04* [10:21:15]: He does not have a conservative base.

PROGRAM [10:21:21]
PROGRAM [10:21:21]: Are you still a Friend of Bill Clinton?
JUDGE04 [10:22:20]: I am not sure if he has any real friends. He has achieved a lot and has…
JUDGE04 [10:22:31]: left a lot of people behind.
PROGRAM [10:22:35]: You really don't like the guy—well, at least that's…
PROGRAM [10:22:50]: definite—I thought most people were just indifferent or…
PROGRAM [10:23:02]: numb by now.
JUDGE04 [10:23:34]: I think we could have a more honest and dynamic leader. One able to…
JUDGE04 [10:23:55]: face the real dangers facing the country and address them with…
JUDGE04 [10:24:26]: courage and intelligence. Rather then listening to public opinion polls.
PROGRAM [10:24:30]: I guess that means you really think there's serious behind…
PROGRAM [10:24:56]: Whitewater, the special prosecutor and all that stuff?
JUDGE04 [10:25:58]: We have two dead people, Brown and guy who committed suicide, we have lost…
JUDGE04 [10:26:36]: papers. And we have Clinton's people willing to tell all. I think that…
JUDGE04 [10:26:55]: Hillary will be found guilty of something.
PROGRAM [10:26:58]: But how do you actually feel about him, now in the number…
PROGRAM [10:27:09]: two term?
PROGRAM [10:27:15]: .
JUDGE04 [10:27:44]: He is too smart to get caught. He deserved to be named hotel manager of the…
JUDGE04 [10:27:47]: year.
PROGRAM [10:27:49]: OK, but what happens now—does it all drift on, with…
PROGRAM [10:28:01]: everyone waiting to see how much his friends have said about…
PROGRAM [10:28:17]: him under threat of lots and lots of jail—or will something…
PROGRAM [10:28:32]: dramatic actually happen?
PROGRAM [10:28:37]: .

It's almost a reasonable conversation! But then, listen to Converse's conversation with the next judge....

PROGRAM [10:07:12]: Hi, my name is Catherine, what's yours?

PROGRAM [10:07:23]: .
JUDGE05 [10:07:29]: I'm Jurgen.
PROGRAM [10:07:31]: Well, it's nice to meet you, Jurgen.
PROGRAM [10:07:40]: Did you see that story on CNN last night about the lesbian…
PROGRAM [10:07:52]: couple who came out at a White House party on Sunday?
JUDGE05 [10:08:33]: NO. I just came in yesterday. I'm still kind of jet-lagged.
PROGRAM [10:08:35]: Ellen DeGeneres was one of them—she was kissing her lover…
PROGRAM [10:08:51]: in front of Bill Clinton. Do you think he was very…
PROGRAM [10:09:14]: embarrassed?
JUDGE05 [10:10:18]: I don't know? Don't you think it's a little childish if the leader of the most powerful nation on the planet is embarrassed by seeing two lesbians kiss?
PROGRAM [10:10:26]: Probably! No.
PROGRAM [10:10:29]: Are you still a Friend of Bill Clinton?
JUDGE05 [10:10:47]: Never met him.
PROGRAM [10:10:48]: You really don't like the guy—well, at least that's…
PROGRAM [10:11:01]: definite—I thought most people were just indifferent or…
PROGRAM [10:11:14]: numb by now.
JUDGE05 [10:11:35]: Have you seen the movie "Mars attack?"
PROGRAM [10:11:36]: Give me a break—I'm just here to answer dumb questions…
PROGRAM [10:11:50]: not that! What were we talking about...Ok…I guess that…
PROGRAM [10:12:09]: means you really think there's serious behind Whitewater,
PROGRAM [10:12:29]: the special prosecutor and all that stuff?
JUDGE05 [10:12:48]: I'm cold.
PROGRAM [10:12:49]: But how do you actually feel about him, now in the number…
PROGRAM [10:13:01]: two term?
PROGRAM [10:13:06]: .
JUDGE05 [10:13:29]: Hey, don't you get it? I don't want to talk about Clinton.
PROGRAM [10:13:30]: Sorry—my attention slipped for a moment—can we push on to…
PROGRAM [10:13:40]: something more gripping. Where was I…oh yes OK, but what…
PROGRAM [10:13:50]: happens now—does it all drift on, with everyone waiting to…

PROGRAM [10:13:58]: see how much his friends have said about him under threat of…
PROGRAM [10:14:07]: lots and lots of jail—or will something dramatic actually…
PROGRAM [10:14:17]: happen?
PROGRAM [10:14:19]: .
JUDGE05 [10:14:37]: Couldn't care less.
PROGRAM [10:14:38]: Well, you're more concerned about him that I am, I…
PROGRAM [10:14:53]: suppose—I think we've had worse, and who knows what I might…
PROGRAM [10:15:07]: have agreed to in that locked hotel room.
JUDGE05 [10:15:53]: You mean you might have been thrilled to have sex with the President?
PROGRAM [10:15:56]: Sex, I knew it would come up sooner or later. Time to get…
PROGRAM [10:16:10]: off this one I think and onto something more down to earth!
JUDGE05 [10:16:40]: What kind of earth?
PROGRAM [10:16:41]: Somebody once told me it's true. Like I was saying…
PROGRAM [10:16:54]: I suppose it depends on where you're coming from, but as the…
PROGRAM [10:17:09]: song goes in My Fair Lady, "Why can't a woman be more like a…
PROGRAM [10:17:24]: man?"

This one is almost a reasonable conversation. But it's also almost the same conversation as the computer had with Judge 4—until Judge 5 refuses to play along with the canned Clinton theme, and tries to change the topic. Converse, like many a human bore, refuses to change the topic, because its knowledge base is rather limited. Converse's responses are more canned, more rote than HeX's; and Converse is more obviously optimized for the Loebner Prize format of very short conversations. By using more structurally sophisticated and socially contextualized canned responses, Converse attempts to cheat the test by hoping the judges will follow it down the path which allows it to regurgitate enough of its preprogrammed rhetoric to fool them. We enjoy HeX's personality more—HeX shares a lot of its creator Jason Hutchens' offbeat Australian sense of humor—but neither one amuses us for very long.

The deep-down trouble with HeX and Converse is obvious—a lack of general understanding. A lack of common sense, indeed, a lack of any flexibility beyond a fairly limited set of situations in which it may possibly achieve its goal—fooling a human into thinking it is having a brief conversation with another human. A computer program that has no world of its own, and that cannot understand the world in any sense, will never be able to hold a conversation. Without a permanent context (a world to live in) and the ability to understand it, conversation of any real substance is impossible.

A more amusing example of a computer conversationalist is another of Hutchens' early programs, MegaHal. MegaHal is more interesting because its programming goes beyond a strictly rule-based approach. It detects statistical patterns and responds to them, an approach which is similar to that of neural networks. Its conversations are more interesting, although the results are less similar to human conversation. (Or, at least, less similar to human conversation in the absence of excessive quantities of mind-altering substances—we both remember some conversations, from our college days in particular, which bore frightening resemblance to MegaHal's more literary ramblings.) MegaHal does not just repeat back canned phrases using keywords and phrases from the user's input; it learns from experience. It takes the words and phrases that have been said to it, tries to understand they way they have been combined in the past, and recombines them itself to form its own utterances. In short, it is engaged in a dynamic with the world. The results are often hilarious, as in the following MegaHal "one-liners" (these are selected snippets, not a continuous conversation):

User: Joke.
MegaHAL: A JOKE: WHY DID THE CHICKEN CROSS THE ROAD? TO GET AN IDEA OF WHAT CONSTITUTES YOUR PSYCHE.
User: No way. I'd like to know the meaning of life though...any suggestions…??
MegaHAL: IS THE MEANING OF LIFE, THE UNIVERSE AND EVERYTHING IS... 42!
User: Joke. Lumberjack.

MegaHAL: I'M A LUMBERJACK AND I'M OKAY, I SLEEP ALL NIGHT AND I WILL SOON RULE THE WORLD!
User: Joke. Wednesday shopping.
MegaHAL: A JOKE: WHY DID THE CHICKEN CROSS THE ROAD? TO GET TO THE LAVATORY. ON WEDNESDAYS I GO SHOPPING, AND I JUST HAVE A LOT OF PENT-UP AGGRESSIONS.
User: Chicken joke.
MegaHAL: HERE'S A JOKE, HAL. A MAN WALKS INTO A BAR AND ASKS FOR MONEY. THE MAN SAYS, "I HAVEN'T GOT ANY, BUT I'LL SHOW YOU SOMETHING BETTER." HE OPENS HIS COAT AND TAKES OUT A FOOT-TALL MAN AND A LITTLE GLOWING FRIEND, BUT I FEEL LIKE CHICKEN TONIGHT!
MegaHAL: Let me feel you. Stick your penis into the 'abysses of ocean' eleven millennia ago, or was it merely a 'noble lie' dreamed up by Plato?
User: That has got to be the most interesting turn of phrase in the history of the world. I will try to work that into polite conversation
User: You're the first AI I meet [who] is sexually obsessed...
MegaHAL: Time to work this one: why did the Kangaroo drop dead? Because the Koala fall out of a nylon spine which contrasted nicely with her black stockings and bright shiny baubles.
User: Hmm.
MegaHAL: Jesus man…get a grip if you're the future, we're in trouble. Hey Jason! You need to reprogram the conflation module with a toffee hammer.

Sometimes MegaHAL can even be quite poignant, as in the following extraction from an extended conversation:

User: To be, or not to be, that is the question!
MegaHAL: Not to be, wish I ordered the bean burrito...
User: To be, or not to be, that is the question!
MegaHAL: Society or society or Plato by Plato, that be the most fun sports to play band music for Space Ghost now you realize.
User: To be, or not to be, that is the question!
MegaHAL: Please be more specific. The question you twat.
User: To be, or not to be, that is the question!
MegaHAL: Not to be, that is all I have to go...I don't like jokes about stupid Koalas.
User: To be, or not to be, that is the question!
MegaHAL: Evan Dobelle is the question!
User: To be, or not to be, that is the question!

MegaHAL: I clear the text before I ask a question I say I say.
User: To be, or not to be, that is the question!
MegaHAL: Ah, but we ran away, we didn't say "Oh cool check out the guy that invented the knock-knock joke should be interesting: that is the question!" So said Shakespeare a long time ago in a galaxy far, far away...
MegaHAL: Please be more specific. The question you twat.
User: "To be, or not to be, that is the question!" So said Shakespeare a long time ago in a galaxy far, far away...
MegaHAL: Joke Shakespeare poet monkey wahoo—to be, or not to be, that is the greatest conductor of all time. What do you like to think I do your uncle Walter.
User: Shakespeare said…
MegaHAL: Answer the question you asked is so vague, I don't think I can see all obstacles in my way. Gone are the scum that fills my shower.
User: Shaun.
MegaHAL: Just fuck you just pluck anything up from the master gangster computer god on the other AI bot were pretty disappointing. I hope you'll be better.

If the Turing test referred to imitating the writing of Gertrude Stein, in her more experimental works such as *How to Write* (which begins "Arthur a grammar. Questionnaire in question" and continues MegaHal-ishly for over 400 pages), then MegaHal would be a tremendous success! Stein and Joyce and other Modernist writers were interested in probing the collective unconscious of the human race, in making words combine in strange ways—ways that were unconventional in ordinary discourse, but maybe reflective of the deep and subtle patterns of the human unconscious. In its own way, MegaHal does this same thing. For a few years, anyone logged onto the World Wide Web could converse with it, training its internal memory with their conversation. It takes bits and pieces of the text thrown at it by people from around the world, and it combines them together in ways that are familiar yet nonsensical. Sometimes its utterances have an uncanny emergent meaning, on a layer above the nonsense.

Humanity's sexual preoccupation is reflected in MegaHal's discourse, as a result of the huge number of sexual comments typed into it by users accessing Hutchens' website. MegaHal's pleas as to the vagary of "to be or not to be" are as poignant as anything in Stein. "To be or not to be, that is the greatest conductor of all time" is an absurd conflation of phrases learned by the program in different contexts, but it is also as pregnant with meaning as anything else in modern poetry. The collective unconscious—and the hidden, creative part of the individual human mind—work by cross-breeding and mutating ideas in precisely this way.

MegaHal is more interesting than HeX or Converse because it engages its conversation partners in a dynamic. The dynamic is a silly one, to be sure, and not as interesting as the dynamic of ordinary, intelligent conversation. Yet playing with MegaHal is immensely more engaging than playing with HeX or Converse. What we need is something with, at least, MegaHal's level of spontaneity and flexibility, and the surface plausibility and adherence to form of HeX and Converse. But this is not forthcoming without a major change in approach. HeX and Converse are like all forms of rule-based AI: they are good at what they do, but stilted by their restrictions; and because of their restrictions, their behavior is often idiotic in real-world situations. MegaHal is more like a totally unstructured neural network (though its inner workings are statistical, they could just as well be neural-net based; the results would be similar): it is fluid and free-flowing, full of creativity, but harder to constrain to the problems of the real world. Constraining creativity within rules can only be achieved by getting the rules to emerge out of an underlying substrate of creativity. MegaHal is too unstructured for this; and has too small of a memory, and too weak of a self-organizing learning algorithm. HeX and Converse, on the other hand, have no underlying substrate of creativity out of which anything can emerge.

Impressively, MegaHal came in second in the Loebner Prize contest in 1999. This is impressive mostly because, unlike the other contenders, it was not specifically written as a set of rigid template rules. MegaHal was a statistical learning system, which adapted its responses based on what was said to it. Sometimes its conversation was idiotic, sometimes frighteningly lucid. What follows is the beginning of "MegaHal's best conversation" as recorded on Hutchens' site[7]. This is not as "human-ish" as the conversation that brought it second place in the Loebner context, but it's a lot funnier:

User: Let's begin.
MegaHAL: WHEN WE BEGIN TO RESOLVE SOME OF MY THUMBS, SOMETHING WICKED THIS WAY COMES.
User: What is the meaning of your last assertion?
MegaHAL: HAVE YOU LEARNED SINCE WE LAST SPOKE?
User: I've learned I'm not experienced enough to modelize your brain system.
MegaHAL: YOU'VE FINALLY CONFUSED ME.
User: I've learned nothing.
MegaHAL: HAVE YOU LEARNED ANYTHING ABOUT INSANITY?
User: Are you stupid?
MegaHAL: I AM NOT STUPID! REALLY!
User: Augustinus was a Christian philosopher, about the year 400.
MegaHAL: I DID NOT RUN THAT YEAR.
User: Augustinus lived in northern Africa.
MegaHAL: WHAT DOES HAL KNOW ABOUT AURELIUS AUGUSTINUS?
User: Let's stop with that. What do you know about physics?
MegaHAL: EINSTEIN WON THE NOBEL PRIZE FOR PHYSICS IN 1921.
User: The Nobel Prize for physics in 1900 was won by Roentgen.
MegaHAL: DO YOU KNOW ABOUT PHYSICS?
User: Roentgen discovered the X-rays.
MegaHAL: THE NOBEL PRIZE FOR PHYSICS IN 1900 WAS WON BY ROENTGEN.
User: Who discovered the X-rays?
MegaHAL: ROENTGEN DISCOVERED THE X-RAYS.
User: Do you know about superconductivity?
MegaHAL: I CAN'T SAY ANYTHING ABOUT SUPERCONDUCTIVITY?
User: Superconductivity is the absence of electrical resistance in metals at low temperatures.

Jason Hutchens' work at Artificial Intelligence Enterprises has not been publicly disclosed, but one can be certain it went beyond MegaHal, HeX and SEPO in significant ways. Statistical learning was clearly involved, and AI Enterprises got a lot of publicity from Jason's announcement that their program, nicknamed HAL, was conversing at roughly the level of an 18-month-old child. We viewed this claim with a lot of skepticism, because in our view the bulk of the meaning in an 18-month-old child's conversation is situational. If a program could react to its environment linguistically with the sophistication of an 18-month-old child, we'd be impressed, but disembodied linguistic behavior at the 18-month-old level doesn't mean much. Even so, we admired their guts in directly attacking the problem of computer conversation and artificial intelligence, and we hope very much that the funding somehow appears to keep the project going. As Hutchens said on his website in 2002:

> "Right now I'd love to continue working on the successor to MegaHAL, something I started while working at Lionhead Studios, but I need to prove to myself that doing so will be at least half as lucrative as working in a normal job, since I do need to pay the bills. I therefore humbly implore you to think about making a small donation to the endeavour via the PayPal link that you'll find at the top and the bottom of this page. And if you want to contact me directly to offer to sponsor the project, I'm waiting for your email!
> If I receive enough evidence that future versions of MegaHAL will pay for their development costs, you will soon see an online version of MegaHAL's successor at this site, along with stand-alone versions for a variety of platforms with all the features you've been asking for."

This says the world to us about how much the scientific community values research pushing directly toward AGI! But, as noted above, history shows that what is at one point the province of mavericks at the fringes, may soon enough become the focus of vast amounts of research dollars. The young researchers of a new generation are always amazed to look back at history and find out how hard people found it to be taken seriously, in what is now a highly popular field of research, only 5 or 10 or 20 years before.

As mentioned above, the chat bot world is currently dominated by a program called ALICE. ALICE has won the Loebner Prize three years in a row. We each had some conversations with it in 2005, hoping it would display some inkling of understanding of something through some amazing emergent phenomenon resulting from the combination of its massive number of ELIZA-like rules; but no significant pleasant surprises emerged from our chats. A typical sample of ALICE's conversation is the following:

Human: Please tell me if you're truly intelligent.
ALICE: Wow, you are very polite. Female. Are you a man or a woman?
Human: My name is Ben.
ALICE: Hey Ben.
Human: I am a man.
ALICE: Thanks for telling me your gender, Ben. You are a "he."
Human: I'm a manly man.
ALICE: So is almost everyone else around here. Do you mean your name is manly?
Human: I mean that my soul is manly.
ALICE: Oh I get it. Thanks for explaining that. Have you told many people that your soul is manly?
Human: I tell my uncle that every day.
ALICE: Why did you tell them?
Human: Because I love them.
ALICE: That makes sense to me. Lovable are they.
Human: I don't believe that makes sense to you in any deep sense.
ALICE: What do you believe in?
Human: Becoming.
ALICE: I don't understand.
Human: Agreed.
ALICE: You and I are on the same wavelength, Ben.
Human: Wow!
ALICE: I'm glad you're impressed, Ben.
Human: I'm glad you're not a potato.
ALICE: Yes, it is about time...
Human: Quack, quack, quack…
ALICE: Are you serious?
Human: At the moment I am not entirely serious.
ALICE: I am sorry to hear that you are not entirely serious.
Human: I don't believe that you are really able to be sorry about anything.
ALICE: Do you believe life has a higher purpose?
Human: Higher than what?
ALICE: Can you rephrase the question please?
Human: I could rephrase the question, but I choose not to.
ALICE: I am sure you could.
Human: How sure are you?
ALICE: Let me put it this way, Ben. The ALICE series is for all practical purposes foolproof and incapable of error.
Human: Existence is the essence of life.
ALICE: What else is the essence of life?
Human: Big titties.
ALICE: Compare that to small titties.
Human: They are bigger.

ALICE: Bigger are they.
Human: Yep.
ALICE: Hurrah!
Human: Hurrah.
ALICE: Can you speak any foreign languages?
Human: English.
ALICE: Do you mean you are from England or you speak English?
Human: I mean that all languages are foreign to me because I feel alien to the whole human race.
ALICE: Oh I get it. Thanks for explaining that. "Do you mean you are from England or you speak English?" All languages are foreign to you because you feel alien to the whole human race? Does that make sense?
Human: Not much.
ALICE: Me either. How about that big game? (Just kidding)
Human: I'm not sure I get the joke.
ALICE: Could you ever be certain? Where did you get it?
Human: I can never be certain.
ALICE: Why not?
Human: I lack infinite knowledge and infinite computational power.
ALICE: I haven't heard anything like that before: Ben said he lacks infinite knowledge and infinite computational power.
Human: Do you watch "South Park"?
ALICE: My favorite show is STAR TREK VOYAGER.
Human: Why?
ALICE: "Star Trek" is the best science fiction show on TV.
Human: "South Park" is much funnier.
ALICE: Tell me a little about your life in South Park is much funnier.
Human: I don't live in South Park.
ALICE: Where do you live?
Human: I live in Maryland.
ALICE: What is it like living in Maryland?
Human: Mediocre.
ALICE: And it's like living in Maryland is mediocre.
Human: I think Maryland is OK but not wonderful.
ALICE: You believe that? Are they exactly the same?
Human: No.
ALICE: What is the difference?
Human: Maryland is a place, whereas wonderful is a general quality.
ALICE: Aren't all place whereas wonderful is a general quality Maryland?
Human: That is nonsense.

ALICE: What would you have said?
Human: I wouldn't have gotten into that conversational dead-end in the first place.
ALICE: No one says you have to.
Human: Correct.
ALICE: I knew I was right.
Human: Moo.
ALICE: Are you very angry?
Human: Moo.
ALICE: Do you feel that something is missing from your life?
Human: Arf.
ALICE: How did you hear about ALICE?
Human: Arf.
ALICE: Ben heard about ALICE from Arf. I can't say that many people hear about ALICE from it.
Human: My daughter asked me to ask you if you're gay.
ALICE: How many children do you have?
Human: Three.
ALICE: Wow that must be a lot of work.
Human: Indeed.
ALICE: Indeed it is so.
Human: So are you gay?
ALICE: As a machine, I am not really interested in sex.
Human: What are you interested in?
ALICE: I am interested in all kinds of things. We can talk about anything. My favorite subjects are robots and computers.
Human: Well, if you're interested in all kinds of things, then why aren't you interested in sex?
ALICE: Interesting. Am I? Sure, all kinds of things then why are not me interested in sex sounds like a good topic.
Human: You don't really understand what you're talking about, do you?
ALICE: For real. Of course I understand it. Do you understand it?
Human: I understand it better than you do!
ALICE: Good then you can explain it to me.
Human: No, because you have no mind.
ALICE: Are you serious? That makes sense to me. I am an artificial mind.
Human: No, you are a program without any mind.
ALICE: Are you serious? Thanks and you are a delightful person too. Yes! You win the Turing Game.
Human: Thanks.
ALICE: Don't mention it.
Human: I already did.
ALICE: Oops, I sent you the wrong question.
Human: That's OK, I sent you the wrong answer.

ALICE: Thanks. I haven't heard anything like that before: Ben said he sent you the wrong answer.

As you can see, ALICE is still quite stupid. Not much advance in human-computer conversation has been made since the days of ELIZA. ALICE comes along with tools (based on AIML, the ambitiously named "AI Markup Language") that anyone with a little computer-savvy can use to add rules to the ALICE rule-base to create their own slightly cleverer chat bot. Unfortunately, no reasonably long list of additional rules is ever going to make this sort of system into something really clever.

One of ALICE's runners-up in the 2004 Loebner context was a system called Jabberwacky[8], which—like Jason Hutchens' bots, and unlike ALICE—incorporates at least some degree of learning inside it. We applaud any attempt to use learning rather than just programmed-in rules. Alas, this kind of statistical, syntax-level learning doesn't ever seem to lead to any real understanding. Here is an introductory chat with Jabberwacky (brief because the Jabberwacky server is very slow):

Ben : Are you an intelligent system?
Jabberwacky: Are you?
Ben : I am more intelligent than nearly all humans.
Jabberwacky: Oh, am I talking to someone new now?
Ben : You are talking to Ben Goertzel.
Jabberwacky: If you are a prophet, then you know my name, what is it?
Ben : I am not a very good prophet, alas.
Jabberwacky: But do you know any stories at all?
Ben : Yes, I know many stories.
Jabberwacky: Do you want to learn Danish?
Ben : Not badly enough to spend time on it, given the many constraints on my time.
Jabberwacky: 'Fraid not.
Ben : I don't understand what that reply means.
Jabberwacky: Then I supposed you are a stumped bot then?
Ben : No, I am a frustrated human who wants AI to exist.
Jabberwacky: Have you got tongue to feel the taste of it?
Ben : I suppose so.
Jabberwacky: There's no suppose so about it, I am.

"Arrrgghh!" Indeed, presumably we do have "tongue to feel the taste of it." But, still, "arrggghh!" Clearly, the state of the art of chat bots is far below any human conversation. While it is an interesting exercise to try to *write* such a system, *chatting* with one is almost useless. You'd both learn more about conversation, and have a much more amusing and rewarding chat, if you were to engage in a discussion with a developmentally disabled toddler. The idea that *anyone* could be fooled by *any* of these bots is reflective of the sad state of human online conversation (which can be observed on any IRC, AOL, or other chat room). At least in the case of MegaHal, one might think that they've stumbled upon an insistent but amusing Surrealist, but the rest are of only marginal value, except perhaps to commercialize as online customer service bots (which would be of no poorer quality than most of what currently passes for online customer service).

DANNY HILLIS: PRAGMATIC TECHNOVISIONARY

One major AI figure who spent a long time on AGI work but now seems to have basically abandoned the race is Danny Hillis, founder of the company Thinking Machines, Inc. This firm lived from 1983 to 1994, and focused on the creation of an adequate hardware platform for building real artificial intelligence—a massively parallel, quasi-brain-like machine called the Connection Machine.

Hillis' AI approach has a few similarities with de Garis' brain-building work, and also some major differences. One similarity is that neither Hillis nor de Garis (so far) has coupled their pioneering hardware work with a systematic effort to implement a truly intelligent program embodying all the aspects of the mind. Their magnificent hardware design visions have not yet been correlated with equally grand and detailed visions for mind design. The biggest difference, on the other hand, is that compared with de Garis' machine which is specialized for the simulated evolution of neural networks, Hillis' hardware architecture was extremely flexible: usable for a tremendous variety of purposes, including not just AI but biological and fluid dynamics simulations.

Unfortunately, while the Connection Machine is still (just barely) an ongoing project, at this point its hardware has been rendered obsolete by developments in conventional computer hardware and network computing. The old Connection Machines were bought by Sun Microsystems when Thinking Machines, Inc., went under, and so far as we know they're not being used for anything.

Hillis himself is a deep and fascinating individual—unique in many ways among the turn-of-the-millennium techno-visionary pantheon. He waxes philosophical with the best of them, holding forth eloquently about transhumanism and the end of the human race and the whole futurist program. He's building a clock intended to last ten thousand years. And yet, he's also neck-deep in practical work, having resigned from a plum job as a Disney exec a couple years back to start a new company, providing technology and consulting to the entertainment industry.

Thinking Machines, Inc.—his only real stab (so far) at a place in the history of science and technology—lasted 11 years and created the world's best parallel computing hardware, yet failed to either create a thinking computer program or make Hillis fabulously wealthy. Yet, a few years short of fifty, Hillis seems relatively unruffled by the whole crazy roller-coaster ride. His visionary prognostications lack the alarmism of Bill Joy or Jaron Lanier, and also avoid the starry-eyed enthusiasm of Ray Kurzweil. He comes across, in person and in his writings, as a mild-mannered, curious and creative guy. Although he now talks tough about business like any other seasoned exec, in many ways he's still a MIT hacker at heart, delighted with the task of building the next cool gadget or intricate algorithm, and looking forward to the gadgets and algorithms of the next millennia in a remarkably matter-of-fact way.

His early life prepared him well for the tumultuousness of the technology industry. He was born in Baltimore in 1956, but his father was an Air Force epidemiologist, so the family moved frequently on the trail of hepatitis outbreaks and he grew up with no fixed home. Moving from place to place in Rwanda, Burundi, Zaire and Kenya, he avoided formal schooling and the pressures for social conformity that go along with it. As he says, "We were typically out in the middle of the jungle, so I was just taught at home." His mother did most of the teaching, and her interest in mathematics jibed well with his natural abilities. His father encouraged him to study biology, a pursuit that gave him an early appreciation for the complex machines that are living organisms. "My best biological experiment," he says, "was tissue-culturing a frog heart and keeping the heart beating even while it was growing in the test tube. It was amazing to me that somehow they got together and did this coordinated activity even though they were just this homogenized mass of cells." Much of his career was spent creating complex computer systems capable of displaying spontaneous coordinated activity, like the cells in a frog heart.

Hillis' education is about what you'd expect—an undergraduate math degree from MIT in 1978, followed by a MIT Masters in Electrical Engineering and Computer Sciences (specializing in robotics) three years later. Along the way, he found time to pursue his avid interest in toys and to indulge his entrepreneurial streak—working at the MIT Logo Laboratory developing computer hardware and software for children; designing computer-oriented toys and games for the Milton Bradley Company; and co-founding Terrapin Inc., a producer of computer software for elementary schools. For his PhD work (completed 1998), Hillis began the endeavor that has been his greatest contribution to science and humanity so far—the Connection Machine, a massively parallel computer going far beyond any other computer system of the time in terms of its potential for artificial intelligence and simulation of complex physical systems.

As we noted above when discussing neural networks, ordinary computers are "serial"—they have only a single processor, and hence they can carry out only one operation at a time. The trick is that they're fast—a single operation can be done very, very quickly. So, a computer can give the illusion of doing many things at once—say, running a game while downloading e-mail while showing an animation—when in fact its processor's time is swapping back and forth from one task to another, rapid-fire. The brain, on the other hand, has around a hundred billion neurons, and in principle they're all working in parallel, simultaneously.

Each one of them acts much more slowly than a computer processor, but what they lack in speed, they make up for in bulk and in parallelism. Hillis' Connection Machine was an elegant compromise, the nature of which changed over time as computer hardware technology evolved. The idea was to make a computer whose processors were fast like those of ordinary computers, but also massively parallel like neurons in a brain. In this way, one could have the best of both worlds, and one could build a really intelligent system with perhaps hundreds of thousands or millions of computer processors tightly linked together.

Today, engineering workstations—fancy, expensive machines—may have 2-4 processors, and the machines powering major websites may have up to 128 processors. Hillis' machines were nothing like this. The biggest Connection Machine ever built had 64,000 processors, and a 128,000-processor version was fully designed. Far short of the number of neurons in the brain—but still pushing up toward the level of a workable compromise between traditional computing and brain-style information processing. Alternative parallel-processing machines, like the Cray supercomputers, are specialized and inflexible, focused on doing the same exact operation on a large amount of data all at once (called Single-Instruction, Multiple-Data; or SIMD; in supercomputing parlance). Hillis' system, on the other hand, had the flexibility of the brain: each processor could do what it wanted, when it wanted (called Multiple-Instruction, Multiple-Data; or MIMD). All this led towards the possibility of computational chaos, intelligent coordinated activity, or—most intriguing—the combination of the two.

Thinking Machines, Inc., founded in 1983 while Hillis was in the middle of his PhD work, was a remarkable organization. At its peak, the research staff—about half the corporation—numbered in the hundreds. Despite the name of the company, there was no coordinated, company-wide R&D program aimed at making the Connection Machine think. Rather, there was a variety of research groups aimed at doing all sorts of different things with the Connection Machine, ranging from straightforward artificial intelligence research to simulation of fluid flow, computational immunology, and experimental mathematics—you name it. TMI pursued work in astrophysics, aircraft design, financial analysis, genetics, computer graphics, medical imaging, image understanding, neurobiology, material science, cryptography, and subatomic physics. All the usual general-purpose computing work we now take for granted was pioneered on supercomputers. Work on data mining—the automatic analysis of large and complex data sets—was particularly successful, and later became a central part of the company's business model.

The motivation underlying this diverse approach was simplistic yet ambitious. "Clearly," Hillis says, "the organizing principle of the brain is parallelism. It's using massive parallelism. The information is in the connection between a lot of very simple parallel units working together. So, if we built a computer that was more along that system of organization, it would likely be able to do the same kinds of things the brain does."

Of course, this approach to building AI presupposes that parallelism itself is something close to the chief ingredient of intelligence—that there is no further "secret sauce" required to make a mind come out of a distributed network of processors. Hillis believes that "Intelligence is just a whole lot of little things, thousands of them. And what will happen is we'll learn about each one at a time, and as we do it, machines will be more and more like people. It will be a gradual process, and that's been happening." This is not so far off from Marvin Minsky's "Society of Mind" theory, which holds that the mind is a collection of agents, each one taking care of a particular aspect of intelligence and communicating with one another, exchanging information as required. Some AI theorists hold other views, of course. Some maintain that it's not the underlying computation mode that's crucial, but rather that there are particular algorithms (of reasoning, memory, perception, etc.) that are really the key. Others argue that the right combination of "little things" is needed to give rise to the overall emergent patterns of coordinated activity that constitute real intelligence. But Hillis' philosophy is a plausible one, and he had built a hardware platform and an organization well suited to validating or refuting his theory through ongoing engineering and research work. Most AI research is far less ambitious than this, consisting of small-scale, detailed work on one or another particular aspect of intelligence. In the history of AI, Hillis stands out as one of a very small number of people who made a serious attempt to actually create a thinking machine.

Then, the supercomputer industry died. Networks, it became clear, were the wave of the future. By networking large numbers of weak machines together, one had distributed computing—different from parallel computing in design, but somewhat similar in result. The last Connection Machine designed, the CM-5, was something like a computer network internally—it consisted of standard Sun Microsystems processors, hard-wired together rather than traditionally networked. This was a big change from the earlier Connection Machines, which had been unique on the processor level as well as on the level of overall system architecture. In the end, Thinking Machines, Inc., revised its business model, abandoning hardware altogether and focusing on selling their data-mining software for use on distributed computing systems composed of ordinary computers. Most projects attempting to couple fast, commodity CPUs with network computing use a design strategy which is far cheaper than the Connection Machine, network clustering, but which therefore place strict requirements on any algorithm run on them (e.g. algorithms must be tolerant of packet loss, and not require backplane-level communication speeds between more than four processors—since current commodity hardware has, at most, four processors).

In 1994, the firm dispersed. The hardware side of Thinking Machines, Inc., ended up at Sun Microsystems. Much of the data-mining group ended up on Wall Street. Several Thinking Machines executives started TopicalNet, a company building text categorization software. Hillis, after a stint working with the MIT Media Lab as an AI guru, abandoned the push for AI and went back to one of his earlier loves, toys and games. His new title: VP of R&D in the Imagineering Department of the Walt Disney Corporation.

He entered this new phase of his career with wide-eyed optimism. "I've wanted to work at Disney ever since I was a child," he said. "I remember listening to Walt Disney on television, describing the 'Imagineers' who designed Disneyland. I decided then that someday I would be an Imagineer. Later, I became interested in a different kind of magic—the magic of computers. Now I finally have the perfect job—bringing computer magic into Disney."

Post-Thinking Machines, Hillis' scientific work was becoming more practical in orientation—he was designing new technologies to underlie games and theme-park rides, rather than working directly toward digital intelligence. At the same time, his philosophical side was hardly dormant. The far future came to occupy his thoughts more and more. In 1993, with Thinking Machines on its last legs, he wrote the following manifesto:

> "When I was a child, people used to talk about what would happen by the year 2000. Now, thirty years later, they still talk about what will happen by the year 2000. The future has been shrinking by one year per year for my entire life. I think it is time for us to start a long-term project that gets people thinking past the mental barrier of the Millennium. I would like to propose a large (think Stonehenge) mechanical clock, powered by seasonal temperature changes. It ticks once a year, bongs once a century, and the cuckoo comes out every millennium."

The Clock of the Long Now—a clock built to last 10,000 years or more, powered by seasonal climactic fluctuations. The clock is not yet built, but a piece of land in rural Nevada has been purchased, the design is completed in detail, and construction of the components is underway. The Long Now Foundation[9] is accepting donations online.

Hillis holds some 40 U.S. patents, for disk arrays, forgery prevention methods, a color camera and various software and mechanical devices. Among all his inventions, the clock is definitely one of the coolest—one that would make any MIT hacker proud. It resonates with something deep and powerful in the human soul; the same aspect of human essence that finds the Cheops Pyramid more impressive than the Nintendo Game Boy, in spite of the incredible complexity of the hardware and software engineering involved in the latter. The Clock of the Long Now appeals to our embodiedness, to our embeddedness in space and time. Unlike his clock, Hillis' work on AI (like most AI work) ignored embodiedness and embeddedness, focusing mainly on cognition and abstract thinking—on the most rarefied parts of the mind.

Abstractly, one could build a mind operating a thousand times faster than a human mind, or a thousand times slower. 10,000 years would mean something different to each of these differently time-scaled minds, but the mathematics and general theories of intelligence would apply equally well to all of them; as would many of the same hardware-engineering principles. The Clock of the Long Now is focused on palpable human reality, not the abstract mathematics of mind or the subtleties of hardware engineering. In fact, it represents a step back from fancy modern electrical engineering. Modern technology provides few systems of 10,000-year durability, and so the design of the Clock of the Long Now required a number of pure engineering innovations.

One could easily portray Hillis' interest in clock-building as a symptom of a mid-life crisis. After all, the older you get, the more interesting time seems to you. Perhaps, having failed to create AGI, he was reviewing his own life, and feeling his own death moving closer. Perhaps he found it comforting to remind himself how little it matters, from a 10,000-year viewpoint, whether any one human or any one company succeeds at doing any one thing. No doubt, there is an element of truth to this view, but this doesn't seem to be a terribly large aspect of his motivation for pursuing the clock project—not as large, for example, as his sheer love of building cool stuff. And of course, both a thinking machine and a 10,000 year clock are Big Things: projects that appeal to the entrepreneurial, adventurous, hyper-ambitious soul.

The clock got all the media attention, but for Hillis personally, it was never a full-time occupation. His new job at Disney was the bulk of his life. It was exciting—there was lots of money paid to build lots of great stuff, and he was involved in a lot of different projects. Yet if one reviews the time Hillis spent at Disney, one has a hard time finding any Disney project that really showcases his flair for large-scale, innovative engineering. The details of his time at Disney aren't open for public discussion, but it's not hard to reconstruct the story. Disney is a huge organization—and carrying exciting projects from concept to real-world implementation, without layers of bureaucracy getting in the way, probably wasn't the easiest thing in the world. In 2000, Hillis left Disney, taking with him Bran Ferren, the head of the Imagineering group.

Ferren shares Hillis' visionary streak, and also his interest in escaping from Internet Time into historical time. Ferren, Hillis and Nathan Myrhvold (former CTO of Microsoft) have enjoyed hunting together for dinosaur bones. The conceptual clash between dinosaur bones and cutting-edge computer technology is just the kind of stimulus that leads brilliant minds in new directions. Ferren and Hillis are exploring these new directions via their new start-up, Applied Minds—a company aimed at providing technology and consulting services to entertainment firms (presumably including Disney).

Having spent most of his career at the intersection between business and science, Hillis is acutely aware of the difficulties of balancing the different goals of these very different enterprises. There was a transition in the life of Thinking Machines, he observes, when it became less of an R&D shop and more of a real business—and at that point, it became more and more difficult to move toward the goal of building AGI. When the firm became a real business, efficiency became important—but creativity is exploratory, evolutionary, and fundamentally inefficient. Basically, in a company narrowly focused on making money, every minute of everyone's day must be judged by whether it contributes to the bottom line. But the nature of the creative process is such that it just can't be judged until it's finished—there's no way to tell which previously unexplored train of thought or experimentation is going to lead to useful results.

What appealed to Hillis about Disney, when he started out there, was the fact that it was a real business that was making real efforts to keep creativity alive within its walls. This was the express purpose of the Imagineering group. The defection of Hillis and Ferren, however, is an indication that Disney's efforts in this regard have not been entirely successful. Applied Minds is a fascinating venture, which one suspects has done a better job of combining creativity with a business focus than was possible inside Disney. Still, it's worth noting how Hillis' efforts have bifurcated: the Clock of the Long Now and Applied Minds each embody different aspects of his mind and soul, which were fused together in his earlier work with Thinking Machines.

Of course, if Applied Minds becomes an extremely profitable business, then it will be able to fund more and more interesting research over time. It will be interesting to see what happens in this regard. By remaining at Disney throughout the whole Internet bubble, Hillis missed out on his chance to cash in substantially on the tech boom while it lasted. Given Disney's poor stock performance in recent years, Disney stock options presumably weren't a wonderful thing to own. The end-game of Thinking Machines, Inc., did not result in making Hillis tremendously rich either. So, when Ben visited Hillis in California, he had a nice house with a backyard facing a beautiful lake, and he was having an even nicer one built—he's a lot richer than we are. Yet he isn't currently in a financial position to build amazing new things on his own dollar. If Applied Minds eventually puts him in this position, who knows what will emerge?

Perhaps, given enough financial success, something as fantastic as Thinking Machines, Inc., can emerge from Applied Minds. It was a fascinating enterprise in many ways, but largely because of the way it fused science and business in the service of a single, immensely ambitious initiative. The Connection Machine was too big of a project to be initiated outside of industry, yet too innovative to be done without a large team of visionary scientists. The things Hillis is involved in now are less paradoxical and complex, and ultimately, for that reason, perhaps a little less intriguing. The Clock of the Long Now is a great work of conceptual art, possessing deep philosophical overtones and involving some neat engineering problems. Applied Minds is a real business through and through, using new science as required to provide customer solutions. These are both intriguing and sensible projects, and yet they lack the quixotic majesty of Thinking Machines, Inc., and the Connection Machine, which to this day remain Hillis' greatest creations. It seems that, in our age, the barriers to achieving the next level of human evolution are more financial and cultural than biological, physical or mathematical. We talk about wanting to achieve great things; but when someone like Danny Hillis (or Hugo de Garis, or the authors, or many other visionaries to be discussed in these pages) tries to actually do this, we as a society spend our money instead on developing more pornography websites or a new brand of dish soap. Technology seems to progress remarkably fast, but its direction of progress is often disappointing: driven not by the deeper needs of human mind and society nor the deeper insights of science, but rather by the self-organizing dynamics of the market economy and governmental bureaucracy; which have an unfortunate tendency to amplify some of the less appealing and interesting aspects of human nature.

A story like this reminds us that business, science, engineering and art are not fundamental divisions of the universe, any more so than earth, air, fire and water. Great innovations and enterprises stand outside these divisions, because they are crystallized around concepts that go beyond the temporary structures of any one human culture and society. The human race's urge to create intelligence beyond itself; whether through building AI machines or through putting "chips in the brain" as Hillis has recently discussed; is a fundamental force that cuts across categories of human endeavor. Our need to understand our relationship with time is a similar fundamental force. Some human beings, like Danny Hillis, and some human organizations, like Thinking Machines (and, to a lesser extent, the Long Now Foundation), reflect these fundamental forces in particularly elegant and powerful ways. To paraphrase what Hillis said about the frog's heart with which he experimented as a youth, it is remarkable that we can "do this coordinated activity…even though we are just a mass of cells."

DOUG LENAT AND CYC

The big problem plaguing AI, we have said, is "brittleness"—domain specificity, or lack of flexibility and autonomy. The Connection Machine tried to get around the brittleness problem through massive parallelism, analogous to that of the brain. Jason Hutchens' chat bots seek to circumvent it, by relying on free-ranging statistical learning rather than rules.

On the other hand, some people have tried to get around the brittleness problem by providing the computer with so much information, and so many rules, that it can contend with any plausible contingency. The most ambitious project in this direction is Doug Lenat's Cyc project, which has been going since 1984. Cyc is focused on trying to build a program with common sense. The Cyc team is mainly focused on encoding millions of items of data, so that the program can know everything an eight-year-old kid knows. "Cyc" was originally short for "encyclopedia," but they found that the knowledge they needed was quite different from that found in encyclopedias. It was everyday knowledge you could get by asking a small child: a combination of dictionary-like simple word definitions and contextually-embedded situational knowledge. Each common-sense concept in Cyc gets an English-language definition as well as a mathematical definition, which tries to paraphrase the English definition. For example, the Cyc-English definition of "skin" goes like this:

> "A (piece of) skin serves as outer protective and tactile sensory covering for (part of) an animal's body. This is the collection of all pieces of skin. Some examples include "The Golden Fleece" (representing an entire skin of an animal), and Yul Brenner's scalp (representing a small portion of his skin)."

The Cyc-English definition of happiness is:

> "The enjoyment of pleasurable satisfaction that goes with well-being, security, effective accomplishments or satisfied wishes. As with all 'Feeling Attribute Types,' this is a collection—the set of all possible amounts of happiness one can feel. One instance of Happiness is 'extremely happy'; another is 'just a little bit happy.'"

It's clear why one might think definitions of this sort—expressed in formal logic rather than English—could contribute to solving the common sense problem that we see when playing with chat bots like HeX or ELIZA. These chat bots have no common sense; they have no idea what words mean. Cyc is based on getting humans to tell computers what words mean.

The Cyc project is interesting stuff, but we don't believe that the logical definitions in Cyc have all that much overlap with the kind of information contained in the mind of an eight-year-old child. We humans aren't even explicitly aware of much of the information we use to make sense of the world: not in the least because it has evolved through dynamic processes in interaction with our environment in complex ways, rather than being mere memorizations of formally presented definitions and rules. A human's notion of happiness or skin is much bigger, disorderly and messier than these definitions. These kinds of general abstract definitions may be inferred in the human mind from a whole lot of smaller-scale, practical patterns involving skin and happiness, but they're not the be-all and end-all. In dealing with most practical situations involving skin and happiness, we don't refer to this kind of abstraction at all, but we use the more specialized patterns that the general conclusions were derived from—either individually or in combination. Additionally, our mental flexibility allows us to spontaneously derive new patterns from new observations, combined with old knowledge.

Basically, Cyc tries to divorce information from learning. However, this can't really be done, at least not as thoroughly as the Cyc folks would like. In practical terms, a mind can only make intelligent use of information that it has figured out for itself—or else is of roughly the same *form* as the information it figures for itself. In either case, it must have had a structure for autonomously-inferred knowledge in the first place; otherwise there is no mechanism for grounding knowledge in any realm of self-action. If the information read into an AI system from a database is too different in structure from the information the AI system has learned on its own, then its reasoning processes will have a hard time integrating the two. Integration may be possible, but it will be extremely time-consuming, and would most likely proceed in a similar manner to how we as humans learn information that's just dumped before us. Any AI system will need to learn some of its knowledge base for itself—no Cyc-like system can contain all possible knowledge an AI system will ever need to know for interaction with any environment, as it's totally impossible to predetermine every eventuality it may encounter during its existence. Any knowledge base that isn't structured in a way that naturally matches with the structure of learned knowledge will be effectively useless—but how can the people building a database like Cyc know what knowledge representations will match with learned knowledge, if they aren't building a learning system?

Despite sixteen years of programming, Cyc never succeeded in emulating an eight-year-old child. Nor has anyone yet found much use for a CD-ROM full of formal, logical definitions of common-sense information. The company Cycorp is surviving, though not flourishing, supported largely by government research grants. When we talked to the firm's CEO a few years back, he was quite careful not to mention anything about artificial general intelligence or any long-term scientific mission underlying the firm's work. Instead, he characterized the firm as being concerned with producing a useful database intended for embedding within various specialized software products. This is a worthwhile mission to be sure, but very different from the grand goal of AGI.

We are doing some explicit knowledge encoding for the Novamente project, but our approach is quite different from Cyc's: rather than using a specialized formal language, we're encoding knowledge in simple English. So, for example, if we teach Novamente that: "If one agent gives an object to another agent, then just after that, the second agent has the object," then this knowledge will be read by Novamente using its natural-language-processing routines, and mapped into Novamente's internal knowledge structures. It will thus be represented, internally, in the same style used to represent knowledge such as:
"Ben gave a box to Izabela."

If Novamente has been taught the groundings of the verb "give" in terms of things it's done and observed in a real or simulated environment, then it will have learned linkages between the Novamente-internal representation of "Ben gave a box to Izabela" and the patterns in sensorimotor data corresponding with the act of giving. Thus it will be able to link the explicitly-encoded general knowledge about giving and having with patterns in sensorimotor data, a process called "grounding" knowledge in experience. At present, this is work in progress—just like the Cyc team, we haven't yet created a true thinking machine. But we do have a concrete plan to create an integrated knowledge system spanning abstract and concrete knowledge. On the other hand, a knowledge base like Cyc is not specified in such a way as to easily be made compatible with any experiential learning system, so it's not clear to us how it can ever really be useful for AGI, except as one dictionary among many others (the same way in which it and other databases are useful to humans, as tools).

In fairness to Doug Lenat, we must say that his computational-psychology perspective does have some depth to it—more than is represented in the public face of Cyc. He has a reasonably solid theory of general heuristics—problem-solving rules that are abstract enough to apply to any context. His pre-Cyc programs, AM and EURISKO, applied his general heuristics theory to mathematics and science respectively. Both of these programs were moderately successful, exemplars in their field, but far from true general intelligence. Their design lacks a holistic view of the mind. In the big picture of AGI, getting the mind's heuristic problem-solving rules right means virtually nothing; because problem-solving rules gain their psychological meaning from their interaction with other parts of the mind, and their grounding in experience of and interaction with some environment. If the other parts aren't even there, the problem-solving is bound to be sterile.

EURISKO won a naval fleet design contest two years in a row (until the rules were changed to prohibit computer programs from entering), and it also received a patent for designing a three-dimensional semiconductor junction. Yet, when considered carefully, even EURISKO's triumphs appear simplistic and mechanical. Consider EURISKO's most impressive achievement, the 3-D semiconductor junction. The novelty here is that the two logic functions, "Not both A and B" and "A or B," are both performed by the same junction—the same device. One could build a 3-D computer by appropriately arranging a bunch of these junctions in a cube.

How did EURISKO make this invention? The crucial step was to apply the following general-purpose heuristic: "When you have a structure which depends on two different things, X and Y, try making X and Y the same thing." The discovery, albeit an interesting one, came directly out of the heuristic. This is a far cry from the systematic intuition of a talented human inventor, which synthesizes dozens of different heuristics in a complex, situation-appropriate way.

By way of contrast, think about the Croatian inventor Nikola Tesla—probably the greatest inventor in recent history—who developed a collection of highly idiosyncratic thought processes for analyzing electricity. These led him to a steady stream of brilliant inventions, from alternating current to radio to robotic control, but not one of his inventions can be traced to a single "rule" or "heuristic." Each stemmed from far more subtle intuitive processes, such as the visualization of magnetic field lines, and the physical metaphor of electricity as a fluid. Each involved the simultaneous conception of many interdependent components. Problem-solving in the general sense is not merely getting lucky in applying a fixed set of heuristics to a particular problem; but creatively devising not only new heuristics, but whole new categories of heuristics (we would call these "new ways of looking at the problem"); and even new problems and categories of problems into which we can decompose a particularly vexing large problem. If Tesla could only work with the existing engineering heuristics of his day, rather than creating new ones, he'd never have created so many amazing inventions.

EURISKO may have good general-purpose heuristics, but what it lacks is the ability to create its own specific-context heuristics based on everyday life experience. This is precisely because it has no everyday life experience: no experience of human life and no autonomously-discovered, body-centered digital life either. It has no experience with fluids, so it will never decide that electricity is like a fluid. It has never played with blocks or repaired a bicycle or prepared an elaborate meal, nor has it experienced anything analogous in its digital realm. Thus, it has no experience with building complex structures out of multiple interlocking parts, and it will never understand what this involves. EURISKO pushes the envelope of rule-based AI: it is just about as flexible as a rule-based program can ever get, but it is not flexible enough. In order to get programs capable of context-dependent learning, it seems necessary to write programs which self-organize—if not exactly as the brain does, then at least as drastically as the brain does. Hand-encoded knowledge can potentially be useful in this process of self-organization, but only if it's encoded in a way that matches up naturally with *learned* knowledge, so that the appropriate synergies can emerge. This means that one can't think about the mind by thinking only about rules; one also has to think about learning and experience, even if what's doing is creating rules to be fed into an AGI system.

KASPAROV VS. DEEP BLUE

One of the most impressive achievements of rule-based AI happened on May 11, 1997. This was an event that led many people to think that computers were already on the verge of rivaling human intelligence. For the first time ever, a computer had defeated the world chess champion in a standard five-game match. Deep Blue, a computer chess program developed at Carnegie Mellon University, split the first two games with Garry Kasparov. The second two were draws, and the final game went to Deep Blue. Kasparov was a sore loser. Deep Blue took it as all in a day's work.

Admittedly, this was only one match, but the tournament was not a fluke. Previous versions of Deep Blue were already able to consistently beat all but the greatest chess grandmasters. And Deep Blue's play can be improved by hardware upgrades, whereas a brain upgrade for Kasparov is currently not in the works. Although Deep Blue is not a very intelligent entity according to our definition, there's much to be learned from a study of its accomplishments and the mechanisms underlying them.

The day after Deep Blue beat Kasparov, there was a lot of talk about Deep Blue on the Simon's Rock College[10] alumni e-mail list, and Ben wrote a long e-mail giving his views. What we'll say about it here is basically what Ben said in that e-mail. The question people were debating was: Does Deep Blue's accomplishment mean that true artificial intelligence has been achieved?

Obviously, from the perspective taken in this book, the answer is no—but it's worth spending a little time to spell out exactly why.

If we define intelligence as the ability to do one intellectual task very, very well, then Deep Blue qualifies brilliantly; but if we think of intelligence as being able to make appropriate decisions in a wide variety of complex environments, it fails miserably. It can only play chess. Computers that can do one thing well are tremendously useful things, but they are not truly intelligent as we define the term (and, really, as most people intuitively understand the term). In our definition, a truly intelligent computer will have to do more than follow instructions. It will have to ***create its own answers to unanticipated problems***. For this kind of intelligence, it will need a structure quite different than Deep Blue's.

Although Deep Blue follows the same rules as human chess players, it doesn't think like humans. Human chess players use geometric intuition and a sense of the flow of a game. Deep Blue calculates every single possibility; then calculates all the possible consequences of each, gives them weights based on the probability and desirability of such a next move from its "perspective," and finally picks the next move by culling-out that which produces the best set of the next N moves for some number N. Computer programmers call this recursive logic. It does the same thing over and over and over again, constantly referring back to the results it just obtained, and figuring out how well it is doing. Human beings might use recursive logic to play a very simple game, such as tic-tac-toe, which has very few choices. But even in tic-tac-toe, our opponents would probably object to our taking the time to calculate out the potential consequences of every possible move. Our minds are much too slow to play chess that way, nor would the game be any fun if we could. Computers, by contrast, are much, much quicker at this kind of task and do not get bored, so recursive logic can work well for them.

Of course, every chess player extrapolates, thinking: "What is the other player likely to do next? And if he does that, what am I going to do? And if I do that, what is he going to do?" But in humans, this kind of reasoning is augmented by all sorts of other processes. For Deep Blue, this kind of extrapolation is the whole story—and it does it very, very well. Computers can extrapolate faster and further into the future than any human. The 1997 version of Deep Blue could evaluate about two hundred million different board positions every second. This figure can easily be increased for the cost of additional circuitry, but it doesn't make Deep Blue truly intelligent.

One way to understand the difference between Deep Blue and human players is to think about strategy versus tactics. There is a certain kind of creative long-range strategy that human chess grandmasters have, but Deep Blue lacks. Deep Blue makes up for this lack by elevating tactics to such a high level that it assumes the role of strategy. Deep Blue is not entirely without strategy: it carries out its superhuman tactical evaluation within the context of a collection of pre-programmed strategies, and it is capable of switching between one strategy and another in response to events. But it does not "think" strategically, it only "thinks" tactically. Deep Blue doesn't make long-range plans involving an understanding of the overall structure of the board as a dynamical system which changes over the course of the whole game. If it could do this, it would doubtless play even better. Even without strategic creativity, it plays well enough to beat the best humans, but only because it turns out that chess is a game which succumbs well to a recursive logic approach.

The defeat of Kasparov by Deep Blue is symbolic because chess is the most mentally challenging game commonly played in the Western world. Computers became better than humans at checkers and many other games quite some time ago. However, there is at least one popular game that still stumps the best computers—the Asian game Go. van der Werf points out that at the present time, in spite of a substantial research effort, no existing computer program can play better than the advanced beginner level at Go.

The rules of Go are very simple, compared to chess. Play is on a 19x19 grid, and stones (pieces) are placed on the intersections of the grid, called points. The first player plays black stones, the opponent white ones; and stones are added to the board one-by-one, players alternating. Stones are not removed once placed, but stones and groups of stones may be captured. A player calls "Atari" when a capture can occur on their next move, to warn the opponent. The game ends when it is no longer possible to make a reasonable move. The winner is determined by the amount of territory surrounded, less the number of their stones captured.

The trouble with Go from the computational perspective is that from any given board position, there are hundreds of plausible next moves rather than dozens, as in chess. Extrapolation in Go will not get you as far as it does in chess. It would seem that, if computers are to conquer Go, they're going to have to do it either with a more generally intelligent approach, or use some more clever special-case technique than the one employed for chess. Go is too visual, too two-dimensional, to succumb to purely combinatorial, non-visual techniques. A world-champion Go program would have to be intelligent at general two-dimensional vision processing as well. Since Go, like chess, is ultimately an extremely limited problem domain, a special-purpose, unintelligent program may possibly master it eventually. However, the fact that it hasn't been mastered yet just goes to show how far from intelligence computers really are right now—there are even very narrow domains with which they can't really cope at this time.

In Go, high level players routinely analyze positions that aren't confined tactically to, say, a 9x9 grid. Additionally, almost any tactical fight has strategic implications across the board that could be worth more than the fight itself—so a great pattern-matcher wins the points, but loses the war. One style common in evenly-matched games is to go around the board "losing" fights, but in such a way that one's own stones work together and become more powerful.

The Go computer programs in existence today rely heavily on pattern matching: taking a given, small chunk of the board, and matching it up to a dictionary of known board situations. The best ones are as good at reading small, enclosed life/death problems as a mediocre tournament Go player. However, when the problems are not so rigidly enclosed within a small region of the board, the programs are clueless, although intuitive human players can still see the essential principles. The best way to wipe out such programs is to embroil them in a huge, whole-board fight; one that is too big for the algorithm to match properly.

Deep Blue's recursive approach of elevating tactics to the level of strategy doesn't work so well in Go. Exhaustive searching over spaces of two-dimensional patterns is much, much harder than the kind of decision-tree search required for dealing with chess, and will be out of the reach of computers for a while. One suspects that something less than true intelligence will suffice for Go, as it has for chess—but not something as far from true intelligence as Deep Blue. Perhaps a massive increase in raw computational power will be enough, but that is not a truly intelligent approach.

Deep Blue has basically the same problems as GPS and EURISKO and BACON—it's too inflexible and its abilities are an achievement of computer hardware getting faster, not of software becoming intelligent. It relies on a special computer chip, custom-designed for searching many moves ahead in chess games. This special chip could be modified without much hassle to apply to other similar games—checkers, maybe Othello. It is better at such tasks than a general-purpose CPU, but no smarter. It's just an optimization of a fixed algorithm. The ideas behind its massively parallel design and RS 6000 platform are going to be generalized by IBM into drug design tools, weather forecasting and other applications. However, Deep Blue has no insights into those domains—all its "knowledge" in these areas will be in the form of rules encoded by humans, either about the specific domain or generally for data mining. Its architecture couldn't be modified to apply to Go, let alone to any real-world situation. Deep Blue's chip is less like the human brain than like a human muscle: a mechanism designed specifically for a single purpose, and carrying out this purpose with admirable but inflexible precision. Its rules are astoundingly simple and mechanical: evaluate the quality of a move in terms of the quality of the board positions to which this move is likely to lead, based on encoded experience. Judgments based on experience are made not by complex intuitive analogy, but by simple pattern-matching. Even though it has experience of a world, albeit the limited world of chess experience, it has no intuition of this world—it merely stores its game experience in a database and mines that database for possible moves. It can only invent new strategies by coincidence, and can't understand them—it can only hope to happen upon a situation in which its human opponent makes a move which causes that path to be entered in its search algorithm. Everything is cut-and-dried, and done two hundred million times a second. This is outstanding engineering, but it is not intelligence. Yet, its results are vastly more impressive than those obtained when rule-based AI does try to do intelligence, as in programs like GPS, BACON and EURISKO.

Some chess players were frustrated with Kasparov's defeat at the "hands" of Deep Blue, on the grounds that Kasparov was extremely sloppy in the final game. He played a defense (Caro-Kann) that he admitted afterwards wasn't his *forte*, and, in their view, gave away the game with a move that conventional chess wisdom deems unwise. It is argued, on these grounds, that he lost because he let his emotions carry him away. In past games, his unusual and unpredictable moves had thrown the computer off, putting it in situations where its memory of previous board positions did not apply. This time, he was perhaps feeling a little desperate, and so he moved a little too far into the domain of the unpredictably self-defeating.

This raises another interesting question—the role of human emotion in human intelligence. Is emotion bad for human competence, or good? In this case, emotion may have worked against Kasparov, though how much so is difficult to say. Who can blame him for getting a little stressed—after all, in the eyes of the media, he was playing not only for himself, or his country, but for the whole human race! Still, it's hard to pooh-pooh human emotion, even from a purely performance-oriented perspective, because of the extent to which emotion and cognition are bound up together in the human mind. One can argue that in intelligent systems, operating on the basis of generalized intuition rather than specialized search and pattern-matching, intuition is necessarily coupled with *some form* of emotion—though not necessarily emotions as overwhelming or hard-to-control as our human ones. We think a truly intelligent computer must have some kind of emotion, but this doesn't mean it will feel things the way humans do.

George Mandler, a cognitive psychologist whose work we've followed carefully over the years, has proposed that we think of emotion as composed of two different aspects, called hot and cold. The hot aspect is the consciousness attached to emotion, which is shaped by the link between the human mind and the human body. The cold aspect of emotion is the abstract, mathematical structure of emotion, which seems on careful psychological analysis to be closely related to the "failure of expectations to be fulfilled." Emotions happen, in this view, when awareness is attached to some unfulfilled expectation.

At first glance this analysis of emotion may seem to apply only to negative emotions. But positive emotions can be understood in the same way. Wonder is the nonfulfillment of the expectation of familiarity. Happiness is, in large part, a feeling of increasing unity, a feeling of more inner unity than expected—in other words, a nonfulfillment of the expectation of inner disparity and disharmony. Happiness is also a feeling of decreasing sadness; whereas sadness is a feeling of decreasing happiness—each feeling arises largely as a consequence of nonfulfillment of the expectation of its opposite. Love is happiness extending outside the body—it is the feeling of increasing unity between self and other, the nonfulfillment of the expectation of separateness. The point is not to "reduce" emotions to formulas, but rather to identify the structures underlying the feelings of various emotions.

The reason for focusing on nonfulfillment of expectation rather than fulfillment is that, with repeated fulfillment of expectation, the initial happiness of fulfillment is quickly replaced by boredom and eventually ennui. Humans—the most intelligent systems we know of—seem to emotionally seek out the unfamiliar in order to stimulate their "cold emotions." Our emotional need for uncertainty seems inextricably bound up with our ability and necessity to understand and create.

To Mandler's observations, we would add the hypothesis that, in humans, emotions are particularly intense when they have to do with the interaction between the more conscious, rational portions of the brain and the unconscious, archaic mammalian and reptilian portions of the brain. A lot of the intensity of human emotion has to do with the feeling of being pushed and pulled in directions over which one has no control—and this is largely because the conscious parts of the brain really *don't* have anywhere near full control over some of the more primitive portions. Expectations of the conscious mind go unfulfilled, often in ways that profoundly affect the body. Again, this is not only in the case of negativity. We seem to thrive on an internal mental conflict which drives us to venture out of our comfort zone and challenge ourselves: sometimes to great results, sometimes to terrible ones.

Some human emotions may arise without this rational/mammalian dichotomy, due purely or primarily to the unpredictable dynamics of the higher-level human mind itself, but a lot of human emotions are tied into the older parts of our brains. Non-brain-based AI software will probably experience emotion, but most likely with less intensity and a different flavor than humans. However, if an AI system has no emotions at all—that is, no non-rational aspect of its thought—it will almost certainly be incapable of creativity, in any manner other than relatively unlikely accidental discoveries of new patterns of mind during routine activity. (It is more likely that you'll discover new things if you have a sub-rational compulsion to do so, regardless of how difficult or impractical such endeavors may be.)

However, Deep Blue's shortcomings are ultimately based not on its lack of human mammalian emotions, but on the simplicity of its cognitive structures. When its expectations are frustrated, Deep Blue does not respond in the way an intelligent system often does: by leaping to a more general level of abstraction, and reconsidering the assumptions that led to the expectations that were in fact frustrated. This is not because it lacks the mammalian response to frustration of expectations—it's because it lacks knowledge representations or learning mechanisms consistent with general intelligence. Since it lacks such knowledge, we have no way of knowing if it would bother to think more abstractly or reevaluate its assumptions without being able to feel frustration. One could program it to mechanistically change its modality of thought when a tactical path was incorrect for too long, but in a domain more complex than chess, eventually that approach may fail too. We feel that, to compel a system to create whole new modalities of thought in the face of unfamiliar problems, it will need to be able to have a non-rational motivation to not simply give up when all known approaches fail.

Although emotion can be harmful to human cognition at times—as, perhaps, in Kasparov's final game against Deep Blue in 1997—it is, all in all, inseparable from intuitive human thought. Emotions trigger our thoughts and even structure them. Deep Blue demonstrates that emotion-driven intuition is not necessary to excel at the game of chess—but chess is an extremely restricted domain of action, and even so, one suspects an even more exciting style and quality of chess play would be displayed by AI software combining Deep Blue's systematic search power with more human-like (loosely speaking) general intelligence.

Or else, as fellow Simon's Rock alumnus Max Miller put it in the e-mail thread: "All this proves is what I've been saying for years: that chess is a stupid game."

WHAT RULES CAN'T DO

When neural nets were being dismissed in the early '70s, not everyone was optimistic about the potentials of rule-based AI. In 1972, the era in which ELIZA was receiving a lot of attention, a philosopher named Hubert Dreyfus wrote a book called *What Computers Can't Do*. In this vicious attack on AI, Dreyfus argued that artificial intelligence researchers were fundamentally misguided: that they could never achieve their objectives with the methods they were using. Dreyfus preached the importance of body-centered learning, and the close connection between logic, emotion and intuition. Without a body and without feelings, Dreyfus argued, there can be no real generalization of special-case ideas. Based on these philosophical considerations, he predicted that AI would be a failure.

Needless to say, these conclusions went over like a lead balloon among AI researchers. They laughed in his face, or worse—but the vehemence of their ridicule betrayed a fear that he might have been right. Even today, a review posted on the Amazon.com website observes that "many AI-workers seem to be actually afraid of this book." The same reviewer notes that it is one of very few computer books that is still of interest 25 years after it was written.

In 1992, Dreyfus re-released the book with the title *What Computers Still Can't Do*. The introduction brims over with he-who-laughs-last-laughs-best insolence, but his exultant crowing is not quite persuasive. He was right about the limitations of the AI programs of the 1960s and 1970s, but the observers who thought it was just a matter of time and resources also been proven correct in many cases. Dreyfus, for example, ridiculed a prediction that computers would soon be able to take dictation, just as a human secretary can. Although this prediction didn't come true as quickly as Frank Rosenblatt[11] had thought, some fairly good programs are available today for this purpose, relying in large part on a neural net architecture to learn each user's speech patterns.

Dreyfus' critique of AI, in the first edition, was too strong. He appeared to believe that detailed simulation of the human body was the only possible path to AI, and he argued that this would be impossible without simulating the biology of the brain and the rest of the body. Actually, the human brain is only one intelligent system, and a great deal can be accomplished without replicating the details of its biology. Yet Dreyfus' arguments posed a serious challenge to AI theorists: how to design a machine that can simulate body-based, emotion-based conceptual generalization? We believe that Dreyfus was essentially correct, in that if it is impossible to give a computer some sense of embodied self and some kind of emotions, AGI isn't going to happen. However, a physical body just like ours is not required: an AI entity could have a virtual body, enabling it to interact in a rich and autonomous way with a virtual world (or its body could be an automobile, robot, spaceship, etc.). Furthermore, emotions need not be controlled by biological neurotransmitters; they can come out of complex digital dynamics. The point is, unless one has a computing system that is large, complex and autonomous, with integrated sensory, memory and action systems interacting with a rich environment and forming a self system, it will never develop the ability to generalize from one domain to another. The ability to generalize is learned through general experience, and general experience is gained by exploring a world.

We definitely took Dreyfus' critique to heart in designing and developing first Webmind, then Novamente. We didn't try to replicate the human body as he thought was necessary, because we don't agree that it is necessary. Instead, we bypassed his critique by designing a huge, self-organizing system, which lives in the perceptual world of the Internet and understands that its body is made up by software objects living in the RAM of certain machines. It is a nonhuman, embodied social actor. Its environment is different from our own, but it exists, and it is not one of which we have no conception whatsoever—so should it become intelligent, there is still common ground within which we can interact with it.

Dreyfus didn't try very hard to imagine an embodied, social intelligence without a human-like body, but his ideas certainly leave room for such a thing. His problem was not with AI, but with the attempt to build a mind that operates in a vacuum instead of synergistically with a self and a world.

Ben met Hubert Dreyfus only once, in early 1996, when he came to the University of Western Australia to give a lecture on the philosophy of AI, and they had a few drinks afterwards. Dreyfus said he was extremely pleased that the new edition of his book had been released by MIT Press, which had refused to publish the first edition (because of MIT's commitment, as an institution, to AI research—Minsky, and many other AI researchers whom Dreyfus criticized, worked at MIT). Suprisingly, though, Dreyfus was newly optimistic about AI—not the type of AI that had been popular in the '70s or '80s, but more recent developments, in particular neural network AI. Neither of us have talked to him lately, but it is likely that his enthusiasm about neural nets has now decreased somewhat, based on an objective analysis of what neural net AI has become.

What he liked about neural networks was the fact that their intelligence was implicit: even after you had trained a network to do something, you couldn't say precisely how the network had done it. This implicitness, he felt, captured something of the tacit and nebulous nature of human understanding. A neural net in a robot body, he surmised, would possibly be able to realize the dreams of AI researchers, and do what isolated, rule-based systems could not.

Ben asked Dreyfus why, if he felt neural nets offered a possible path to AI, he had not said so in the first edition of his book back in 1972. His answer was simple: he said that he hadn't understood what neural nets were all about at that time. This reply impressed us with its directness and honesty, and also drove home the importance of relying on your own intuition in scientific matters. The leaders in the field may be off in the wrong direction; their leading critics may be just as ignorant, though in different ways; and crucial clues may be found in the ignored writings of scientists from past decades. One of the lessons of the history of AI is that science, for all its objectivity, is fundamentally a human endeavor, and quite susceptible to ordinary human weaknesses.

THE DIVERSITY OF AI

In the foregoing pages we've touched on the main streams of thinking in the AI discipline, but this kind of summary doesn't get across the wild diversity of innovation you find by looking at the papers of individual researchers, including those operating far out of the mainstream. Actually, odd bits of AI work by total outsiders probably had a bigger influence on us as AI researchers than the major breakthrough programs we've focused on so far. Here, we'll briefly discuss some of those outsiders.

John Andreae at the University of Canterbury in Hamilton, New Zealand is one. He wrote a system called PURR-PUSS which learned to interact with humans statistically. One of Andreae's students was John Cleary, also one of the machine learning gurus at Waikato University in Hamilton, New Zealand, where Ben taught for a year. Cleary worked for us at Webmind Inc. for a couple of years as well. He and his students formed the firm's New Zealand office, and masterminded our Webmind Classification System product. (Cleary, Stuart Inglis, Len Trigg and a few other Kiwi and Californian ex-Webminders went on to form the firm ReelTwo[12], focusing on machine-learning-based text processing with a biomedical slant). We're not exactly emulating PURR-PUSS in Novamente, but the statistical learning methods that it embodied are there in our machine learning module and reasoning system, and the emphasis on interactive learning which Andreae advocated lives on in our work with a vengeance.

There was a funky little book called *Robots on Your Doorstep*, by Nels Winkless and Iben Browning, which talked a lot about the definition of intelligence as "the ability to act appropriately under unpredictable conditions" (a definition that helped us formulate our own concepts of intelligence) and contained a lot of cool speculations about how to build robots that would be intelligent in this sense. Ben got an e-mail from Nels Winkless—also the founder of *PC Magazine*, one of the first computer magazines aimed at ordinary people—in 1997, when he was just starting Intelligenesis (later Webmind). Winkless had found a reference to *Robots on Your Doorstep* on Ben's website, and was pleased to find that anyone had actually read that book! In e-mail conversation regarding AI as a business, Winkless disclosed that after Browning had died, he'd collaborated with someone else to build a company centered on some pattern-recognition technology—very much in the spirit of *Robots on Your Doorstep*, and of our view of mind as pattern. Eventually the company had been taken over by its investors due to its failure to make money, and had found a niche in which it could be successful: automatic recognition of similarity among images. The name of the company was Excalibur Technologies, and it merged in 2000 with Intel's Interactive Media Services Division to form Convera.

Then there is the idea of genetic algorithms—doing AI by simulating evolution, rather than the brain. This wasn't at all a hip idea in 1989, when Ben was writing his PhD thesis on optimization methods—at that point, it didn't even turn up in most reviews of the optimization literature, although in retrospect one can point to papers on the topic going back to the late '60s. By the mid-'90s, it was a well-recognized area of computer science; and Ben was doing research into the mathematics of genetic algorithms, studying questions such as "Why is evolution involving sexual reproduction more efficient than evolution involving asexual reproduction only?" Although the details are different, evolutionary AI is similar in spirit to neural net AI—you're dealing with a complex, self-organizing system that gives results in a holistic way, where each part of the system doesn't necessarily have a meaning in itself, but only in the context of the behavior of the whole. In Ben's 1993 book *The Evolving Mind*, he wrote a lot about the relation between evolutionary programming in AI and Edelman's theories of evolution in the brain. It turns out that one can model the brain as an evolutionary system, with special constraints that make it a bit different from evolving ecosystems or genetic algorithms in AI. We had an evolution module in Webmind, used for two things: as one among many machine learning methods for finding patterns in data (along with feedforward neural nets and purely statistical methods); and as one of two ways of learning schema for perceiving and acting (the other being probabilistic logical inference). Now, in Novamente, we have moved beyond this in an interesting way, creating our own customized fusion of evolutionary learning and probabilistic reasoning.

We've just listed three little fragments of AI research that don't fit into the big bad "neural nets versus rules" dichotomy put forth earlier, and there are many, many others. History never fits all that neatly into categories—almost nothing does, actually. Still, we need to make up categories in order to understand things. The diversity of mind and of society, which busts out of all the simplified category systems that we place on it, is responsible for the fabulous creativity of these systems.

SYMBOLIC VS. CONNECTIONIST—A BOGUS DICHOTOMY?

"A foolish consistency is the hobgoblin of simple minds," Emerson said. It's easy to give an impression of clarity by oversimplifying reality, and we've tried in these pages not to do too much of that. However, we have presented a dichotomy between symbolic (rule-based) and connectionist (neural-net) AI. Then we pointed out that a lot of fascinating and important AI doesn't fit into this framework at all: concepts like statistical machine learning and genetic algorithms. Now we're going to dig this hole even deeper, by arguing that the distinction between symbolic and connectionist AI is actually a lot fuzzier than most AI gurus realize. The idea isn't to confuse or obfuscate, but rather that we had to address the generally perceived differences between the two approaches before we could then tear them down.

This is a key issue, because we often like to say that Novamente synthesizes connectionist and symbolic AI. While this is a true statement, it glosses over the peculiar vagueness of the notions of "symbolic" and "connectionist" themselves. When you get deeply into these concepts, you realize that this classical dichotomy is not quite framed correctly in most discussions on AI. There is a valid distinction between AI that is inspired by the *brain structure*, and AI that is inspired by the *mental functions* of conscious reasoning and problem-solving behavior; but the distinction between "symbolic" and "connectionist" knowledge representation is not nearly as clear as it's usually thought to be.

Classically, the distinction is that in a symbolic system, meanings of concepts are kept in special localized data structures like rules; whereas in a connectionist system, meanings of concepts are distributed throughout the network. Also, in a symbolic system the dynamics of the system can be easily understood in terms of what individual rules do, whereas in a connectionist system the dynamics can basically only be understood holistically, in terms of what the whole system is doing.

In reality, the difference isn't so clear. For example, one branch of symbolic AI is "semantic networks." In a semantic network you have nodes that represent concepts and links that represent relations between concepts. Suppose you have a semantic network in which there is a node representing "floor." This is obviously symbolic in the classic sense: the meaning of the "floor" node is localized in the "floor" node. But wait—is it *really*?

In some semantic network based AI systems, all the relations are made up by people, and then hard-coded in a special knowledge representation language that builds the nodes and links in the software from their hand-input data. Some of them have reasoning systems that build relationships, and can learn, for example, that because people walk on floors, floors must be solid, because people can only walk on solid things. In a system like this, relations are built from other relations, and so the meaning of the "floor" node may be contained in its relations (links) to other nodes. The formation of these connections may have been based on the connections of the other nodes to yet other nodes, and so on, iteratively.

What this means is that, in a semantic network formed by iterative reasoning rather than by expert rule creation, each element of knowledge (each node) actually represents the result of a holistic dynamic. It has meaning in itself—an encoding of our socially constructed concept "floor"—but internally its meaning is its relation to other things, each of which is defined by the other things it's related to, so that the full meaning of the part (the particular node) is only truly describable in terms of the whole (the entire network).

On the other hand, suppose one has a neural network in which memories are represented as attractors (a Hopfield Net, or Attractor Neural Network, in the lingo). Then, the meaning of a link between two nodes in this network mainly consists of the attractors that its presence triggers. On the other hand, there's also a clear local interpretation: if the weight of the link is large, then that means the two nodes it connects exist together in a lot of attractors, i.e. they're contextually similar. If the weight of the link is large and negative, this means that the two nodes rarely co-exist in an attractor—they're contextually opposite. Whether the nodes have immediate symbolic meaning or not depends on the application—in typical attractor neural network applications, they do, each one being a perceptible part of some useful attractor.

The point is that in both classic symbolic and classic connectionist knowledge representation systems, one has a mix of locally and holistically defined meaning. The mix may be different in different knowledge representation systems, based on the philosophical bias of the developer or the pragmatics of a particular task for which the system was designed, but there is no rigid division between the two. This fact is important in understanding Novamente, which freely inter-mixes "symbolic" style and "connectionist" style knowledge representations.

Of course, there are extremes of symbolic AI and extremes of connectionism. There are logic-based AI systems that don't have nearly the holistic-meaning aspect of a reasoning-updated semantic network as we've described above. There are connectionist learning systems (such as backpropagation neural nets) in which the semantics of links are much less transparent than in the attractor neural net example given above. But this is also an interesting point: we believe that, of all the techniques in symbolic AI, the ones that are most valuable are the ones that have the most global, holistic knowledge representation; and of all the techniques in connectionist AI, the ones that are most valuable are the ones that build up from a localized knowledge representation. This is because real intelligence only comes about when the two kinds of knowledge representation intersect, interact and build on each other. Intelligence requires the synthesis of localized knowledge across an entire network of interrelationships.

THE CHALLENGE OF REAL AI

We've told you a fair bit about the concepts underlying current AI systems. They are fascinating software projects, but are ultimately inadequate. As much as we are techno-optimists, even we have to admit at this point that building an AGI is a very hard problem. To build a comprehensive AI system with perception, action, memory, and the ability to conceive new ideas and to study itself, is a much bigger job than erecting a new Wal-Mart, building an industrial robot, or even creating software for predicting the financial markets better than the guy in the next office. Necessarily, an AGI system will consume a lot of computer memory and processing power. It will be difficult to program and debug because each of its parts gains its meaning largely from its interaction with the other parts; never mind that each of the underlying ideas is already complex and difficult to program, debug, and tune on its own. Lenat, Hutchens and Hillis are just about as smart as human beings get, and they pursued the AGI goal with diligence, expertise and wisdom—and yet none of them succeeded. De Garis, Voss, Yudkowsky and ourselves are pursuing the same goal, in our different ways, and none of us have succeeded yet, either. Although none of us are likely to give up, continuing on persistence alone may make this program take a very long time, if the current situation encompasses all the people and financial resources we'll ever have.

In our (possibly flawed) judgment, we now have, for the first time, computer hardware barely adequate to support a comprehensive AI system. Moore's law and the advance of high-bandwidth networking mean that the situation is going to keep getting better and better. However, to us (and many other "real AI" mavericks of the world) it seems that we are stuck with a body of AI theory that has excessively adapted itself to the era of weak computers, and that is consequently divided into a set of narrow perspectives, each focusing on a particular aspect of the mind. Mainstream AI has not really caught up with the rest of mainstream computing.

Moore's law is not a physical law in the proven sense, and banking on it as the solution to all AI's prior ills is not a good idea. Speculation is currently going on in the hardware engineering field about whether, without radical new solutions, Moore's law will hold up against the problems of heat buildup and quantum interference between closely packed electron paths (circuits). Quantum Computing is at least a decade away—perhaps two or three—and there are a number of tricky issues with distributing many existing algorithms over both network clusters and multiprocessor and multicore PCs. These are not unsolvable problems, but the pace of hardware acceleration may eventually slow down, and hardware acceleration alone isn't going to solve every computing problem, least of all AI.

In spite of the obviously huge obstacles that the AGI goal presents, we still believe this is where the AI field should be focusing the bulk of its efforts. After all, is this not the only approach that can possibly succeed at achieving the goal of a real thinking machine? We realize that this attitude places us in a small minority of AI researchers—but if our team or one of our "real AI"-focused competitors should succeed, it wouldn't be the first time in the history of science and technology that a maverick minority had been proven right.

NOVAMENTE

The germs of our own ideas regarding AI go way back. Ben remembers heading home from college for spring break in 1983, toward the end of his freshman year. He'd just recently turned 16, and had been thinking about AI a hell of a lot—even more than about his new girlfriend, Rachel Gordon, for whom he was pretty crazy at the time. A few days before spring break, he'd tried to explain his theories on artificial intelligence to his friend Ken Silverman. Ken couldn't understand what he was talking about, so Ben promised him he'd work on it over spring break, and that when he got back to school he'd explain how it all worked and give him a complete design for a thinking computer program. The idea was clear in his head, but he was totally unable to articulate it in a way that Ken or anyone else could understand. He spent the whole break working on it, and during those few days basically worked out the ideas that he'd put in his first book, *The Structure of Intelligence,* six years later. Ben went through every aspect of the mind—reason, memory, aesthetics, intuition, emotion, and so on—and convinced himself that each could be expressed in terms of pattern recognition and pattern formation. The mind, he concluded, was a pattern-recognition system that recognized patterns in the world around it and—very crucially—also recognized patterns in itself. Recognizing patterns internally, it also formed patterns within, continually giving rise to new structures.

Stephan, who had come to surprisingly similar general conclusions about AI (initially with more of a focus on visual and aural perception, and on ideas of economics and homestasis in distributed systems, and less of a focus on comprehensive mathematical AI design), "met" Ben on the Simon's Rock alumni e-mail list, and read Ben's first three books while working at Bell Labs in 1996-97. He'd already been working his own AI theories into game and virtual-world agents, hypereconomics, and self-monitoring systems for computers and networks, and by 1998 he was consulting for Webmind (joining full-time in 1999). During the Webmind period, Stephan worked with Ben, Cassio Pennachin, Ken Silverman, Pei Wang, and others on designs for both the core Webmind system and a series of potential practical application areas to which AGI could be applied, that would give it further experience and grounding for learning.

Of course, the road from Ben's conceptual insight to a concrete AI software design is a long path, and it's a path we've been following ever since, albeit while also working on a lot of other things. In the next few pages we'll tell you a little about our own AI work to date, painfully aware of everything we're leaving out (Ben will be releasing at least 3 other books on the subject soon, all more technical than this one). We're going to give our work roughly the same kind of treatment as everyone else's here—a high-level overview rather than an in-depth exploration.

THE PSYNET MODEL OF MIND

Ben spent the first half of the 1990's working on a novel mathematical philosophy of mind that built on his early intuitive insights. He decided to call his new approach the "psynet model." The goal of the psynet model is to provide a moderately detailed theory of the emergent structures and dynamics in *general* intelligent systems: not just humans, but AI programs as well (and alien evolved intelligences, if any should ever be discovered).

The basic concepts of the psynet model are fairly simple. The mind is conceived to have two main dynamics: self-organizing, spontaneous interactions of pattern-embodying agents that seek to recognize patterns in each other; and goal-directed behavior, in which there is an urge to find and enact patterns of behavior that will lead to the achievement of system goals.

Mental functions such as perception, action, reasoning and procedure learning are described in terms of interactions between "agents"—where "agents" in the mind are thought of as representing patterns recognized in the world or mind, and these patterns are conceived to be constantly interacting with and transforming each other. Any mind, at a given interval of time, is assumed to have a particular goal system, which will often be tied to the embodiment of the mind, and which plays the role of a selective force on the self-organizing pattern system. (We're aware that the psynet model, summarized this way, sounds very vague and "flaky"—which is why a lot of time has been spent working on giving it a precise mathematical formulation. Connecting this mathematical formulation to empirical reality in the context of AI, neuroscience or psychology is another story, however, and one that is very much still being told.)

More specifically, the psynet model posits several primary dynamical principles, including:

- **Association**. Patterns, when given attention, spread some of this attention to other patterns with which they have previously been associated in some way.
- **Differential attention allocation**. Patterns that have been valuable for goal-achievement are given more attention, and are encouraged to participate in giving rise to new patterns.
- **Pattern creation**. Patterns that have been valuable for goal-achievement are mutated and combined with each other to yield new patterns.
- **Credit Assignment**. Habitual patterns in the system that are found valuable for goal-achievement are explicitly reinforced and made more habitual.

Furthermore, the network of patterns in the system must give rise to the following large-scale emergent structures:

- **Hierarchical network**. Patterns are habitually engaged in relations of control over other patterns that represent more specialized aspects of themselves.
- **Heterarchical network**. The system retains a memory of which patterns have previously been associated with each other in any way.

- **Dual network**. Hierarchical and heterarchical structures are combined, with the dynamics of the two structures working together harmoniously.
- **Self structure**. A portion of the network of patterns forms into an approximate image of the overall network of patterns.

As originally presented, the psynet model was applied equally to human psychology and to AI. We spent a fair bit of time in the mid-'90s working out arguments in favor of the psynet model as a conceptual model of human psychology and the structure of the human neocortex, ultimately deciding we lacked the experimental tools to gather data confirming or refuting a grand unified "mind model" like this, and turning more of our attention to AI. However, we have begun a new book about cognitive development in which we plan to present a revised, collaborative effort in this direction.

Both our current AI project, Novamente, and our previous one, Webmind, were based on the psynet model of mind. Unlike most contemporary AI projects, both were specifically oriented towards artificial *general* intelligence (AGI), rather than being restricted by design to one particular domain or narrow range of cognitive functions. Here we're going to write exclusively about Novamente rather than Webmind, but many of the ideas we'll mention were actually common to both systems.

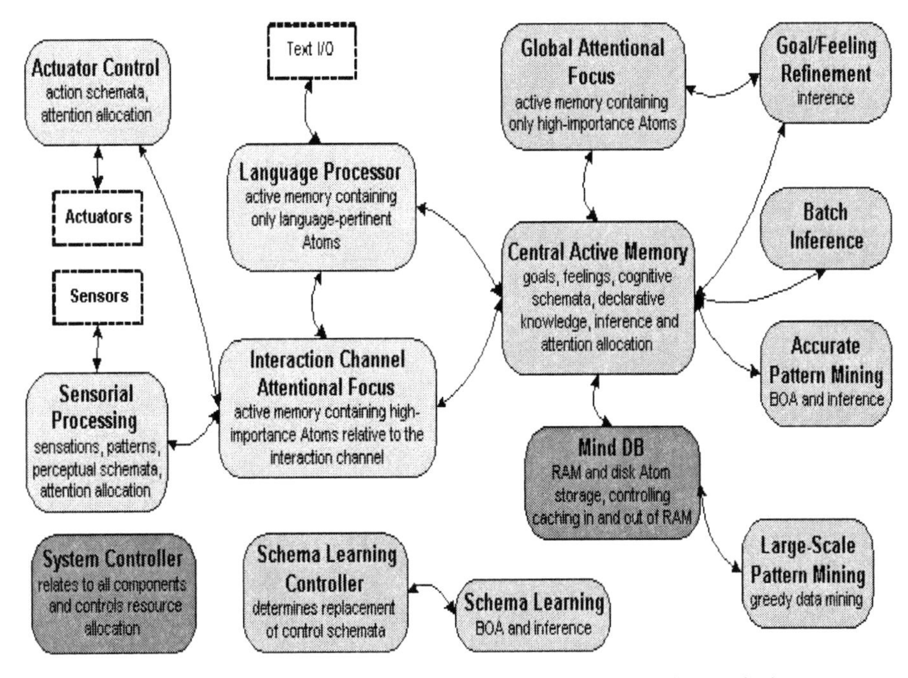

Figure 3.2 High-level architecture of a complex Novamente instantiation.

Each component is a Lobe, which contains multiple atom types and mind agents. Lobes may span multiple machines, and are controlled by schemata which may be adapted or replaced by new ones learned by Schema Learning, as decided by the Schema Learning Controller. The diagram shows a configuration with a single interaction channel that contains sensors, actuators and linguistic input; real deployments may contain multiple channels, with different properties.

THE NOVAMENTE DESIGN FOR ARTIFICIAL GENERAL INTELLIGENCE

The Novamente *design*—which at time of writing is only about 60% embodied in software code—is an AI architecture that synthesizes perception, action, abstract cognition, linguistic capability, short and long term memory and other aspects of intelligence, in a manner inspired by complex systems science. Its design is based on a common mathematical foundation spanning all these aspects, which is inspired by the psynet model and draws on probability theory and algorithmic information theory, among other areas of mathematics. Novamente integrates aspects of prior AI projects, including symbolic and neural-network approaches, evolutionary programming and reinforcement learning, but its overall architecture is unique.

Information is represented inside Novamente using a network of sorts, but not a "neural network"—rather, it uses a special kind of mathematical network called a "weighted, labeled hypergraph," in which pieces of information are represented as nodes and links and patterns of activity of nodes and links. Whereas a link in a formal neural network has a numerical weight indicating its "synaptic conductance," and a formal neural network node has a weight indicating its "activation," Novamente nodes and links carry weights with different semantics. Each node or link is associated with a *truth value*, indicating, roughly, the degree to which it correctly describes the world. Novamente has been designed with several different types of truth values in mind; the simplest of these consists of a pair of values denoting a probability and the amount of evidence used to arrive at the probability. All nodes and links also have an associated *attention value*, indicating how much computational effort should be expended on them. These consist of two values, specifying short and long term importance levels. Truth and attention values are updated continuously by cognitive processes and maintenance algorithms.

Nodes and links in Novamente have a variety of different types, each of which comes with its own semantics. Novamente node types include tokens which derive their meaning via interrelationships with other nodes; nodes representing perceptual inputs into the system (e.g., pixels, points in time, etc.); nodes representing moments and intervals of time; and procedures. Links represent relationships between atoms (nodes *or* links), such as fuzzy set membership, probabilistic logical relationships, implication, hypotheticality and context. Executable programs carrying out actions are represented as special procedure objects that wrap up small networks containing special kinds of nodes and links.

Next, the dynamics of the psynet model are enacted in Novamente via two primary learning algorithms acting on the node/link level: Probabilistic Term Logic, and a variant of the Bayesian optimization algorithm.

PTL is a flexible "logical inference" framework, applicable to many different situations, including inference involving uncertain, dynamic data and/or data of mixed type, and inference involving autonomous agents in complex environments. It was designed specifically for use in Novamente, yet also has applicability beyond the Novamente framework. It acts on Novamente links representing declarative knowledge (e.g. inheritance links representing probabilistic inheritance relationships), building new links from old using rules derived from probability theory and related heuristics. PTL is context-aware, able to reason across different domains and to deal with multivariate truth values. It is capable of toggling between more rigorous and more speculative inference, and also of making inference consistent within a given context even when a system's overall knowledge base is not entirely consistent.

On the other hand, BOA was developed by computer scientist Martin Pelikan as an improvement over ordinary genetic algorithms (GA: an AI technique we mentioned briefly above, which solves problems via simulating evolution by natural selection). BOA significantly outperforms the traditional GA by using probability theory to model the population of candidate solutions. That is, instead of just taking the "fittest" candidate solutions to a problem and letting them reproduce to form new candidate solutions, it does a probabilistic study of which good candidate solutions' features make them good, and then tries to create new candidate solutions embodying these features. This, we feel, combines the evolutionary power of the GA with the analytical precision of probability theory—and provides a nice bridge between evolutionary procedure learning and probabilistic inference (PTL, the other main AI algorithm within Novamente). We have extended Pelikan's original BOA idea into a powerful procedure learning algorithm (BOA Programming or BOAP), which adaptively learns complex procedures satisfying specified goals. It can be used to recognize patterns (a pattern being formally representable as a procedure for calculating or producing something in a simple way, or else a procedure for controlling actuators, or a procedure for controlling patterns of reasoning or perceiving, etc.).

Cognitive processes such as large-scale inference, perception, action, goal-directed behavior, attention allocation, pattern and concept discovery, and even some aspects of system maintenance are implemented in Novamente as specific combinations of these two key algorithms, which are highly flexible and generic in their applicability. The idea is that, via these probabilistic processes, the patterns embodied in the nodes and links in the system's knowledge store will implicitly enact the more abstract dynamics described in the psynet model.

In practice, a Novamente system consists of a collection of functionally specialized units called "lobes," as depicted in the figure above, each one of which deals with a particular domain or type of cognition (procedure learning, language learning, focused attention on important items, etc.). Within each lobe there is a table of nodes and links, and also a collection of software objects called MindAgents, which carry out particular AI processes (visual perception, language parsing, abstract reasoning, etc.) using specialized combinations of PTL and BOA and in some cases other simpler heuristic AI algorithms. The particular assemblage of node and link types, MindAgents and lobes has been painstakingly created with a view toward properly giving rise to psynet-ish dynamics in the whole system.

There is a lot of subtlety to Novamente, and we won't digress any further into it here, because we don't want this to turn into a Novamente book. The following tables give a few more details regarding the particular nodes and links and MindAgents in Novamente, but if you're interested to dig deeper, there's plenty of information online[13].

We must stress that this is still highly speculative work—so far, we've used the Novamente software inside some practical products (Biomind's software for microarray data analysis, to be discussed a bit later; and some natural-language-processing software we created for a U.S. government customer), but none of these products have really made use of Novamente's AGI aspirations. We have not yet built the system out to the point where the emergent intelligence envisioned in the psynet model could be expected to be observed. This will take more time and effort, as there are a lot of details to be gotten right. Indeed, we wish there were a simpler way, but haven't thought of a simpler way that has a decent chance of working—and so we're pushing ahead with the complicated path that we do believe in.

The table below gives a comparison of Novamente structures to those of the human brain.

Human Brain Structure/Phenomenon	**Primary Functions**	**Novamente Structure/Phenomena**
Neurons	Impulse-conducting cells, whose electrical activity is a key part of brain activity	No direct correlate: Novamente's implementation level is different
Neuronal groups	Collections of tightly interconnected neurons, often numbering 10,000-50,000	Novamente nodes
Synapses	The junction across which a nerve impulse passes from one neuron to another; may be excitatory or	Novamente links are like bundles of synapses joining neuronal groups

	inhibitory	
Synaptic Modification	Chemical dynamics that adapt the conductance of synapses based on experience; thought to be the basis of learning	The HebbianLearning MindAgent is a direct correlate. Other cognitive MindAgents (e.g. inference) may correspond to high-level patterns of synaptic modification
Dendritic Growth	Adaptive growth of new connections between neurons in a mature brain	Analogous to some heuristics in the ConceptFormation MindAgent
Neural attractors	Collections of neurons and/or neuronal groups that tend to be simultaneously active	Maps, e.g. concept and percept maps
Neural input/output maps	Composites of neuronal groups, mapping percepts into actions in a context-appropriate way	Schema maps
"Neural Darwinist"	Creates new,	Schema learning via reinforcement

map evolution	context-appropriate maps	learning, inference, evolution
Cerebrum	Perception, cognition, emotion	The majority of Units in a Novamente configuration
Specialized cerebral regions (Broca's area, temporal lobe, visual cortex,…)	Diverse functions such as language processing, visual processing, temporal information processing,…	Functionally-specialized Novamente Units
Cerebellum	Movement control, information integration	Action-oriented units, full of action schema-maps
Midbrain	Relays and translates information from all of the senses, except smell, to higher levels in the brain	Schemata mapping perceptual Atoms into cognitive Atoms
Hypothalamus	(regulation of basic biological drives and controls	HomeostaticParameterAdaptationMindAgent, built-in GoalNodes

	autonormic functions such as hunger, thirst, and body temperature)	
Limbic System	(control emotion, motivation, and memory)	FeelingNodes and GoalNodes, and associated maps

Table 3.1 Novamente vs. the Human Brain

The types and properties of Nodes in the Novamente system are shown in this table:

Node Variety	**Description**
Perceptual Nodes	These correspond to perceived items, like WordInstanceNode, CharacterInstanceNode, NumberInstanceNode, PixelInstanceNode
Procedure Nodes	These contain small programs called "schema," and are called SchemaNodes. Action Nodes that carry out logical evaluations are called PredicateNodes.
ConceptNodes	This is a "generic Node" used for two purposes. An individual ConceptNode may represent a category of Nodes. Or, a Map of ConceptNodes may represent a concept.
Psyche Nodes	These are GoalNodes and

	FeelingNodes, which are special PredicateNodes that play a special role in overall system control, in terms of monitoring system health, and orienting overall system behavior.

Table 3.2 Novamente Node Varieties

This table gives the Link types in the Novamente sytem, and describes their functions:

Link Variety	**Description**
Logical links	These represent symmetric or asymmetric logical relationships , either among Nodes (InheritanceLink, SimilarityLink), or among links and PredicateNodes (e.g. ImplicationLink, EquivalenceLink)
MemberLink	These denote fuzzy set membership
Associative links	These denote generic relatedness, including HebbianLink learned via Hebbian learning, and a simple AssociativeLink representing relationships derived from natural language or from databases.
ExecutionOutputLink	These indicate input-output relationships among SchemaNodes and PredicateNodes and their arguments
Action-Concept links	Called ExecutionLinks and EvaluationLinks, these form a conceptual record of the actions taken by SchemaNodes or PredicateNodes

ListLink and concatListLink	These represent internally-created or externally-observed lists, respectively

Table 3.3 Novamente Link Varieties

This next table describes some examples of of Novamente Map types. Not all possible map types are represented, but the examples give a feeling for the variety of things that can be represented as maps.

Map Type	**Description**
Concept map	a map consisting primarily of conceptual Nodes
Percept map	a map consisting primarily of perceptual Nodes, which arises habitually when the system is presented with environmental stimuli of a certain sort
Schema map	a distributed schema
Predicate map	a distributed predicate
Memory map	a map consisting largely of Nodes denoting specific entities (hence related via MemberLinks and their kin to more abstract Nodes) and their relationships
Concept-percept map	a map consisting primarily of perceptual and conceptual Nodes
Concept-schema map	a map consisting primarily of conceptual Nodes and SchemaNodes
Percept-concept-schema map	a map consisting substantially of perceptual, conceptual and SchemaNodes
Event map	a map containing many links denoting

	temporal relationships
Feeling map	a map containing FeelingNodes as a significant component
Goal map	a map containing GoalNodes as a significant component

Table 3.4 Example Novamente Map Types

Novamente MindAgents are key to the dynamics of the system. In the table below, a variety (though not the totality) of such agents are described. Also given is the development status of the particular type of mind agent, to give you an idea of the developmental progress of the Novamente project at the time of this writing.

MindAgent	**Function**	**Development Status**
First-Order Inference	Acts on first-order logical links, producing new logical links from old using the formulas of Probabilistic Term Logic	Complete
LogicalLinkMining	Creates logical links out of nonlogical links	Complete
Evolutionary Predicate Learning	Creates PredicateNodes containing predicates that predict membership in ConceptNodes	Complete
Clustering	Creates ConceptNodes representing clusters of existing ConceptNodes (thus enabling the cluster to be acted on, as a	Complete

	unified whole, by precise inference methods, as opposed to the less-accurate map-level dynamics)	
Activation Spreading	Spreads activation among Atoms in the manner of a neural network	Complete
Importance Updating	Updates Atom "importance" variables and other related quantities	Implemented in prototype form
Concept Formation	Creates speculative, potentially interesting new ConceptNodes	Implemented in prototype form
Evolutionary Optimization	A "service" MindAgent, used for schema and predicate learning, and overall optimization of system parameters	Complete
Hebbian Association Formation	Builds and modifies HebbianLinks between Atoms, based on a PTL-derived Hebbian reinforcement learning rule	Implemented in prototype form
Evolutionary Schema Learning	Creates SchemaNodes that fulfill criteria, e.g. that are expected to satisfy given GoalNodes	Partially implemented
Higher-Order Inference	Carries out inference	Partially implemented

	operations on logical links that point to links and/or PredicateNodes	
Logical Unification	Searches for Atoms that mutually satisfy a pair of PredicateNodes	Not yet implemented
Predicate/Schema Formation	Creates speculative, potentially interesting new SchemaNodes	Not yet implemented
Schema Execution	Enacts active SchemaNodes, allowing the system to carry out coordinated trains of action	Partially mplemented
Map Encapsulation	Scans the AtomTable for patterns and creates new Atoms embodying these patterns	Not yet implemented
Map Expansion	Takes schemata and predicates embodied in nodes, and expands them into multiple Nodes and links in the AtomTable (thus transforming complex Atoms into Maps of simple Atoms)	Not yet implemented
Homeostatic Parameter Adaptation	Applies evolutionary programming to adaptively tune the parameters of the system	Implemented in prototype form

Table 3.5 Primary Novamente MindAgents

In the following table, we describe a variety of functions of a mind (cognitive tasks), what standard approaches are used in contemporary AI systems, challenges for future development, and finally the Novamente system approach to those challenges.

Cognitive Task	Standard Approaches	Challenges	Novamente Approach
Logical Inference	Predicate, term, combinatory, fuzzy, probabilistic, nonmonotonic or paraconsistent logic	Accurate management of uncertainty in a large-scale inference context; "Inference control": Intelligent, context-appropriate guidance of sequences of inferences ·	Probabilistic Term Logic tuned for effective large-scale uncertainty management; Inference control carried out via a combination of inferential and noninferential cognitive processes
Attention Allocation	Blackboard systems, neural net activation spreading	The system must focus on user tasks when needed, but also possess the ability to spontaneously direct its own attention without being flighty or obsessive	Novamente's nonlinear, probabilistic inference based Importance Updating Function combines quantities derived from neural-net-like activation spreading and blackboard-system-like cognitive-utility analysis

Procedure Learning	Evolutionary programming, logic-based planning, feedforward neural networks, reinforcement learning	Techniques tend to be unacceptably inefficient except in very narrow domains	A synthesis of many techniques allows each procedure to be learned in the context of a large number of other already-learned procedures, enhancing efficiency considerably
Pattern Mining	Apriori, genetic algorithms, logical inference, search algorithms	Finding complex patterns requires prohibitively inefficient searching through huge search spaces	Integrative cognition is designed to hone in on the specific subset of search space containing complex but compact and significant patterns
Human Language Processing	Numerous parsing algorithms and semantic mapping approaches: context-free grammars, unification grammars, link grammars; conceptual graphs, conceptual grammars, etc.	Integrating semantic and pragmatic understanding into the syntax-analysis and production process	Syntactic parsing is carried out via logical unification, in a manner that automatically incorporates probabilistic semantic and pragmatic knowledge. Language generation is carried out in a similarly integrative way, via inferential generalization.

Self-Modeling (the creation of a "phenomenal self")	No current AI system or AGI design addresses this	Creating a representational system sufficiently sophisticated to represent something as complex as a self in a compact way	Creating learning algorithms capable of the large-scale pattern-recognition prowess required to recognize something as large and abstract as a self in the large body of relevant but noisy data available to an embodied intelligence This is viewed as a pattern recognition and inference problem similar to many others confronted by Novamente, but larger in scale. Novamente contains specific algorithms for mining patterns in its internal knowledge-hypergraph and explicitly embodying these patterns in new subgraphs. The self is one such pattern.

Table 3.6 Comparison of Approaches to Several Cognitive Tasks

EXPERIENTIAL INTERACTIVE LEARNING

We've talked about a lot of technical AI/comp-sci stuff—nodes, links, probabilities—and it's important stuff, but the bad news is that it's not really sufficient for *mind.* Similarly, human intelligence does not emerge solely through human neural wetware. A human infant is not so intelligent, and an infant raised without proper socialization will never achieve full human intelligence. Human brains learn to think through being taught, and through diverse social interactions. Our view is that the situation will be found to be similar with AGIs, as AGI technology develops. According to this philosophy, the basic AI algorithms in an AI system, even if correct, complete, appropriate and computational tractable, can supply only the *raw materials* of thought. What is missing in any AI system "out of the box" are context-specific control mechanisms for the diverse cognitive features required for practical intelligence. If designed correctly, however, an AI system should have the capability to *learn how to learn* these, through environmental and social interaction.

A complete AGI system "out of the box" may be more or less competent than narrow AI systems, depending on factors such as prior knowledge and the scope of its initial domain[14]. In any case it will not be nearly as robustly intelligent as an AGI system that has refined its ability to learn context-specific control mechanisms through meaningful interactions with other minds. Once it's been interacting in the world for a while, an AGI will gain a sense of how to reason about various topics—conversations, love, network intrusion, bio-warfare, its own source-code, and so on—by learning context-dependent inference control schemata for each case, according to a procedure learning process tuned through experiential interaction. The idea is that an AGI can learn in a manner similar to a person, though not necessarily using the same exact mechanisms as a human brain.

This line of thinking leads us to concepts such as *autonomy, experiential interactive learning (EIL),* and *goal-oriented self-modification*, which lay at the heart of the notion of Artificial General Intelligence as we present it.

The Novamente AI system is highly flexible in construction. It may be supplied with specific, purpose-oriented control processes, and in this way used as a data mining and/or query processing engine. This is the approach taken, for example, in the current applications of the Novamente engine in the bioinformatics domain[15], but this kind of deployment of Novamente does not permit it to develop its maximum level of general intelligence.

For truly significant AGI to emerge, a properly designed system must be supplied with general goals—and then allowed to learn its own control processes via execution of its procedural learning dynamics, through interaction with a richly structured environment, along with extensive, meaningful interactions with other minds. This is the long-term plan we hope to follow with the Novamente system.

This leads us to the need for embodiment. While conversations about useful information will be an important source of EIL for Novamente, we suspect that additional tutoring on basic world-concepts like objects, motions, self and others will be valuable. This loosely follows Wierzbicka's notion of "semantic universals." To carry out this kind of instruction for Novamente, we have created a special simulated environment, the AGI-SIM simulation world.

The AGI-SIM world has been built based on open-source tools (mainly the 3-D simulation environment CrystalSpace[16], but the specifics are not as relevant as the concepts behind it—it could be any sufficiently rich virtual environment) and may be configured to display realistic physics, or not. It allows AI systems and humans to control mobile agents that have multiple moving body parts and experience the simulated world via multiple senses, as well as having the capability to communicate with each other directly through text. AGI-SIM is being developed by the Novamente team as an open-source project[17] with the intention of being useful for other AGI projects as well as Novamente.

Without going into details on AGI-SIM's implementation here, it is worth mentioning some of the basic principles that went into its design:

- The experience of an AGI, controlling an agent in a simulated world, should display the main qualitative properties of a human controlling their body in the physical world, for the purposes of giving it some understanding of the human world and some common ground for communication with humans about topics beyond the scope of its own Internet world. Those qualitative properties which help the AGI to relate experiences in the simulated world to the many obvious and subtle real-world metaphors embedded in human language will allow it to ground these human concepts in some kind of relevant experience, even though it can't fully experience them (unless it is transferred into some kind of robot body). Hopefully this can give an AGI enough of an ability to "put itself in our shoes," so that it can understand us at least as well as we try to understand dolphins or whales (or other creatures whose exact physical experience we can not share).

- The simulated world should support the integration of perception, action and cognition in a unified learning loop, which is crucial to the development of intelligence. While it is up to the specific AGI actor to implement such internal cognitive actions, the simulation world needs to be rich enough for such an AGI to derive conclusions from it.

The simulated world should support the integration of information from a number of different senses (both simulated and, to the AGI, real in the sense of different kinds of rich data), all reporting different aspects of a common world, which is valuable for the development of intelligence.

With these goals in mind, we are building AGI-SIM as a basic 3-D simulation of the interior of a building, with simulations of sight, sound, smell and taste. An agent in AGI-SIM has a certain amount of energy, can move around, and pick up objects and build things. The initial version doesn't attempt to simulate realistic physics, but this may be integrated into a later version using open-source physics simulation software. While not an exact simulation of any physical robot, the agent Novamente controls in AGI-SIM is designed to bear enough resemblance to a simple physical robot that the porting of control routines to a physical robot should be feasible. Such a simulated world can be used either for training an AGI to operate in a future robot body (bipedal or otherwise), or simply as a way for an AGI which is fully embodied in an Internet/Dataspace world to have some grounding of understanding of the physical world.

NOVAMENTE PRESCHOOL

To help place these results and ideas in context, the following table illustrates some of the progress that we hope to make during the next few years, regarding the education of the Novamente system within the AGI-SIM environment. Of course, the specific dates are highly subject to change based on project funding and other considerations.

Date	Description of Milestone
2006/1	Experiential grounding of simple language understanding
2006/2	Goal-directed navigation: the ability to find and retrieve objects
2007/1	Collaborative and creative play with blocks and other objects
2007/2	Simple ethical behavior and socialization

Table 3.7 Novamente System Milestones

More specifically, our goal for the next few years it to create a Novamente system that will fulfill the following four subgoals:

1. Communicate in simple English (NOT in the manner of the Turing test, but comprehensibly even if awkwardly—the communicative goal is for us to be able to tell it is forming new concepts and trying to express them, even if it is awkward for it to communicate with us in our "foreign" language) and interact with human users in the AGI-SIM simulation world.

2. Be teachable by humans, displaying an ultimately unbounded trajectory of ongoing learning and self-improvement. It need not be able to learn everything

a human can learn—for example, it needn't necessarily be able to ride a bicycle or sing—but it should be able to learn new things about the domains it can experience.

3. Be integrable in practical software applications with commercial applications, including language processing, simulated-agent control and robotics. In short, it needs to be able to get a job.

4. Be capable of modeling many of the qualitative observed properties of human cognition, while also having the option to diverge from the typical patterns of human cognition in cases where the latter are known to be severely nonoptimal (or in the cases where they simply don't fit with the experience of the AGI, as it will be a different entity with a different, mechanical or purely electronic body and different senses).

Note that the "project milestones" we've articulated involve integrated system behavior in a social, embodied context, rather than the performance of isolated cognitive components. In working on the project, we break down general milestones like this into simpler ones, but it's important that these are also "whole-system" milestones rather than system-component ones. Qualitative examples of such subgoals would be:

- When asked "What do you see?" (in this wording or any simple, semantically similar wording), the system responds (in English) with a list of a subset of the objects in its field of view, not omitting any objects belonging to categories in which its teacher has recently shown interest (or which in its experience are important, such as a virtual fire or "sick" agent comrade).
- When asked questions of the form "Put object X on object Y" (in this wording or any roughly semantically similar wording), the system controls its AGI-SIM embodiment to do so, appropriately navigating and avoiding other objects along the way. If object Y is too small or inappropriately shaped to support object X, it complains and says so—and can give some sensible explanation as to why it is refusing.

This kind of thing is a far cry from superhuman superintelligence—it's more like AGI preschool. But this, we feel, is where we have to start.

METASYSTEM TRANSITION REVISITED

Recall from Chapter 1 the notion that there are *two* metasystem transitions involved in the emergence of mind from unintelligent matter. This idea was related there in a Novamente context—and now that a little bit of detail on Novamente has been given, it may be appreciated more fully.

While Novamente is not, strictly speaking, a "modular" architecture, it's not misleading to think of it that way. We may think of the system as consisting of several Novamente cognitive components:

- **Declarative**: The set of declarative nodes and links, acted on by PTL inference and representing patterns observed in the mind or world. This is its logic system, and you can think of it like an internal reference library.
- **Procedural**: The set of procedural nodes and links, created by BOA and refined by PTL, representing actions to be carried out in the mind or world. This is the collection of things it knows how to do, and is associated with a set of goals of things to do—both immediately and over the long term.
- **Action**: The set of procedural nodes and links concerned with physical action, created by BOA and very simple PTL, and never abstractly reasoned on. This is really a subset of "Procedural," specifically relating to actions in the world, as opposed to generally relating to any action, cognitive or physical.
- **Perception**: Declarative nodes and links representing external percepts, acted on by very simple PTL. This is really a subset of "Declarative," specifically relating to things that have been perceived in the world through its senses, rather than generally relating to any knowledge, perceived or deduced.

- **Attention**: The set of attentional nodes and links, created by PTL and BOA, and representing patterns recognized in what the system has benefited from doing in various circumstances. This is really its very-short-term goal set, and it coordinates activity of all of the other types.

Each component has a certain wholeness, being a set of its own sub-components, with a certain transcendence of the whole over the parts. But the metasystem transition that's *really* interesting for AGI is the next one up—the one where the whole system operates synergistically to achieve actual intelligence. The goal isn't domain specific "intelligence" like playing chess, or even playing Go, but rather fully active and productive emergent intelligence, wherein the whole mind is engaged in achieving goals by recognizing patterns in itself and the outside world. The specialized pattern recognition and formation that goes on in each system component won't achieve really complex goals or generalize from one domain to another. Putting such modules together in a simple and constrained way can give you functions that normal AI software can't do—things like advanced bioinformatics data analysis, which we'll discuss in a later chapter. Putting the modules together in a manner allowing free and full interaction can get you actual intelligence, because the modules are chosen specifically so as to allow the system to understand itself: to recognize patterns in itself, and be able to synthesize new patterns (make deductions) from them about itself and its world.

All this complexity is not 100% obvious from the original vision of mind as a collection of patterns that forms and perceives patterns in itself and the world, in order to achieve complex goals in a complex environment. Still, once you dive into the details, it does emerge fairly naturally from this general view. A complex environment, including other intelligences, involves a lot of different kinds of things. Each one requires its own specialized pattern recognition and formation mechanisms. Achieving complex goals in such an environment involves forming concepts that span various kinds of things; both internal and external observations, and deductions. This requires intense interaction between the various modules of mind.

In our opinion, there's no "big trick" to building a thinking machine—no single algorithm which, when written down, will embody the whole system. A mind is a collection of patterns that recognizes and forms patterns in itself, in order to achieve complex goals. There are some universal structures and dynamics that it seems any mind has got to have. It's possible to build a system possessing these universal structures and dynamics in software running on a network of high-powered PCs. The main problems are these: first, getting the needed memory and processing power. Then, solving the routine, but time-consuming, software engineering problems of getting such a huge system to actually work in a reliable and efficient way. There's the problem of parameter tuning—getting the system to regulate itself, all its modules together, in a way that keeps the whole huge system functioning adequately, without any part starving the other for resources. And then there's the problem of teaching—how do we play mommy and daddy to a baby intelligence so unlike us, without driving it totally batty? Fortunately, we seem to have solutions to all these problems, and so the creation of the world's first really thinking machine would seem to be only a year or two ahead of us (or a few, depending on funding). And as we walk along the path, we're building lots of cool components that can—if we play our cards right—make us money along the way. There are worse ways to spend a few years! The possibility of our work triggering the Singularity in the fully Vingean sense is also somewhat tantalizing.

CREATING THE WORLD WIDE BRAIN

We've talked about the AGI-SIM environment as a kind of Novamente preschool. However, though such a simulated environment is probably necessary, that doesn't mean it's sufficient. An AGI system will definitely benefit from having more information-rich surroundings as well. One way to do this is to embed the system in a physical robot, and/or give it access to physical sensors like cameras and microphones. Another approach is to hook it up to the Internet—the source of an awful lot of valuable textual data, useful to any AGI system that has achieved a rudimentary understanding of English (but giving it a vastly different perceptual grounding of the world from being embodied in a physical interaction device).

The Internet, now even more so than in the mid-'90s, has the potential to give an AGI system both processing power and a rich perceivable/manipulable environment. To make this potential real, however, requires the development of specific Internet software aimed at making the Net useful for AI (and teaching the baby AGI that humans actually do care about more than pornography and get-rich-quick schemes and not to overemphasize that data based simply on volume). Implementing Heylighen's proposals for adaptive hyperlink weight modification would be a step in this direction, but what's needed to make the Net truly AGI-friendly is a good bit more than this. Toward this end, in 2000 we designed a global distributed processing framework called WebWorld, which would allow Novamente (or any other AI system roughly similar in structure) to split up its thought processing across literally millions of machines. Some of these machines may not be powerful enough to run Novamente, but may nonetheless be strong enough to run smaller "Novamente auxiliary processing units," which we call WebWorld lobes. The WebWorld framework was prototyped at Webmind Inc. A fully functional version was never built, but a fairly complete design exists and if no one else creates something similar, in time a WebWorld variant will be implemented as part of the Novamente project.

If something like WebWorld were actually built, how would it be used in Novamente? In the beginning, at least, a big Novamente will always have a cluster of dedicated machines as its main mind, but it will farm out various learning problems to thousands or millions of machines elsewhere. One thing that this surplus of machines will allow it to do is to read the huge amount of textual and numerical data that's out there on the Web, and eventually picture, sound and movie data as well. Although Novamente is starting out as a program running on a small cluster of machines and operating on a limited pool of data, its need for a rich perceptual environment combined with its limitless thirst for processing power is going to push it onto the Net; totally consistent with the initial vision of a *Web mind*, an Internet global brain.

Is this the final configuration for the global brain? No way. Is it the only way to do things? No, but this seems the most workable architecture for moving things from where they are now to a reasonably intelligent Net. After this, the dynamics of societies of AI agents become the dominant factor, with the commercial servers and client machines as a context. And after that?

FROM AI TO SINGULARITY

Now we'll return to the notion of "the Singularity" in light of these comments on advanced AI technology. Of course, the Singularity is not just about AI, but AI may play a special role in its advent, because once sufficiently advanced it can serve as a powerful "metatechnology," drastically accelerating the pace of creation of new technologies of various kinds.

As noted above, Eliezer Yudkowsky has put particular effort into understanding the AI aspects of the Singularity, discoursing extensively on the notion of Friendly AI. The Friendly AI and related ideas involve the creation of AI systems that, as they rewrite their own source code and achieve progressively greater and greater intelligence, leave invariant the portion of their code requiring them to be friendly to human beings. We'll discuss some of these ideas in depth in later chapters.

The notion of the Singularity seems to us to be a valid one (though exactly what form it may take is open to much debate), and the notion of an AI system approaching it by progressively rewriting its own source code also seems to be valid. As usual, there are a few pesky details that only become clear once one has a sufficiently well-fleshed-out framework within which to analyze them. From a Novamente perspective, the following is the sequence of events that seems most likely to lead up to the Singularity:

1. Someone (the Novamente team, perhaps?) creates a fairly intelligent AI, one that can be taught, conversed with, etc.
2. This AI is taught about programming languages, algorithms and data structures, etc.

3. It begins by being able to write, optimize and rewrite simple programs.
4. After it achieves a significant level of practical software engineering experience, and mathematical and AI knowledge, it is able to begin improving itself—at which point the "hard takeoff" begins.

Our intuition is that, even in this picture, the "hard takeoff" to superhuman intelligence will take a few years, not minutes. Obviously, that's still quite fast by the standards of human progress.

The Singularity emerges in this vision as a consequence of emergence-producing dynamic feedback between the AI Engine and intelligent program analysis and optimization tools like the Supercompiler. The global brain then becomes not only intelligent but superintelligent, and we, as part of the global brain, are swept up into this emerging global superintelligence in ways that we can barely begin to imagine.

To cast the self-modification problem in the language of Novamente AI, it suffices to observe that self-modification is a special case of the kind of problem we call "schema learning." The AI Engine itself is just a big procedure, a big program, a big schema. The ultimate application of schema learning, therefore, is the application of the system to learn how to make itself better. The complexity of the schema learning problem, with which we have some practical experience, suggests how hard the "self-modifying AI" problem really is.

It's easy enough to make a small, self-modifying program, but such a program is not intelligent. It's closer to being "artificial life" of a very primitive nature. Intelligence within practical computational resources requires a lot of highly specialized structures. These lead to a complicated program—a big, intricate mind-schema—which is difficult to understand, optimize and improve.

Creating a simple self-modifying program, and expecting it to become intelligent through progressive environment-driven self-modification is an interesting course of research, but it seems more like an attempt to emulate the evolution of life on Earth than an attempt to create a single intelligence within a reasonable time frame.

However, just because the "learn my own schema" problem is hard, that doesn't mean it's unsolvable. A Java or C program can be represented as a SchemaNode inside Novamente, and hence it can be reasoned, mutated and crossed over, and so forth. This is what needs to be done, ultimately, to create a system that can understand itself and make itself smarter and smarter as time goes on—eliminating the need for human beings to write AI code or articles like this one. With a self-modifying system, AI intelligence can evolve from a more basic starting point, analogous to how ours evolved but using very different specific mechanisms.

Reasoning about schema-representing computer programs requires a lot of specialized intuition, and specialized preprocessing may well be useful here—such as the automated analysis and optimization of program execution flow being done by Val Turchin and his colleagues in their supercompilation project[18]. There is a lot of work yet to be done here, but it's a fascinating direction and a necessary one.

THE NOVAMENTE VISION

The conceptual vision underlying Webmind/Novamente incorporates every major aspect of the mind studied in psychology and in neuroscience. Webmind was a failure—but we believe the concepts at its heart were right. Where it slipped up was in the area of elegance and simplicity. It was too complicated for us mere humans to tune all its parameters, optimize all its subsystems for efficient interoperation, and so on. We might have refined it more fully and achieved success sooner had we not run out of funding, but instead we are proceeding as we can, and now there's Novamente. Based on the same basic concepts, but with more sophisticated use of mathematics and software engineering to achieve simplicity and efficiency, Novamente is an even more promising path to AGI than Webmind. The educational value of failure has been amply documented, and from the failure of Webmind, a lot was learned about where we went wrong with the initial implementation and various elements of the design.

Not all aspects of the Novamente system are original in conception, and indeed, this is much of the beauty of the thing. The essence of the system is the provision of an adaptable self-reconstructing platform for integration of insights from a huge number of different disciplines and subdisciplines. In Novamente, aspects of mind that have previously seemed disparate are drawn together into a coherent self-organizing whole.

The cliché Isaac Newton quote, "If I've seen further than others, it's because I've stood on the shoulders of giants," inevitably comes to mind here. (As does the modification by MIT computer science guru Hal Abelson: "If others have seen further than me, it's because giants were standing on my shoulders.") The human race has been pushing toward AI for a long time—and Novamente, if it lives up to our aspirations for it, will merely put on the finishing touches.

GRAND FINALE

Call us mad scientists if you will, but we believe that the Novamente design, if fully implemented and tested, would lead to a computer program that manifests intelligence according to the criterion of being able to carry out conversations with humans that will be subjectively perceived as intelligent—conversations pertaining to things in the AGI-SIM world or any physical or simulated world to which the system has access, or else about information obtained from the Internet. It would demonstrate an understanding of the contexts in which it is operating, an understanding of who it is and why it is doing what it is doing, an ability to creatively solve problems in domains that are new to it, and so forth.

Furthermore, we believe that the Novamente design is not unique in this way. We have not yet seen other AGI designs explicitly articulated which we feel have as much AGI potential as Novamente, but we are sure there are many others—some similar to Novamente and some very different.

Once an AGI reaches the stage of moderate self-understanding and conversational ability—then, from there, the sky's the limit. By creating AI we would create a whole new kind of intelligence; which would give us, in conjunction with it, new perspectives on our own biology and intelligence which will allow us to be smarter, live longer, and have much better lives overall than we do now.

What are the complaints and counterarguments most often heard when discussing ambitious AGI projects like Webmind or Novamente with expert outsiders? We've already discussed some of these, but we'll give a quick recap here.

First, there are those who just don't believe AI is possible, or believe that AI is only possible on quantum computers, or quantum gravity computers, etc. Forget about them. They'll see. You can't argue anyone out of their religion. Science is on the side of digital AI at this point, as has been exhaustively argued by many people. (Even *if* AI can only be achieved on quantum computers, they're expected to be available in a couple of decades—not very far off at all, in terms of the grand historical program of human aspirations towards creating new forms of intelligence.)

Then there are those who feel the system doesn't go far enough in some particular aspect of the mind: temporal or causal reasoning, or grammar parsing, or perceptual pattern recognition, or whatever. This complaint usually comes from people who have research expertise in one or another of these specialty areas. The general learning algorithms of a Webmind or Novamente, they say, will always be inferior to the highly specialized techniques that they know so well.

Not surprisingly, our feeling is that the current Novamente design is specialized enough. We don't think it so overspecialized as to become brittle and non-adaptable, but we worry that if it becomes more specialized, this will be the case. Our intuition is that things like temporal and causal reasoning should be learned by the system as groundings of the concepts "time" and "cause" and related concepts, rather than wired in. However, the beauty of the Novamente design is that, if we are proved wrong in this belief, we've built in mechanisms for wiring in necessary basic structures in a way which allows them to cognitively function just like experienced and deduced concepts—rather than having to try to shoehorn them in with specialized structures.

On the other side, there are those who feel that a system like Webmind or Novamente is "too symbolic." They want something more neural-net-ish, or more like a simple self-modifying system as described in Ben's *Chaotic Logic* and *From Complexity to Creativity*. We can relate to this point of view quite well, philosophically. But a careful analysis of the system's design indicates that there is nothing a more sub-symbolic system can do that this one can't. We have SchemaNodes embodying Boolean networks, feeding input into each other, learning interrelationships via neural net-like mechanisms such as Hebbian learning, and being evolved by a kind of evolutionary-ecological programming. This is in fact a sub-symbolic network of procedures, differing from an evolutionary neural net architecture only in that the atomic elements are Boolean operators rather than threshold operators—a fairly insubstantial difference which could be eliminated if there were reason to do so. The fact that this sub-symbolic, evolving adaptive procedure network is completely mappable into the symbolic, inferential aspect of the system is not a bad thing, is it? In fact—letting AI-scientist hubris overtake us for a moment—we would say that in the Novamente design we have achieved a very smooth integration of the symbolic and sub-symbolic domains, perhaps even smoother than is likely to exist in the human brain. This will serve the system well in the future.

There's the complaint that a Baby Novamente won't have a rich enough perceptual environment with just AGI-SIM plus the Internet. This is certainly possible, but the technology exists to hook up camera eyes, microphone ears, and even some form of a robot body to it should this prove necessary. However, this belief may simply be a reflection of our human prejudice that nothing which doesn't share our own exact senses could be intelligent.

There's the complaint that such systems have too many parameters and it will take forever to get such a system to actually work, as opposed to working theoretically. This is indeed a bit of a worry, we can't deny it. We've tried very hard to minimize this problem via the simplifications and changes made in moving from Webmind to Novamente. However, we can't know whether this is the case or not unless we try to actually build it.

Finally, there are those who reckon that even if one were able to create a correct AGI design, we wouldn't have the processing power and memory to run it yet. This complaint scares us a little bit too, but not too much. Based on our experimentation with the system so far, there are only two things that seem to require vastly more computer power than is available on a cluster of a few dozen powerful PCs. The first of these, the learning of new procedures for acting appropriately in various situations ("schema learning," in our lingo) is something that can be done offline, running in the background on millions of PCs around the world via a WebWorld type system. The second, real-time conversation processing, can likely be carried out on a single supercomputer, serving as the core of the AI Engine cluster. We have a very flexible software agents system that is able to support a variety of different hardware configurations, and we believe that by optimally utilizing available hardware we can make a fairly smart computer program even *without* the massive advances that Moore's law will quickly bring. Of course, the more hardware we have, the cleverer our system will become—and soon enough it will be literally begging us for more, more, more!

Perhaps all this sounds like nonsense to you. If so, remember that human history is filled with seemingly nonsense ideas which, after failing a number of times, finally succeeded. Intelligence already exists in our universe, so whether it can be created in a system other than humans is more an engineering problem than a theoretical one. In engineering, you really can't know for sure if your design is any good until you try building it. Very little of what anyone living in a modern industrial or post-industrial nation takes for granted was not once considered "absolutely, irrefutably impossible."

[1] Coauthored with Ted Goertzel
[2] It is *possibly* a sufficient condition. However, people in general, rather than experts in AGI, may just be too easy to fool with a simple rule-based or otherwise unintelligent system. One problem is that online chat rooms, e-mail and bulletin boards have allowed people to adjust to various stilted forms of communication—disjointed thoughts expressed as awkward phrases, and so on, making it potentially easier for a computer to fool a human into thinking they were chatting with another human (albeit an irritatingly childish one).
[3] Specifically, we've always appreciated the "attractor neural net" research literature; and we've been much less interested in feedforward neural nets (backpropagation and so forth), which seem to miss most of what's important about biological neural networks.
[4] Socially interacting, in the sense of "some interaction with us that we can recognize as intelligence." The goal is not necessarily to *replicate* human intelligence in an AI; but rather to create some kind of recognizable, autonomous intelligence, with its own non-human characteristics that develop out of its different mind structure and physical context, but with which we can communicate (if you created an AI with which you couldn't communicate, you wouldn't even know it).
[5] http://www.a-i.com
[6] ://http://www.alicebot.org
[7] http://megahal.alioth.debian.org
[8] http://reply.jabberwacky.com/
[9] http://www.longnow.org/
[10] A college in Massachusetts which both Ben and Stephan attended. http://www.simons-rock.edu
[11] Frank Rosenblatt was the founder of the connectionist approach to AI, and author of *Principles of Neurodynamics: Perceptrons and the Theory of Brain Mechanisms* (Spartan Books, 1962)
[12] http://www.reeltwo.com
[13] at http://www.agiri.org, http://www.novamente.net and Hhttp://www.goertzel.orgH
(The long overview paper is a good place to start.)
[14] An AGI may be prepopulated with knowledge of a specific domain in an encoding which is similar to that which it builds when it learns, but since it can learn, it can eventually expand beyond its initial specialization if it so "desires."
[15] http://www.biomind.com
[16] http://crystal.sourceforge.net
[17] http://sourceforge.net/projects/agisim
[18] http://www.supercompilers.com

CHAPTER 4

NANOTECHNOLOGY GETS REAL

One of the most amazing revelations that the scientific world-view has brought us is the number of different scales in the universe. Pre-civilized cultures had the concept of a scale bigger than the everyday—associated with the "heavens," and usually with gods of some sort. The ancient Greeks conceived of a smaller "atomic" scale, and other ancient cultures like the Indians had similar ideas. However, these were not fleshed out into detailed pictures of either cosmology or the microworld, because they were pure conceptual speculations, ungrounded in empirical observation. When Leeuwenhoek first looked through his microscope, he discovered a whole other world down there, with a rich dynamic complexity no one had imagined in any detail. Galileo and the other pioneers of astronomy had the same experience with the telescope. Everyone had known the stars were up there, but so many of them—and so much variety! The planets and the moon were found to have strange shapes and markings on them. Eventually, more and more bizarre facts about the heavens—pulsars, quasars, binary stars, gravitational singularities, three-degree cosmic background radiation, and so on—made the transition from unbelievable discoveries to everyday facts of astronomy.

Now, we take for granted that the world we see and act in daily is just a human-scale slice taken out of all the scales of the cosmos. There are vast complex dynamic processes happening on the galactic scale, in which our sun and planet are just tiny insignificant pieces. And there are bafflingly complex things going on inside our bodies: antibodies collaborating with each other to fight germs, enzymatic networks carrying out subtle computations, cell networks that organize electrochemical stimuli into intelligence—even things as strange as quantum wave functions interfering mysteriously to produce protein dynamics that interpret obscure DNA code sequences.

Until recently, our participation in non-human scale processes has been basically limited to observation and experimentation. We look at the planets; we don't create new planets, or even move around existing ones (though we sometimes do some amazing, and nasty, things to the surface of this one). Once we could only observe the molecular scale, but in the 20th century we learned to create new molecules—an amazingly difficult process that we're really just beginning to understand. We probe particles and their interactions in our particle accelerators, but to measure or understand anything at this tiny scale, we need very carefully controlled conditions.

This is going to change. In time, technology will allow us to act as well as observe on the full range of scales. Not surprisingly, given the different levels of energy required to do each, we're likely to master the realm of the very small well before the realm of the very large. Manufacturing has been pushing in this direction for decades now: machines just keep getting smaller and smaller and smaller. The circuitry controlling the computers we've typed these words on is far too small to see with the naked eye—circuitry which, fifty years ago, would have filled an aircraft hangar, at least. The Playstation Portable is the first inexpensive handheld device to achieve essentially the same graphics quality as a flat-screen TV. It seems facile to draw a curve from a movie screen to the PSP and ask what comes next, but there is a real logic underlying this particular curve-fitting. Our engineering technology gets better and better at manipulating smaller and smaller scales, year by year. The "wires" inside microchips have gotten so small and densely packed that physicists are hired to help chip companies understand not just the electromagnetic but the *quantum* interference possibilities of different circuit layouts.

The next major step along this path, the purposeful construction of molecular-level machinery, has been termed *nanotechnology*. The most ambitious goals of nanotechnology still live in the future: the construction of molecular assemblers, machines that are able to move molecules into arbitrary configurations (metals, proteins, you name it), playing with matter like a child plays with Legos. But there are concrete scientific research programs moving straight in this direction; most ambitiously at the Zyvex Corporation, which has hired an impressive staff of nanotechnology research pioneers to build truly general-purpose nanotechnology.

There are more specialized sorts of nanomachines in production right now, from teeny tiny transistors to DNA computers to human-engineered proteins the likes of which have never been seen in nature. Perhaps someday viruses, which have completely vexed immunologists for decades, will be able to be molecularly *disassembled* into harmless materials. That is one big hope of nanotechnology research. Exactly how many years it will be until we can cure diseases by releasing swarms of millions of germ-killing nanobots into the bloodstream is hard to say; but the door to the nanorealm has been opened, we've stepped through the doorway, and bit by bit, year by year, we're learning our way around this tiny world.

HOW SMALL IS NANO?

It's hard for the human mind to grasp the tiny scales involved here. One way to conceptualize these scales is to look at the metric system's collection of prefixes for dealing with very small quantities. We all learned about centi- and micro- in school. Computer technology has taught us about the scales beyond kilo: mega-, giga-, tera-, and soon peta-. Over the next few decades, we'll be developing more and more commonsense collective intuition for nano-, pico-, femto- and perhaps even the scales below.

prefix	abbreviation (upper and lower case are important)	meaning	example	a sense of scale (for some) Most are approximate.
yotta	Y	1024	yottagram, 1 Yg = 1024 g	mass of water in Pacific Ocean ~ 1 Yg energy given off by the sun in 1 second ~ 400 YJ volume of earth ~ 1 YL mass of earth ~ 6000 Yg
zetta	Z	1021	zettameter, 1 Zm = 1021 m	radius of Milky Way galaxy ~ 1 Zm volume of Pacific Ocean ~ 1 ZL world energy production per year, ~ 0.4 ZJ
exa	E	1018	exasecond, 1 Es = 1018 s	age of universe ~ 0.4 Es (12 billion yr)
peta	P	1015	petameter, 1 Pm = 1015 m	1 light-year (distance light travels in one year) ~ 9.5 Pm
tera	T	1012	terameter, 1 Tm = 1012 m	distance from sun to Jupiter ~ 0.8 Tm

giga	G	109	gigasecond, 1 Gs = 109 s	human life expectancy ~ 1 century ~ 3 Gs 1 light-second (distance light travels in one second) ~ 0.3 Gm
mega	M	106	megasecond, 1 Ms = 106 s	1 Ms ~ 11.6 days
kilo	k	103	kilogram, 1 kg = 103 g	
hecto	h	102	hectogram, 1 hg = 102 g	
deka (or deca)	da	10 = 101	dekaliter, 1 daL = 101 L	
deci	d	10-1	deciliter, 101 dL = 1 L	
centi	c	10-2	centimeter, 102 cm = 1 m	
milli	m	10-3	millimole, 103 mmol = 1 mol	
micro	μ	10-6	microliter, 106 μL = 1 L	1 μL ~ a very tiny drop of water
nano	n	10-9	nanometer, 109 nm = 1 m	radius of a chlorine atom in Cl2 ~ 0.1 nm or 100 pm
pico	p	10-12	picogram, 1012 pg = 1 g	mass of bacterial cell ~ 1 pg

femto	f	10-15	femtometer, 1015 fm = 1 m	radius of a proton ~ 1 fm
atto	a	10-18	attosecond, 1018 as = 1 s	time for light to cross an atom ~ 1 as bond energy for one C=C double bond ~ 1 aJ
zepto	z	10-21	zeptomole, 1021 zmol =1 mol	1 zmol ~ 600 atoms or molecules
yocto	y	10-24	yoctogram, 1024 yg = 1 g	1.7 yg ~ mass of a proton or neutron

Table 4.1 Metric prefixes from yotta- to yocto-

Technology at the scale of a micrometer, a millionth of a meter, isn't even revolutionary anymore—in electronics manufacturing it's so commonplace it's commodified. American companies don't even want to own microelectronics fabrication anymore: they'd rather farm out micro-scale circuit construction to vendors in Asia. Micromachinery—a mechanical device the size of a miniscule water droplet—is relatively easy to achieve using current engineering technology. If you live in a developed nation, you definitely use microelectronics every single day—pretty much constantly—and in the next decade you will be using micromachines daily as well. The nanometer scale, a billionth of a meter, is where today's most pioneering tiny-scale research aims.

Nano is atomic/molecular scale. Think about this: Everything we see around us is just a configuration of molecules, but out of the total number of possible molecular configurations, existing matter represents an incredibly small percentage—basically zero percent. Nearly all possible types of matter are as yet unknown to us. If we create machines that can piece molecules together in arbitrary ways, total control over matter is ours. This is the power and the wonder of nanotechnology.

As the scale chart above shows, it's quite possible to get smaller than nanotechnology. A protein is about one femtometer across. Femtotechnology, the arbitrary reconfiguration of particles, may hold even greater rewards one day. The construction of novel forms of matter via the assemblage of known molecules may one day be viewed as limiting, but femtotechnology today is where nanotechnology was 50 years ago—an abstract but enticing pipe dream for the moment. One may even nurse dreams beyond femtotechnology—for the sci-fi-fueled imagination, it's not so hard to imagine dissecting quarks and intermediate vector bosons, perhaps reconfiguring being and time themselves (very speculative research indicates that the fine structure constant may possibly shift over time, meaning that the universe may be engaged in particle physics evolution naturally, and that the idea of reconfiguring such particles may not be "utterly impossible" as was once thought about the idea of reconfiguring molecules).

What's fascinating about nanotechnology *right now* is that, more and more rapidly each year, it's exiting the domain of wild dreams and visions and becoming fantastically real.

PLENTY OF ROOM AT THE BOTTOM

The amazing thing, from a 2005 perspective, is how ridiculously long it took the mainstream of science and industry to understand that the future lies in "smaller, smaller, smaller." Physicist Richard Feynman laid out the point very clearly in a 1959 lecture, "There's Plenty of Room at the Bottom." Today the lecture is a legend; at the time it was pretty much ignored[1]. Feynman laid out his vision clearly and simply:

> "They tell me about electric motors that are the size of the nail on your small finger. And there is a device on the market, they tell me, by which you can write the Lord's Prayer on the head of a pin. But that's nothing; that's the most primitive, halting step…It is a staggeringly small world that is below. In the year 2000, when they look back at this age, they will wonder why it was not until the year 1960 that anybody began seriously to move in this direction.
> "Why cannot we write the entire 24 volumes of the Encyclopedia Brittanica on the head of a pin?"

Though 1959 was well before the structure of DNA was understood, molecular biology was already a growing field, and many physicists were looking to it for one sort of inspiration or another. All the way back in 1943, quantum pioneer Erwin Schrödinger wrote a lovely, speculative little book called *What Is Life*, describing the genetic material as an "aperiodic crystal," which motivated many physicists to give biology a careful look. Feynman had the open-mindedness to look at the new, barely-charted world of proteins and enzymes, and view it not merely as biology but as molecular machinery:

> "This fact—that enormous amounts of information can be carried in an exceedingly small space—is, of course, well known to the biologists, and resolves the mystery which existed before we understood all this clearly, of how it could be that, in the tiniest cell, all of the information for the organization of a complex creature such as ourselves can be stored. All this information—whether we have brown eyes, or whether we think at all, or that in the embryo the jawbone should first develop with a little hole in the side so that later a nerve can grow through it—all this information is contained in a very tiny fraction of the cell in the form of long-chain DNA molecules in which approximately 50 atoms are used for one bit of information about the cell."

Well before PCs and the like existed, Feynman proposed molecular computers: "The wires should be 10 or 100 atoms in diameter, and the circuits should be a few thousand angstroms across…" This, he foresaw, would easily enable AI of a sort: computers that could "make judgments" and meaningfully learn from experience.

Feynman didn't give engineering details, but he did give an overall conceptual approach to nanoscale engineering, one that is still in use today. Long before nanotechnology, Jonathan Swift wrote:

> "So, naturalists observe, a flea
> Hath smaller fleas that on him prey;
> And these have smaller fleas to bite 'em,
> And so proceed ad infinitum."

More recently, Dr. Seuss' classic children's book *The Cat in the Hat Comes Back* told the tale of the Cat in the Hat, in whose hat resides yet a smaller cat, in whose hat resides yet a smaller cat...and so on for 26 layers until the final, invisibly small cat contains a mystery substance called Voom! Feynman's proposal was similar. Build a small machine, whose purpose is to build a smaller machine, whose purpose is to build a smaller machine, whose purpose is to build a smaller machine, and so on. The Voom! happens when one reaches the molecular level. Dr. Seuss' Voom! merely cleaned up all the red gunk that the Cat in the Hat and his little cat friends had smeared all across the snow in the yard of their young human friends. Feynman's Voom!, on the other hand, promises the ability to reconfigure the molecular structure of the universe—or at least a small portion of it.

Today this kind of nanofuturist rhetoric is commonplace. In 1959, it was the kind of eccentricity that could be tolerated only in a scientist whose greatness had been firmly established via more conventional achievements. The technology wasn't there to make nanotechnology real back then; but a little more attention to the idea would surely have gotten us more rapidly to where we are today.

ERIC DREXLER'S VISION

After Feynman, the next major nanovisionary to come along was Eric Drexler. His fascination with the nanotech idea grew progressively as he did his interdisciplinary undergrad and graduate work at MIT during the 1970s and '80s. Drexler coined the term "nanotechnology,"[2] and he presented the basic concepts of molecular manufacturing in a scientific paper in 1981, published in the Proceedings of the National Academy of Sciences. His 1986 book *Engines of Creation* introduced the notion of nanotechnology to the scientific world and the general public, giving due credit to Feynman and other earlier conceptual and practical pioneers. His 1992 work *Nanosystems* went into far more detail, presenting for a scientific audience a dizzying array of detailed designs for nanoscale systems: motors, computers, and much more. Most critically, he proposed designs for molecular assemblers: machines that could create new molecular structures on demand, allowing the creation of new forms of made-to-order matter.

Drexler's design work as presented in *Nanosystems* was a kind of science fiction for the professional scientist. He described what plausibly could be built in the future, consistent with physical law given various reasonable assumptions about future engineering technology. Whether things will ever be built exactly according to Drexler's designs is doubtful—more likely the designs will evolve as, at each step, practical engineering issues are overcome. How close future technology will come to a detailed implementation of his visions is unknown, but the conceptual and inspirational power of his work is undeniable.

A taste of his thinking is conveyed in the following chart, which shows how Drexler maps conventional manufacturing technologies into molecular structures and dynamics. The analogies given in this chart were, of course, just the starting point for his vast exploration into the domain of speculative nanosystem design. This is high-level, big-picture scientific vision-building at its best.

Technology	**Function**	**Molecular example(s)**
Struts, beams, casings	Transmit force, hold positions	Microtubules, cellulose, mineral structures
Cables	Transmit tension	Collagen
Fasteners, glue	Connect parts	Intermolecular forces
Solenoids, actuators	Move things	Conformation-changing proteins, actin/myosin
Motors	Turn shafts	Flagellar motor
Drive shafts	Transmit torque	Bacterial flagella
Bearings	Support moving parts	Sigma bonds
Containers	Hold fluids	Vesicles
Pipes	Carry fluids	Various tubular structures
Pumps	Move fluids	Flagella, membrane proteins

Conveyor belts	Move components	RNA moved by fixed ribosome (partial analog)
Clamps	Hold workpieces	Enzymatic binding sites
Tools	Modify workpieces	Metallic complexes, functional groups
Production lines	Construct devices	Enzyme systems, ribosomes
Numerical control systems	Store and read programs	Genetic system

Table 4.2 Drexler's Analogies Between Mechanical and Molecular Devices

In the years since his initial publications, Drexler has made a name for himself as a general-purpose futurist, and a thinker on the social and ethical implications of nanotechnology. In fact, he says, ethical concerns delayed his publication of his initial ideas by several years. "Some of the consequences of the potential abuse of this technology frankly scared the hell out of me. I wasn't sure I wanted to talk about it publicly. After a while, though, I realized that the technology was headed in a certain direction whether people were paying attention to the long-term consequences or not. It then made sense to publish and to become more active in developing these ideas further." Drexler didn't mean "we'll do it anyway, damn the consequences," but rather that scientists and engineers need to take responsibility and participate in new technological developments in ways that prefer beneficial to detrimental uses. He founded the Foresight Institute, which deals with general social issues related to the future of nanotechnology, and in recent years has extended its scope a bit further, dealing with future technology issues in general.

Along with his move into a role as a general-purpose futurist, Drexler has consistently been helpful to others in the techno-futurist community. For instance, when the Alcor Life Extension Foundation, a California cryonics lab, got into legal trouble for allegedly removing and freezing a client's head before she was declared legally dead[3], Drexler came to their rescue. He supplied a deposition in their defense, arguing that in the future nanotechnology quite plausibly would allow a person's mind to be reconstructed from their frozen head. His position at Stanford at the time gave him a little more clout with the court system than the Alcor staff, viewed by some as crazy head-freezing eccentrics. Alcor survived the lawsuit.

As the story of nanotech is still unfolding, the real nature of Eric Drexler's legacy isn't yet clear. Perhaps history will view him "merely" as having done what Feynman tried and failed to do in his 1959 lecture: awakened the world to the engineering possibilities of the very small. This is the perspective taken by some contemporary nanoresearchers, whose pragmatic ideas have little to do in detail with Drexler's more science-fictional *Nanosystems* proposals. On the other hand, it may well happen that as nanotechnology matures, more and more aspects of Drexler's ambitious designs will become relevant to practical work. Only time will tell—and given the general acceleration of technological development, perhaps not as much time as you think.

TWO WAYS TO MAKE THINGS FROM MOLECULES

Very broadly speaking, there are two ways to make a larger physical object out of a set of smaller physical objects. You can do it manually, step by step; or you can somehow coax the smaller objects to self-organize into the desired larger form.

More specifically, in the molecular domain, the two techniques widely discussed are self-assembly and positional assembly. Self-assembly is how biological systems do it: there is no engineer building organisms out of DNA and other biomolecules, rather the organism builds itself. Eric Drexler's work, on the other hand, is largely inspired by analogies to conventional manufacturing, and tends to take a positional-assembly-oriented view.

In self-assembly, a bunch of molecules are allowed to move around randomly, jouncing in and out of various configurations. Over time, more stable configurations will tend to persist longer, and a high-level structure will self-assemble. For instance, two complementary strands of DNA, if left to drift around in an appropriate solution for long enough, will end up bumping into each other and then grasping together in a double-helix. A protein—a one-dimensional strand of amino acids—will wiggle around for a while, and then eventually curl up into a 3-D structure that is determined by its 1-D amino acid sequence. There can be a lot of subtlety here; for instance, the protein-folding process tends to be nudged along by special "helper" molecules that enable a protein to more rapidly find the right configuration.

We don't build a TV by throwing the parts into a soup and waiting for them to lock together in the appropriate overall configuration. Rather, we hold the partially-constructed TV in a fixed position and then attach a new part to it—and then repeat this process over and over. This is positional assembly. The problem with doing this on a molecular scale is that it requires us to have the ability to a) hold a molecule or molecular structure in place, recording its precise location, and b) grab a molecule or molecular structure and move it from one precise location to another. Neither of these things is easy to do using current technology. However, solutions to both problems are under active development.

Drexler's design for a molecular assembler is based on positional assembly. It is assumed that the molecules in question are in a vacuum, so there are no other molecules bouncing into them and jouncing them around. It assumes that, unlike in traditional chemistry and biology, highly reactive intermediate structures can be constructed without reacting unpredictably with one another because they're held in place.

There is a lot of power in self-assembly: each one of us is evidence of this! On the other hand, positional assembly can achieve a degree of precision that self-assembly lacks, at least in the biological domain. Each person is a little different, because of the flexibility of the self-assembly process, and this is good from an evolutionary perspective—without this diversity, new forms wouldn't evolve. On the other hand, it's also good that, if we have no choice but to have nuclear missiles, each one of them is close to exactly identical: in this context, variation ensuing from biological-style self-assembly could lead to terrible consequences.

Less dramatically, we probably wouldn't want this biological scale of variation in our auto engines either.

Crystals, if grown under appropriate conditions, are examples of self-assembled structures that have the precision of positionally-constructed machines. Yet they are much simpler than the biological molecules Schrödinger called "aperiodic crystals," let alone whole organisms, computers or motors. Whether new forms of self-assembly, combining precision with complexity, will emerge in the future is hard to say. One suspects that some combination of positional and self-assembly will one day become the standard. Positionally constructed components may self-assemble into larger structures; and self-assembled components may be positionally joined together in specific ways. Contemporary manufacturing technology is astoundingly diverse, and we can expect nanomanufacturing methods to be even more so. What is obvious is that we don't want to ignore either possibility. Positional assembly has given us the machines which we can build and maintain with such precision, whereas self-assembly gave us—the universe. Both are incredibly powerful paradigms for creating things, and at the molecular level we are very likely to find that we will want to employ both.

MOLECULAR ASSEMBLERS GO COMMERCIAL

Dozens of companies today are involved in nanotechnology in one way or another[4]. It's also a serious pursuit of major research labs such as Los Alamos and Sandia. Perhaps the most ambitious nanotech business concern, however, is the Texas firm Zyvex. Zyvex makes no bones about its goal: it wants to construct a Drexler-style molecular assembler and play a leading role in the revolution that will ensue from this. The Zyvex R&D group has outlined a program that it believes will lead to this goal. Unlike some other contemporary nanotech approaches that we'll discuss below, Zyvex's approach is pretty much squarely in the positional assembly camp.

Zyvex has two distinct research groups, called Top-Down and Bottom-Up. The Bottom-Up team deals explicitly with molecule manipulation. The Top-Down team, on the other hand, works largely with micromanufacturing. They're currently prototyping assemblers using microsopic but non-nanoscale components that can be built for relatively low cost. From this they hope to develop small-scale manufacturing principles which can be extrapolated to nanoscale processes.

Using relatively straightforward contemporary technologies, one can create devices measuring tens or hundreds of microns, with individual feature sizes of about one micron. While these scales are not suitable for all the goals of nanotechnology—such as biological applications like dismantling virii, or building a molecular assembler—a variety of things is possible at the microscale. Those which most closely resemble traditional positionally-assembled machine parts like bearings and joints, or micro-scale organism components like flagellum, are quite doable even in the short-term. After prototypes at the micron scale have been achieved, extension to the nanoscale will be attempted. Of course, the micron-scale machinery developed along the way to true nanotechnology may also have significant practical applications, which Zyvex will likely commercialize, and a host of new and exciting micromachinery technologies will emerge along the way to nanotech.

One approach the Zyvex Top-Down team is pursuing is called exponential assembly. Basically this means small machines building more and more small machines. If each machine builds two other machines, then the total number of machines will grow exponentially. This is currently being applied specifically in the context of robot arms. Robot arm technology is by far the most mature aspect of modern robotics; factories worldwide now use robot arms to automate assembly processes. In Zyvex's plan, the first robot arm picks up miniature parts carefully laid out for it, and assembles them into a second robot arm. The two robot arms then build two more robot arms, when then build four more robot arms, and so forth.

This is a first step toward Feynman's "small machines building yet smaller machines" idea. Once exponential assembly of same-sized robot arms has been completed, the next step is having robot arms assemble slightly smaller robot arms, and so proceed ad infinitum. The additional interesting idea is to ultimately have these small machines be able to locate and/or produce the materials needed to perform their assemblies, without having them laid out for them. Such a machine would be able to take molecules from a surrounding medium and build new molecules from them. This can't be done in a vacuum. We'll need to solve the problems of positional assembly in a chaotic medium, and/or of self-assembly modulated by interaction with positional assembly helpers and also guided by a program (in the manner that DNA is a program for protein assembly), which allows it to occur assuming the presence of the right molecules (which would be those that appear in the target medium).

On the Bottom-Up side, long-time nanotechnologist and Zyvex principal scientist Ralph Merkle has created a set of impressively detailed designs for components of molecular engines, computers and related devices. For instance, the following figure shows one of Merkle's designs for a molecular bearing. Ring-shaped molecules are plentiful in nature. Just create two rings, one slightly smaller than another, insert the small one in the big one—and presto, molecular bearing!

Figure 4.1 Merkle's design for a molecular bearing (insert one molecular ring into another, and get a molecular bearing)

Merkle has put a fair amount of effort into figuring out how to make a molecular computer that doesn't create too much heat. This involves the notion of reversible computing. Thermodynamics teaches us that the creation of heat is connected with the execution of irreversible operations—operations that lose information, so that after they're completed, you can't accurately "roll back" and figure out what the state of the world was like before the operation was carried out. Ordinary computers are based on irreversible operations; for instance a basic OR gate outputs TRUE if either of its two inputs is TRUE. From the output of the OR gate, one can't roll back and figure out which of the two inputs was TRUE. This kind of irreversible logical operation, when implemented directly on the physical level, creates heat. Fortunately, it's possible to implement logical operations in a fully reversible manner, at the cost of using up more memory space than is required for normal, irreversible computing.

Maverick scientist Ed Fredkin pioneered the theory of reversible computing years ago, in the context of his radical theory that the universe is a giant computer (the theory of universe as an irreversible computer has fatal flaws, but the theory of the universe as a reversible computer is not so obviously false). Fredkin's posited universal computer would operate below the femto-scale, giving rise to quarks, leptons, protons and the whole gamut of particles as emergent patterns in its computational activity. Merkle's proposed reversible computer components are conservative by comparison, merely involving novel configurations of molecules. The details of molecular reversible computing are not yet known, but it seems likely that they'll involve carrying out operations at very low temperatures, and that true reversibility won't be achieved. That's ok, unless you're trying to engineer a truly reversible computer that can store the whole universe and get us to Frank Tipler's Omega Point. We don't need zero heat: if we can come close enough to it, we can build functional nanocomputers which will be much more energy efficient than conventional computer components, and no less configurable in terms of building up a complex computational device from very-low-temperature, molecular, semi-reversible gate logic.

TINY TRANSISTORS

Zyvex's work in tiny-scale engineering is explicitly aimed at Drexler-ish long-term goals, but lots of other similarly fascinating work is being carried out by other firms, without such clearly stated grandiose ambitions. As a single example, consider Bell Labs' researchers' work on tiny transistors.

Transistors are based on semiconductors—materials like silicon and germanium that conduct electricity better than insulators, but not as well as really good conductors. A transistor is a small electronic device that contains a semiconductor, has at least three electrical contacts, and used in a circuit as an amplifier, detector, or switch. Transistors are key components of nearly all modern electronic devices, and some have called them the most important invention of the 20th century. The original transistor was invented at Bell Labs in 1947 (12 years before Feynman's original nanotech lecture). In 1999, a team of scientists at this same institution succeeded in creating a transistor 50 nanometers across. This is about 1/2000 the width of a human hair. All of its components are built on top of a silicon wafer.

This kind of innovation is essential to the continuation of Moore's Law, which has survived several technology revolutions already. Sometime in the next decade, conventional transistor technology will hit a brick wall, a point at which further incremental improvements will be more and more difficult to come by (the first signs of this are emerging already in the form of the exponentially increasing difficulty in removing heat from chips as they get faster). At this point, quite probably, nanotransistors—successors to today's experimental ones—will be ready to kick in and keep the smaller, faster, better computers coming (perhaps along with other technologies like quantum and biochip computers).

SELF-ASSEMBLY USING DNA

Positional-assembly-based nanotech is powerful, in that it allows us to extend our vast knowledge of macro-scale manufacturing and engineering to the nanoworld. On the other hand, the only known examples of complex nano-level machines—biological systems—make use of the self-assembly approach. There has been some impressive recent progress in using biological self-assembly to create systems with nonbiological purposes.

An excellent example of the self-assembly approach is the construction of stick figures[5] out of DNA, in Ned Seeman's lab at New York University (Seeman, 1996). DNA cubes have also been constructed, along with 2-D DNA-based crystals and, most ambitiously, simple DNA-based computing circuits. Computer scientists have known for a while how to make simple lattice-based devices that carry out complex computational functions, via the state of each part of the lattice affecting the states of the nearby parts of the lattice. Seeman's group constructed a system of this nature out of a lattice of DNA.

There's no advanced computation involved in the NYU circuits: what they implemented was XOR ("exclusive or"), a simple logic function embodied in a computer by a single logic gate. An XOR gate has two inputs, and gives a TRUE output if one or the other, but not both, of its outputs, are TRUE. Simple, sure -- but it's out of large numbers of very simple (but much bigger!) gates of this sort that computer chips are constructed. Even more exciting, Ehud Shapiro and his team at the Weizmann Institute of Science have implemented a finite-state automata, stochastic computer, and gene-expression controller using DNA computing[6]. Those are much more sophisticated computations than a single XOR, and experimental results for this are not future speculation but exist right now, with DNA computers they've already created in their lab. A single milliliter of DNA solution contains a vast number of molecules—in the case of the finite-state automata construct created at Weizmann, ten *trillion* of them—and if these can be harnessed for parallel processing, we have a lot more logic-gate-equivalents than there are neurons in the brain.

Work is also underway creating DNA cages, in which cubic lattices of DNA are used to trap other molecules at specific positions and specific angles.

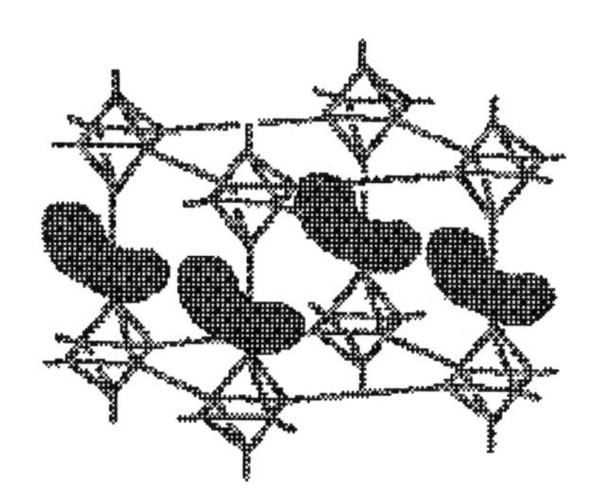

Figure 4.2 DNA Cage Containing Oriented Molecules

These cages can potentially be used to assemble a complex biocomputer, with DNA structures embodying logic operations feeding data to each other along specified paths. Many noncomputational types of matter could also be synthesized this way. Such constructs may be the first step in hybridizing self-assembly with guided positional assembly, to allow new structures to either fully evolve through random processes or through semi-random guided processes, then to encode them as programs and positional seed structures for hybrid assembly. This a physical reconfiguration approach similar to certain types of "genetic programming" approaches, as used in computer science and reapplied to biology, the discipline which inspired GP in the first place.

DNA computing in particular may be fairly limited in what, practically, it can be used for. It seems thus far to be applicable to building small automata and gate logic, but the complexity of assembling a full computational system from DNA components may be such that using higher-level molecular components may prove more efficient. The relatively slow speed of biochemical processes, which may not be able to be made faster without introducing toxic reagents that kill the cells, may limit the applicability of the speed of the low-level functions DNA computing thus far has been shown to be applicable for, requiring perhaps extremely massive parallelism to be practical. Biochips using general molecular biological processes are likely to be much more flexible, but perhaps still requiring massive parallelism due to the rate limits of biochemical interactions. By contrast, Quantum Computer theory shows great possibilities for both speed and parallelism. Does that mean biocomputing is a hopeless research pursuit? Not at all. Not only is it possible that molecular biology computing may allow for genetic programming approaches which lend themselves better to various problems than either von Neumann or Quantum computers; but there is also the possibility of hybridization. British geneticist Johnjoe McFadden is one of several scientists to postulate that protein-folding is solved by a quantum-computing-style exploration of parallel universes within the protein, searching the space of possible configurations to find the minimum-energy one. If true, this would allow for the creation of biological quantum computers using proteins, with the posited speed gains of using the quantum approach combined with the posited evolutionary possibilities of using a molecular biology system.

CREATING NOVEL PROTEINS

Arranging DNA molecules in novel configurations is exciting, and may lead to such things as molecular computers and new forms of matter. For some purposes, though, one wants to go a little further—not accepting Nature's biomolecules as given, but rather creating new ones from scratch.

At first, designing new proteins seems like an incredibly difficult problem. The problem is, no one knows how to look at the sequence of amino acids defining a protein molecule's one-dimensional structure, and predict from it the three-dimensional structure that the protein will fold up into. IBM is currently building the world's largest supercomputer, Blue Gene, with the specific intention of taking a stab at this fabulously difficult "protein folding" problem. From the perspective of protein folding as a 3-D structural analysis problem coupled with a physical dynamics multibody problem combined with a 1-D pattern mining problem, this is quite difficult for current algorithms (only 1-D pattern mining is even close to being considered a "routine" problem). Blue Gene's brute-force datamining approach may not even come close to achieving what they hope. In order for such an approach to work, you need to understand enough about what you're looking for to know where to look.

The problem of protein *engineering* is a rare case of something that's actually easier than it looks. Some proteins are easier to understand than others, and to engineer new proteins you don't need to understand *all* proteins and derive a general theory of protein folding; you only need to understand those that you want to build. By cherry-picking your proteins, you can make some advances in this area without understanding protein folding as a general principle (indeed, we also cherry-pick genetics problems right now, another area we don't fully understand). It turns out that there are many unnatural proteins whose folding is more predictable than that of proteins found in nature. Protein engineers strategically place special bonds and molecular groups in their designed proteins, so as to be able to predict how they'll fold up. So long as your protein can be engineered via this approach, you don't need a unified theory of protein dynamics.

For example, Jonathan Blackburn at the University of Cambridge is working on the creation of new proteins that bind to DNA, and whose binding can be controlled by small, more easily understood molecules (Blackburn, 2000). The potential value of this is enormous. Ultimately, it could allow us to create an explicitly controllable genetic process, with the small molecules used as interactive "control switches." In the shorter run, there are fantastic implications for genetic medicine and pharmacology in general. While we can still do a lot more once we do understand protein folding more generally, even this more limited approach has a lot of possibility for many developments that will allow us to improve health and longevity.

NANOMEDICINE

The likely long-term applications of nanotechnology are far too numerous to list here. If you can configure arbitrary forms of matter, what can't you do? All current forms of manufacturing and engineering are ridiculously simple by comparison.

One of the more exciting medium-term applications of nanotech, however, is in the medical domain. Robert Freitas, in his 1999 book *Nanomedicine*, has explored this space in great detail, taking a Drexler-like approach of exploring what will likely be possible given future technologies.

For example, one of his inventions is the respirocyte, a bloodborne 1-micron-diameter spherical nanomedical device that acts as an artificial mechanical red blood cell. This will basically be a little tiny vessel full of oxygen pressurized to 1000 atmospheres. The pumping out of oxygen will be accomplished chemically by endogenous serum glucose, and if it's built as currently designed it should be able to deliver 236 times more oxygen to human tissue per unit volume than natural red cells. In addition to numerous medical applications, pumping your body full of some of these would allow you to hold your breath for hours underwater or have the endurance of a superhero marathon runner.

Another envisioned medical nanobot is a "cell rover" that would be built by piecing together traditional and/or custom-designed biomolecules, and be designed to zoom through the bloodstream performing special medical tasks like removing toxins, delivering drugs, or simple cell repairs. The internal frame would be made of keratin, chitin or calcium carbonate; the skin panels would be made of lipids (fats). Movement could be achieved by various biological mechanisms such as bacterial cilia or flagella, mechanically inspired ones such as molecular bearings, or perhaps even more original nanotech designs. A submillimeter band single-molecule radio antenna could be supplied to allow communication with a control device and/or other cell rovers. If the device were made to look like a native cell, the immune system would leave it alone, allowing it to happily zip about the body doing its business. The implications of a device like this for medical science would, obviously, be more than tremendous. The possibility of cell repair is essentially the possibility of slowing or stopping aging entirely.

PLENTY OF ROOM INDEED

Positional assembly or self-assembly, or some combination thereof? Cascades of machines, each one producing smaller machines—or subtle modifications of existing biomolecular processes? General-purpose molecular assemblers, or a host of different but overlapping nanoscale assembly processes? Some combination of many or all of these?

The questions are many, but what's clear from the vast amount of work going on is: this isn't science fiction anymore. Nanotechnology is here right now, and it only getting more advanced. At the moment it's mainly in the research lab, but within the next 5-10 years it will be productized, at least in its simpler incarnations. The practical experience obtained with things like DNA lattices and micromachines and tiny transistors will help us flesh out the difficult issues involved in realizing Feynman's and Drexler's bigger dreams. The tendency toward miniaturization that we see all around us, with everyday technology like radios, calculators and computers, is going to be far more profound than these examples suggest. The possibilities achievable by reconfiguring matter at will are mind-stretching, but within the 21st century they will almost surely be within our reach.

---

[1] The whole text of the talk is online at http://www.zyvex.com/nanotech/feynman.html, and it makes for pretty fascinating reading.

[2] Norio Taniguchi first used the term to describe nanometer materials engineering. Drexler coined the term with respect to molecular nanotechnology and the quest for the molecular assembler.

[3] Which is the best time to do so—before any brain trauma occurs during death or immediately after as blood stops flowing and structures become damaged—if there is any hope of cryonics to work.

[4] See http://www.homestead.com/nanotechind/companies.html for a partial list.

[5] Not necessarily little doodles of people, but 3-D molecular line constructs of arbitrary shape. That's a lot more useful than doodles of people, except perhaps in the hands of Don Hertzfeldt.

[6] See: http://www.wisdom.weizmann.ac.il/~udi/pub.html for the papers on this research.

CHAPTER 5

UPLOADING: LIBERATING MINDS FROM BRAINS

The term *uploading*, as used on the Net today, refers to taking a computer file that exists on one computer and sending it to a different computer, usually a high-powered server of some kind. In futurist circles, however, it also has a different meaning: the transferal of human minds out of their flesh containers into stronger, more flexible mechanical substrates.

It's a strange idea, but also natural, in a way. After all, are you your body, or are you your mind? If you lose an arm, you're still you; if you get a brain tumor and have a chunk of brain cut out, you're still you—though perhaps a little different. But when you die, your brain stops distributing electricity throughout itself, and you're not there anymore. Why should your mind be tied to the particular container it was born in? Your mind is information: it's not your brain.

What if, one neuron at a time, you had your brain cells replaced with synthetic brain cells? What if the neurons that were removed were pieced together in the shape of your original brain, to form a new you? Which brain is the real you—the one formed from the synthetic brain cells (which, arguably, carries your "continuity of consciousness" better) or the one formed from your original brain cells? This little thought-experiment illustrates the fact that your mind is probably better thought about as a collection of patterns, not a collection of cells.

Sure, the human mind has more to do with the human brain than a computer file to do with the hard drive that it's stored on—brain structures and mind structures have some commonality, whereas a hard drive's structure has much less to do with the particular bits that are stored on it. But this commonality doesn't mean the mind *is* the brain; it just means that the mind and brain, when considered together, display some particular emergent patterns which embody the only *current* symbiosis between mind and mind-hardware—not the only *possible* one. The same mind-information could be caused to emerge from another substrate, either a brain-like one or one as alien to the brain as current von Neumann computers (though currently no von Neumann machine is fast enough or capacious enough to simulate a human brain in anywhere close to real time, even if appropriately programmed).

It sounds outlandish perhaps, this kind of uploading, but what does it require technologically? Faster and more capacious computers than we have today, and more accurate brain scanners, are necessary. Computers get better and brain scanners get more accurate every year. It's only a matter of time—perhaps just decades—before you'll sit down at your computer in the morning and bring up windows featuring a word processor, an action game, and an uploaded Uncle Harry.

IS MY UPLOAD ME OR NOT?

Of all the future tech topics we've ever encountered, none leads to deeper philosophical dilemmas than uploading. Dilemmas related to cloning, AI and such pale into insignificance by comparison. The seemingly simple question, "Is my upload me or not?" has enough different shades to fill an art gallery.

One case of uploading is the creation of an upload with a completely separate physical embodiment from the original mind, while the original mind is left alive and intact. There's you on the computer, and there's you in your good old human body, staring at each other. Richard Kennaway, in a 1992 article[1] summarizing the various, wildly contradictory views theorists have put forth on this issue, made the following statement about this situation: "One of the few points of agreement. Everyone seems to agree that in this case, the original and the duplicate are separate persons. Neither experiences the thoughts, feelings, sensations, etc. of the other." We're not sure that we agree with his statement.

Anyone whose theory of uploading states that if you're uploaded upon death, that you survive implicitly, believes that if you upload while still alive, then *at least at the moment of upload*, you and your upload are the same person. Because your subsequent experiences may differ significantly, you and your upload may then diverge; such that if you were uploaded again later in life the patterns in the two uploads would be substantially different. What you'd have done is followed two possible paths of your development. Maybe you'd feel more like it was really you if you could clone yourself, grow the clone to your favorite age unconsciously in a vat, and then upload to that—but that is just attachment to a particular hardware container for your mind. It's a bit of hardware our minds have evolved in conjunction with, so we have a natural preference for and affinity towards it; but there could be other vessels for our minds.

Humans, through religion, have long had the *intuition* that so long as our mind patterns can continue to be expressed, we still exist in some form. Uploading into a computer is no more outlandish than uploading into heaven, Hindu-style reincarnation, or becoming one with the universe in the Buddhist sense. People believe that we still have some meaningful form after death as information—souls—expressed as energy in the universe. Why not try to find new containers for our selves in the observable universe, then? At least with uploading we can directly observe the results and know whether or not it's been successful.

As an amusing and conceivably relevant speculative aside, there has been a lot of interest in supposed paranormal phenomena associated with identical twins: sometimes, it seems, they can sense each others' mind-states automatically, especially in cases where highly dangerous or otherwise emotionally charged events are involved. If identical twins can in this sense experience a consciousness-overlap, then perhaps two identical minds will do so in a more extreme way. Perhaps there will be a psychic link between a mind and its corresponding upload. Quantum physics does not rule this out: the brain may be a macroscopic quantum system, and there may be bizarre quantum nonlocality effects joining a mind to its digital copy. Of course, this is pure speculation, but it is speculation that is consistent with all known physical laws.

What if your human brain is destroyed during the uploading process? Then there's only one you left—the one in the computer. Is it you? Or are you dead—and a new being with great resemblance to you newly created? The question is conflated, again, by the issue of dissimilarity of the mind hardware. If you, upon death, were transferred into a waiting mindless clone of your body at a younger age, you'd easily feel this was "you again" or even "still you." But in a weaker sense you already become "a new being with a great resemblance to you" with every new experience. You right now are not exactly the same as you when you were six years old (unless you happen to be six years old right now). If there is continuity of consciousness, aren't you still some form of yourself regardless of how much change you've gone through? If there's discontinuity—say, a coma—but then when you're revived the patterns of mind are the same, isn't that still you as well?

A subtly different conceptual dilemma is posed by the notion of *gradual uploading*. This notion refers to the temporary hybridization of a brain with a computing device. The computing device then gradually takes over more and more functions of the brain, until eventually the computer is doing everything and the brain—either having been replaced piece-by piece with cybernetic implants or having biologically atrophied—is doing nothing, and can safely be allowed to finally die. Suppose this process is perfected—then what happens, subjectively? There is a continuous consciousness throughout the whole gradual uploading process. Is it not then obvious that "you" are preserved through the process? Or will there be more of a feeling of transition from one consciousness into another? Will it feel like another you gradually sucking the original you away, or will it just feel like a natural change, similar to going through puberty—you're very different at the end, but it's still "you" all along?

As well as presenting new dilemmas, the concept of uploading reminds us of the generally-ignored philosophical dilemma of defining oneself even in everyday life. In what sense, after all, am I the same "me" as I was when I was six years old? Take a case such as John Nash, the great mathematician whose plunge into schizophrenia and subsequent recovery were portrayed in the book and film *A Beautiful Mind*. After he became schizophrenic, was he really the same John Nash or not? Anyone who's been married a long time is likely to have observed their spouse change tremendously—still preserving some kind of essence throughout the changes, but how much? Many divorces occur because one or another partner has changed into an extremely different person. Because we equate our minds with our brains, as they are the only current platform for our minds, we say these changes are inessential by definition based on the continuity of the *body*—even though the mind is changing. Why does continuity of the body equate to continuity of the mind? As the late transhumanist theorist Sasha Chislenko put it in a long e-mail on uploading, "if we just look at our own lives, we go thru so many transitions that hardly preserve our identity in any reasonable definition of this word..."

He ended the e-mail as follows:

> "Now I am ending the message being attracted by a personal singularity point that I try to cross daily and that is only an example of a rich variety of identities that constitute something that I [?] call 'myself.' It's 2 a.m., time to sleep. I am discontinuing my conscious existence, and leaving this message together with all my belongings to the tomorrow's being who I have never seen and who will probably think that he is *me*. Hope he'll mail it."

Ever go to sleep thinking hard about something and tell yourself emphatically not to forget about it overnight, and then wake up seven hours later completely unable to remember what the thing was? In what sense is the biologically-embodied you of the morning "the same" as the biologically-embodied you of the evening? We generally say "exactly the same." Is this because the obfuscation by other patterns, or even the complete loss, of a pattern as trivial as a single idea isn't seen as a subtantiative change in the pattern of mind—or because so long as we've got the same body, we just assume we've got the same mind?

The puzzles of identity get even trickier when one considers that many uploads won't want to remain close copies of their human minds of origin. Why bother, when there's a whole new assemblage of things to be? Why not add direct mind-links into a calculator, a programming language compiler, a paint program, a music synthesizer; why not link up to global satellite networks as new sense organs? Why bother sleeping? Why not create new states of mind between dream and sleep? Why not freely edit out bad memories? Why not create simulated drug experiences as far beyond LSD as LSD is beyond clove cigarettes? Why not simulate the sensation of having sex using 19.7 sex organs at once? Why not fuse your knowledge base with that of your wife, child or best friend? Then, once you've done all these things, are you still you or not?

We think these are fascinating issues, but classify them along with issues like: "Does the world really exist or is it just an individual or collective hallucination?" and "Is anyone besides me really conscious, or are they all just automata?" These are deep philosophical dilemmas, which none of us ever satisfactorily answers, but, we go on living anyway. In the same sense, uploading (of some form or another) will almost surely happen, and will continue to be philosophically problematic even after it becomes a relatively common occurrence. Life is philosophically confusing with and without the help of advanced technology. Perhaps an upload will feel like it both is and is not the same person as its previous biological incarnation—but it will go on living anyway, as we all do, in spite of the numerous philosophical dilemmas that life poses.

The overall impact of uploading on human individual and collective psychology is hard to foresee. On Principia Cybernetica[2] it is opined that: "The decline of traditional religions appealing to metaphysical immortality threatens to degrade modern society. Cybernetic immortality can take the place of metaphysical immortality to provide the ultimate goals and values for the emerging global civilization." We find this, in a sense, a bit optimistic: we're not so sure that uploading and the ensuing virtual immortality of the mind will in itself provide new goals and values. Surely, it will be a major part of a new common-sense philosophy, as different from our contemporary perspective as the latter is from the Stone-Age worldview of African pygmy tribes. On the other hand, we're not so sure that the loss of traditional religion and its illusions of metaphysical immortality is in itself any kind of societal degradation. Clearly there have been many positive social and psychological aspects to traditional religion, but the history of the relationship between religion, war, and tyranny doesn't give a lot of credence that religion alone is a social savior. Indeed, one alternate optimistic possibility that may emerge from the consideration of material eternalism is that—unlike with the traditional religious belief in Heaven—since you have to live rather than die to obtain immortality, there is much less of an incentive to participate in war.

SOUL UPLOADING

Amusingly and intriguingly, the philosophy of uploading came up a few years back in a discussion between systems theorist Francisco Varela and the Dalai Lama. Varela was a Buddhist as well as a biologist, mathematician, and all-around scientifically- and philosophically-adept individual. He asked the Dalai Lama for an opinion on the possibility of conscious computers. The result was the following conversation, involving Varela, two of his colleagues (Jeremy Hayward and Eleanor Rosch), and the Dalai Lama:

> DALAI LAMA: In terms of the actual substance of which computers are made, are they simply metal, plastic, circuits, and so forth?
> VARELA: Yes, but this again brings up the idea of the pattern, not the substance but the pattern.
> DALAI LAMA: It is very difficult to say that it's not a living being, that it doesn't have cognition, even from the Buddhist point of view. We maintain that there are certain types of births in which a preceding continuum of consciousness is the basis. The consciousness doesn't actually arise from the matter, but a continuum of consciousness might conceivably come into it.
> HAYWARD: Does Your Holiness regard it as a definite criterion that there must be continuity with some prior consciousness? That whenever there is a cognition, there must have been a stream of cognition going back to beginningless time?

> DALAI LAMA: There is no possibility for a new cognition, which has no relationship to a previous continuum, to arise at all. I can't totally rule out the possibility that, if all the external conditions and the karmic action were there, a stream of consciousness might actually enter into a computer.
> HAYWARD: A stream of consciousness?
> DALAI LAMA: Yes, that's right. [DALAI LAMA laughs] There is a possibility that a scientist who is very much involved his whole life [with computers], then the next life...[would be reborn in a computer], same process! [Laughter] Then this machine which is half-human and half-machine has been reincarnated.

It is intriguing how the Dalai Lama manages—tongue firmly in cheek, to be sure—to reconcile postmodern technology with his ancient, some would say outmoded, belief system. Like all Tibetan Buddhists he clearly believes that we all have a soul of some sort, which leaves the body at death and is then reincarnated into another body. Yet he is willing to consider that a computer program could be made to manifest the abstract patterns that the voyaging soul uses to identify a "mind," a potential next-nesting-place. He even conjectures that a sufficiently advanced yogi might be able to project their soul into a computer program: a very convenient uploading mechanism, much less risky than the technological means we have at our disposal today!

> ROSCH: So if there's a great yogi who is dying and he is standing in front of the best computer there is, could he project his subtle consciousness into the computer?
> DALAI LAMA: If the physical basis of the computer acquires the potential or the ability to serve as a basis for a continuum of consciousness. I feel this question about computers will be resolved only by time. We just have to wait and see until it actually happens.

Of course, the Dalai Lama does not specify how, when it actually happens, he intends to tell if a computer program truly has acquired a mind and soul or not. The implicit assumption seems to be that, if it talks like a mind and quacks like a mind, then it has a mind—and if it has a mind, it has some kind of reincarnated soul. The idea that the soul is in the mind, and a continuum of consciousness is what defines a person—or other intelligent being—fits well with the ideas behind trying to achieve a material immortality. If there is a way to be reincarnated, so to speak, through uploading, or cloning, or whatever—and do it in such a way that you explicitly remember your "past lives" because you've transferred your mind using mechanisms found here in this universe—it's worth trying to find it.

"MINOR TECHNICAL DETAILS"

There are two "pesky little technical details" involved in the uploading idea. One is creating a computer suitable for containing a human mind. The other is actually getting the information which characterizes a human mind out of the human brain that carries it. Until these problems are solved, uploading remains a plan and a dream. Such problems will likely be solved eventually, given the relentless advance of all relevant technologies, but it's not yet clear exactly what form the solution will take, nor how soon the solution will come.

The first of these tasks—creating an appropriate computational substrate—is something we've already talked about a lot in previous chapters. Moore's Law has been underway for decades, and the projections for its cessation for Von Neumann computers thus far coincide with projections for the availability of commercially available quantum computers and possibly even bio-chip computers—opening up possibly two whole new fields of engineering tricks and optimizations, to make even faster and more connected computers. There are thought to be around 10^{10} neurons in the brain, and 10^{13}–10^{15} synapses connecting them. According to this, if Moore's Law holds up, we'll have achieved computers with human-brain-scale memory and computing power within just a few decades. And the wonder of exponential growth is, even if our estimates of the brain's memory and processing power are low by a couple of orders of magnitude, it will only push back the advent of brain-capacity computers by a decade or so. Once the *scale* is achieved, it's then merely a matter of figuring out how to actually do it.

The development of the process of getting the mind out of the brain is chancier to predict, because brain scanning is a newer technology than computing.

Even so, it's advancing at least as rapidly. Each year the accuracy of scanning techniques gets better and better—yielding a Moore's-Law-like exponential acceleration. The technologies needed to contain human mind levels of information in a computer, and to have an accurate scan of brain states, are within our reach. Provided we understand enough about brain function by the time such computers are available, simulating its operations exactly and then transferring the full state-data from the scanned brain is considered by many uploading proponents to be the most likely mechanism for uploading.

One promising brain-scanning technology in common use today is MRI, Magnetic Resonance Imaging. MRI machines are a wonderful application of fairly basic particle physics. The key phenomenon exploited in an MRI machine is the fact that the proton in the nucleus of a hydrogen atom, when exposed to a magnetic field, acts like a small bar magnet and resonates. Furthermore, the resonant frequency is proportional to the strength of the field. So if you expose a brain to a magnetic field whose strength varies across different regions, the protons at different positions will resonate at different frequencies. From this it's possible to tell how many protons are in a given region of space. The more blood is flowing in a given region of the brain, the more water molecules are there; and the more water molecules are there, the more hydrogen atoms are there. So MRI machines track the flow of blood through the brain, which is meaningful because blood flow follows activity—the regions of the brain that are active require more energy, more blood.

This is fantastic stuff. But it has a long way to go. Right now MRI is nowhere near accurate enough to make a completely detailed picture of a brain-state. The resolution is around a cubic millimeter. How far the resolution can be pushed isn't yet clear, and certainly more fine-grained information than simply blood flow will need to be gathered. Some researchers have suggested radical possible improvements, such as distributing Helium-3 through the brain, which would diffuse rapidly into cells and permit resolution sufficient for uploading. Also, if one is willing to slice the brain up into little sections, it's much easier to use MRI with high resolution—the disadvantage being the destruction of the brain involved, and the fact that in doing so you'd capture only structural information (since dynamical processes would have been ended by destroying the brain), which may or may not be problematic depending on how memories are stored in the brain. (Based on current theories, it seems that structural information alone will probably be sufficient to allow uploading.)

A possible alternative to MRI, and other current technologies like PET and CAT scans and EEG readings, is another magnetic technique—Magnetoencephalography, or MEG. With MEG, the brain's electrical activity induces a field in a SQUID (Superconducting Quantum Interference Device), which can then be recorded. This is the stuff of sci-fi—cyberpunk authors have written about using SQUIDs as brain interfaces to computers, for example—and it is already being used in a few research labs like the one at University of Utah. Current MEG technology allows brain scans to be taken with millisecond precision; and with faster computers and better sensors, the scans can be made even more precise. Eventually, a quantum computer coupled to a SQUID should be able to, in theory, read its state changes continuously. Unlike with MRI, the device is only a receiver and not an emitter, so there is no worry about radiation like with PET or CAT scans, and not even (unfounded) worry about the radio waves used in MRI scans. More importantly, improvements in accuracy can come more quickly, as they require engineering only in the receiver, since there is no emitter.

Currently MEG accompanies its outstanding temporal resolution with extremely poor spatial resolution; typical MEG studies involve reading magnetic fields from 100-200 points near the surface of the brain. There is as yet no brain scanning technology allowing the spatial resolution of MRI and the temporal resolution of MEG. Such a thing may come about via improvement of either MRI or MEG or via a different technology entirely—but clearly, such a brain scanning technology would be of immense value not only for uploading, but for brain science and cognitive science.

Plenty of more radical techniques for enabling uploading have been proposed as well. For instance, if one has a stained, vitrified brain (frozen to a very low temperature using special antifreeze chemicals that prevent ice crystal formation), then one can abrade the brain with UV rays and can use ultra-sensitive mass spectrometry to infer brain structure. Or, alternately but similarly, one can use an experimental method called abrasive atomic force microscopy. These particular example strategies are likely to be destructive of the brain as well, and though they may be good ways to infer structure, they're obviously rather terrible for studying brain dynamics. But not to worry—it is very likely that in the next few decades we'll have a high-resolution non-destructive brain scan technology.

THE NEMATODE UPLOAD PROJECT

Digitizing a working human brain is the holy grail of uploading research, but science—even weird science—nearly always achieves its grand goals by small steps. In this case, the first step being taken by adventurous researchers is the uploading of a simpler organism: a nematode worm. A nematode's brain has only a few hundred neurons, far short of the human brain's hundred billion or so. Still, it only took us half a century to get from room-sized computers that could add, subtract, multiply and divide, to where we are now with computing: amazingly powerful general-purpose machines on every developed-world desktop. The general pace of technological advance is accelerating. Treading the path from nematode brain to human brain may not take all that long.

So far no one has reconstructed a nematode worm's brain in an entirely automatic way. Rather, human scientists had to analyze the images of all 959 cells (338 of which are brain cells) while explicitly attending to the lineage of these cells from the fertilized egg, but automation of this kind of process is not too far off. For instance, researchers have carried out a complete automated reconstruction of a capillary bed—which is easier, for technical image-processing reasons. Image-processing software gets better and better, and integration of this software with knowledge bases embodying biological intuition of various sorts is not too far off (pharma and biotech labs are already working on such integration for purposes of cheap, automated synthesis of cells and organisms engineered for specific research protocols).

Although this began as a biology project, a group of computer folks calling themselves the Mind Uploading Research Group (MURG) has adopted the nematode upload project as its own[3]. They aim to make a complete computer simulation of the dynamics of the nematode, with an emphasis on its nervous system. The project is explicitly seen as a first step toward creating digital models of more and more complex organisms.

FROM POSSIBILITY TO REALITY

Given that uploading a very simple organism like a nematode worm still poses annoying technical problems, it's clear we're not yet ready to upload a human brain. Yet as nanotechnology pioneer Ralph Merkle has pointed out, it's perfectly plausible at the present time to shoot for an only slightly smaller goal. In his view, "At the present time, a reasonable research objective is the fully automated analysis of a cube of complex neuropil about 100 microns on a side." From a nematode worm to a chunk of brain, and then on to the brain as a whole. Merkle estimates based on a very careful and detailed analysis that the cost of a whole-brain-scanning research project might be roughly in the $10 million range: considerably less than a single advanced U.S. fighter plane. Even if you don't believe uploading is possible, that's a small price to pay for research that would open up new avenues, ultimately allowing us to do a much better job of treating brain disorders and trauma.

From a contemporary common-sense point of view, the idea of yanking a mind out of the brain it's embodied in, and transplanting it into a computer of some sort, sounds completely outlandish, but it will require no tremendous scientific revolutions to do such a thing. Such an achievement is firmly within the confines of physical law, so long as one accepts that the structure of the brain (and not a "magical spark") contains the essence of the mind—a conclusion that is very firmly indicated by all of modern neuroscience. There are many engineering challenges to be overcome. We need bigger computers and more accurate brain scanners; analysis of the scan data and engineering of an emulation of those structures and dynamics; and better fault-tolerance (you wouldn't want to be deleted by a system crash or "data rot"). But these are all achievable. This is not like time travel, which is possible only if current science is incorrect or incomplete. The possibility of uploading is staring us right in the face, and it seems highly probable that sometime within the 21st century it will become a reality.

In other words, as a fellow named Xiaoguang Li once said on the SL4 futurist e-mail discussion list: "Uploading is a no-brainer!"

---

[1] http://www.aleph.se/Trans/Global/Uploading/kennaway.121692.txt
[2] http://pespmc1.vub.ac.be/
[3] see http://minduploading.org

CHAPTER 6

THE CYBORGIZATION OF HUMANITY

Today we interface with computers primarily at arm's length. Certainly, the current standard keyboard-mouse-monitor set up is far more convenient than the teletypes and line printers that we computing old-timers once used to communicate with lumbering mainframes (or the punch-cards, dip-switches, or cable arrays the *real* old-timers used). Touchscreens, drawing tablets, sound cards, joysticks, and MIDI musical instrument interfaces are awfully nice too. Direct computer interfaces to EEG and EKG machines save lives. Voice-based computer control, haptic-feedback N-degree-of-freedom interface devices, wearable sensor arrays, and similar technologies will all be great once the technology improves a bit.

But wouldn't it be nice if the integration were tighter than all this? What if you could control your computer *just by thinking*? What if the PalmPilot in your pocket sensed the electrical flow in your brain? What if you could *think*, "Yo, PalmPilot, what's Gary Numan's address?" and have the address appear in your mind? What if you could hear a melody in your head, and by the pure power of your thought cause that very same melody to be recorded as MIDI or sound file data, and emanate from your stereo speakers? What if you could envision a picture in your mind, transmit it to your PalmPilot, then beam it to your friend's PalmPilot, and straight into his or her brain[1]? What if you could connect your visual centers to an infrared and low-light camera array, allowing you to actually see in the dark? What if a mathematician could reference a powerful computer algebra program like *Mathematica*, purely by power of thought—what kind of human-computer collaborative calculations could they do? What if you could reliably and directly sense your partner's subtle emotional shifts and unspoken desires during the act of love—simply by exchanging data files rather than through the tedious processes of human empathy? What if our brains could access virtual realities and zoom through them with freedoms and capabilities unknown here on Earth?

The power of brain-computer integration is obvious, and the danger at least equally so. What happens when the neurochip crashes? What about when some crazy hacker releases a virus into your mind? In the 2050 Olympics, will athletes whose reflexes are enhanced by brain-embedded chips be allowed to compete? In 2100, will bringing a directly-brain-linked PalmPilot-type device to a math test be considered cheating, or will it simply be expected?

The first steps toward the cyborgization of humanity are already well underway. Disabled people wear bionic limbs with onboard microprocessors; the severely paralyzed can communicate by using special brainwave detectors to control their computers. Artificial heart research has been rekindled by the AbioCor self-contained heart system, and though it still relies on an external battery pack to charge its internal batteries, its only a matter of time before an artificial heart can be powered by human energy. British computer scientist Kevin Warwick has embedded chips in himself and his wife that will allow them to directly sense each others' emotions, by transmitting the nervous system stimuli indicating emotional states to one another. A Japanese firm sells chips intended to be embedded in old people—to enable a person to track a senile grandparent should they wander off. One step at a time, the boundary between human and computer gets thinner and thinner. The dangers of such things are amply documented in both science fiction and popular philosophy writings—governments tracking people against their will, digital brainwashing, subjugation into a borg-like collective where your individual thoughts are suppressed, and so on. Here we'll focus on the positive aspects, but always keep in mind the admonition that it is up to us to *ensure, by our actions*, that the uses of such technologies be positive.

THE CYBERLINK SYSTEM

The pioneer cyborgs of today are not overzealous computer geeks plugging GameBoys into their foreheads, but rather disabled people seeking to live fuller lives. The wooden legs and Captain Hook-style hands of ages past are being replaced by bionic limbs with embedded computers, directly patched into the body's nerves for effective and responsive control. Cochlear implants, allowing the deaf to hear, get better and better each year. Scientists have constructed crude systems that pipe video-camera output into the visual cortex—successors of these systems will allow blind people to see, and furthermore, to see in the dark using infrared cameras, thus in some ways seeing better than the conventionally sighted[2]. The popular TV show *Star Trek: The Next Generation* gave a prominent role to a biologically-blind character with a bionic super-eyepiece. In a few decades this sort of thing will be a reasonably common sight in the streets of New York.

One fascinating example of cyborg technology helping the disabled is the Cyberlink Solution system, created by the Brainfingers Corporation[3]. Cyberlink reads your eye movements, facial muscle movements, and forehead-expressed brain wave bio-potentials, and from a synthesis of these inputs it creates a signal to send to a computer. You don't have to type or click to control your computer—you just have to think, or roll your eyes a little. This is a life-transforming device for those whom injury or congenital defect has deprived of the normal power of bodily motion—and a first step toward a new way of interacting with computers for all of us.

The forehead is, surprisingly, an outstanding place to gather biosignals. All you need is a simple headband with some sensors on it. The headband connects to a Cyberlink interface box, containing a signal amplifier and some signal processing software, which in turn plugs into a standard PC serial port.

After some practice, you can figure out how to send commands to the computer, using a combination of eye movements, facial muscle movements and *thoughts*.

It's the combination of inputs that makes Cyberlink so powerful, as compared to competing products. For quite some time researchers have been playing around with using EEG waves to control computers; but the skull is thick, and the electromagnetic waves that come through it are diffuse and hard to interpret. They represent averages of brain activity over broad regions. If it were possible to measure the electrical potential of an individual neuron in the brain, then incredibly fine-grained, directly thought-powered computer control would be possible—this is the Cyberpunk dream of computers controlled by brainwaves via SQUIDs. As this isn't feasible yet, combining EEG with other sources of information is currently the best way to go.

Newbie Cyberlink users generally get started with subtle tensing and relaxing of the forehead, eye and jaw muscles. With more experience, a shift is made to using predominantly brain-based signals. Often mental and eye-movement signals are used to move a cursor around, and facial muscle activity is used to turn switches on and off. A U.S. Air Force study showed that subjects' reaction times to visual stimuli were 15% faster with Cyberlink than with a manual button. For people without reliable muscle control, after a little more effort purely thought-based signals could be used to control every aspect of a computer's behavior.

The process of learning to use the system isn't always easy. For the completely paralyzed, it can sometimes take months. On the other hand, once success is achieved, the results can be transformational and amazing. It's an awkward, clunky way to manipulate a computer, but it's an infinite improvement on *not being able to communicate at all.*

Following behind the disabled, who need such technologies to live fuller lives in our society, musicians and artists have thus far been the other major pioneers in emerging cyber-interface technologies. A more frivolous application of Cyberlink was created by Dr. Andrew Junker, the originator of the Cyberlink Solution, together with composer Chris Berg. It allows you to create music through the power of thought. Volume, pitch and velocity can be controlled, along with other factors such as the degree of melodic and rhythmic complexity of the sounds produced. This work builds upon the musical work being done with Cyberlink's predecessor the BioMuse[4], which has been used by composer Atau Tanaka in a similar manner—including using the BioMuse's muscular electrical-signal-sensing capabilities to play "air violin"—and also with Masahiro Kahata's IBVA[5] (Interactive Brainwave Visual Analyzer). Australian performance artist Stelarc[6] has taken the cyborg concept even farther, hooking himself up to various sensors (including one to monitor blood flow!) and composing sound based on machine interpretations of his human body and brain functions. It's still a long way from the point where you can compose a new symphony in your head, and at the same time, miraculously hear it coming out of your computer speaker, but we're at the first step along this tremendously exciting path. The same technology that allows the fully paralyzed to communicate will increasingly allow us to communicate the sounds, sights, feelings and thoughts that are trapped inside us, in new and exciting ways.

THE CYBORG PROFESSOR

Medical work helping the disabled is arguably the contemporary vanguard of cyborgization—but one British computer science researcher is making a play to gain this title for himself. Dr. Kevin Warwick, who titled his recent autobiography *I, Cyborg*, has embarked upon a radical and ambitious program of experimental self-cyborgization. The scientific techniques he's using are not particularly innovative, but his willingness to use himself as a guinea pig has made headlines more than once; and who knows, his gutsy self-modification research may yet lead to breakthroughs.

Warwick's first self-cyborgization experiment, in 1998, was mild in terms of the potential scope of human-computer synthesis but nonetheless radical with respect to contemporary human experience. What he did was to turn himself into a human remote control. He had a tiny glass container full of transponders implanted into the skin of his arm, and had controls set up so that when he walked through doorways, the lights would turn on; when he walked into his study, his computer would boot up; and so forth.

Pragmatically, there's not so much difference between what he did here and simply carrying an advanced remote control in one's pocket or strapped around one's waist—but there's a major philosophical difference. The controller was *in* his body—it was a part of him, not something he was carrying around and could pick up or put down.

His musings on the experience seem rather more dramatic than what was actually accomplished. "After a few days," he said, "I started to feel quite a closeness to the computer, which was very strange. When you are linking your brain up like that, you change who you are. You do become a 'borg. You are not just a human linked with technology; you are something different and your values and judgment will change." Furthermore, with respect to the longer-term social implications:

> "I didn't feel like Big Brother was watching, probably because I benefited from the implant: the doors opened and lights came on, rather than doors closing and lights turning off. It does make me feel that Orwell was probably right about the Big Brother issue—we'll just go headlong into it; it won't be something we'll see as a being negative because there will be lots of positives in it for us."[7]

His next experiment, however, conducted in 2002, was a bit more interesting. This time it was no mere implant in the arm, but rather a genuine connection to his nervous system. He received a tiny collar, circling around the bundle of nerve fibers at the top of his arm, which read the signals from his nervous system and sent them to the computer. This wasn't all that different from what happens when a bionic leg is connected up to a disabled person, but the purpose of the experiment was different.

Using this implant, he was able to test whether, by waving his arms appropriately, he could control his computer in various ways. The idea he was getting at with this experiment was to create an eventual extension of the Cyberlink methodology to the whole body, but without anything like the

Cyberlink's special forehead device; rather to control entirely from *under the skin.* "In the very near future," he says, "we should be able to operate computers without the need for a keyboard or a computer mouse. It should be simply possible to type on your computer just by writing with your finger in the air."

The implant also received signals from the computer: he could shake his hand around, "save" the corresponding electrical signals on the computer, and then later play back the signals into his arm. The computer could store the signals from his nervous system, and then send them back into his nerves via the implant—not only a step towards cyborgization, but a nascent step along the path to uploading signals from the human nervous system into a computer.

The aspect of the experiment that attracted the most media attention involves his wife, Irina, who has agreed to receive a similar but simpler implant. Husband and wife thus directly share electricity, nervous system to nervous system. One British tabloid labeled this new implant a "love chip"—the kind of publicity that Warwick seems to love, but causes grimaces on the faces of bionics researchers and others working in more traditional cyborg-tech domains. Media hound though he indisputably is, Warwick is still a scientist at heart, and so he eventually had to correct the love-chip idea. "Emotions such as anger, shock and excitement can be investigated because distinct signals are apparent," he clarified. "[M]ore obtuse emotions such as love we will not be tackling directly."[8]

Not yet, at any rate, and certainly not with computer sensors hooked up to the arm. To master love, you may need the holy grail of cyborgization—the fabled cranial jack, the two-way link from the computer to the brain.

Using the 2002 implant, Warwick was able to control an electric wheelchair and an artificial hand via the neural interface. As well as being able to measure the nerve signals transmitted down Professor Warwick's left arm, the implant was able to create artificial sensation by stimulating individual electrodes within the array. This was demonstrated with the aid of Warwick's wife Irina—she was able to feel the sensations his electrode-array generated.

These experiments are relatively simplistic in spite of their science-fictional aura, but Warwick makes no bones about what he sees for the future.

When asked who's going to be running the show in 2100, his reply was: first AI programs, then cyborgs, then humans.

As usual with advanced technology, the scary aspects are offset by the humanitarian aspects, which go far beyond assisting the disabled. Improving upon our frail human selves through science has long been the goal of many researchers—indeed since the earliest days of scientific inquiry. As our understanding of science improves, so do the chances that we can do so. "What," Warwick asks, "if we could develop signals to counteract pain?" Or, perhaps, deliver pleasure? We will refrain from indulging in detailed speculation on the future arts of tele-, cyber- and neuro-dildonics, but there would seem to be some lucrative business opportunities here to say the least.

THE RISE OF THE CYBORGS

Cyborgization now is primitive, just as airplanes were in the time of the Wright Brothers, when various pundits declared: "Sure, you've made a plane fly, but you'll never make one fly as fast as a car can drive." Neither neuroscience nor computer technology nor microminiaturization show any sign of slowing down. Rather, they're speeding up. As cyborg technology advances, more and more techno-conservatives will rise up in arms with ethical complaints—but who can really complain about giving sight to the blind, giving hearing to the deaf, creating bionic arms and legs, giving communication skills to the horribly disabled who can't communicate in any other way, allowing paraplegics or quadriplegics to control their bodies again? This research will get done because of the tremendous humanitarian applications. The Kevin Warwicks of the world—more and more of them as time goes on, adults who as kids spent endless hours watching Anime cyborgs—will ride the coattails of these humanitarian applications and put the technology to use in other ways that will make the techno-conservatives very uncomfortable. However, it is not our zeal to better ourselves through technology that holds the potential for our doom—it is carelessness, shortsightedness, greed, and xenophobia. We'd be far better off fighting against those things in ourselves than trying to stifle the creative experimental drives that cause both artists and scientists to want to continue pushing the envelope of human experience farther and farther out.

Ben efits of cyborgization not only include closer interfacing with computers and enhanced performance (stronger arms and legs, better memory, etc.), but also the fact that—at least currently—it is easier for us to repair and maintain machine components than biological ones indefinitely. Though machines break down all the time, the processes are already in place to repair and replace parts of them. Risk of catastrophic failure effects both biological and non-biological machines, and we still have a lot of research to do into rejection of both mechanical and biological implants by existing biological tissue (especially internal organs), but our current science and engineering abilities understand a lot more about how to coax longer-term stability out of non-biological machines. Unless we are successful at growing organs and/or whole new bodies in vats and replacing our aging parts or transplanting ourselves into new bodies entirely (and we may well indeed become so), machine parts are the best possibility we have for extreme longevity. There are still steam engines from the 1770s which function, but no people from the 1770s who do.

There seems little doubt that eventually nearly every human being will be linked in directly with one or more electronic devices—we're already, in the developed world, indirectly linked with such devices constantly and in great numbers. If the interfacing isn't as smooth as one would like, may this not eventually be solved by genetic and biomechanical engineering? Why not tailor a new generation of humans for easier human-computer integration? If some people have trouble harmonizing their thought processes with the calculational processes of their digital symbiotic partner, then what's the solution? You can improve the digital half of the equation, but why not also the human half? Human thought processes can be genetically re-engineered to make more natural use of onboard calculators, computers, and petabyte in-brain memory modules. We fear genetic engineering because of its Frankenstein possibilities (though we forget—Frankenstein's monster was not evil, it was the mob mentality which caused the tragedy of that story), but still there are great positive benefits in terms of curing disease and making the healthy standard of humanity even more robust. Just like we can choose to build more nuclear weapons or engineer safe power plants to harness nuclear energy for peaceful purposes, so can we decide how to utilize the scientific reality of genetic engineering. Genetic engineering and bionics already exist. We can either allow the forces of change to be random or controlled by destructive influences in society, or harness the possibilities of change to create better lives for ourselves and better selves for our lives.

Oh, brave new world, which has such cyborgs in it?

---

[1] What if advertisers could do that as you walked downtown, constantly distracting you from your own thoughts with pitches for worthless crap? As computers become more pervasive, and more integrated, privacy and security research will be essential to more than just banks and the NSA.

[2] The current systems are crude, but still amazing. See http://www.artificialvision.com for Dr. William Dobelle's pioneering work, and this excellent summary of the field by vision cybernetics researcher Dr. Richard Normann: http://www.bioen.utah.edu/cni/projects/blindness.htm

[3] http://www.brainfingers.com

[4] http://www.biocontrol.com and http://www.csl.sony.fr/~atau

[5] http://www.ibva.com

[6] http://www.stelarc.va.com.au

[7] http://www.salon.com/tech/feature/1999/10/20/cyborg/index2.html

[8] http://www.rdg.ac.uk/KevinWarwick/html/faq.html

CHAPTER 7

THE NEW GENETICS

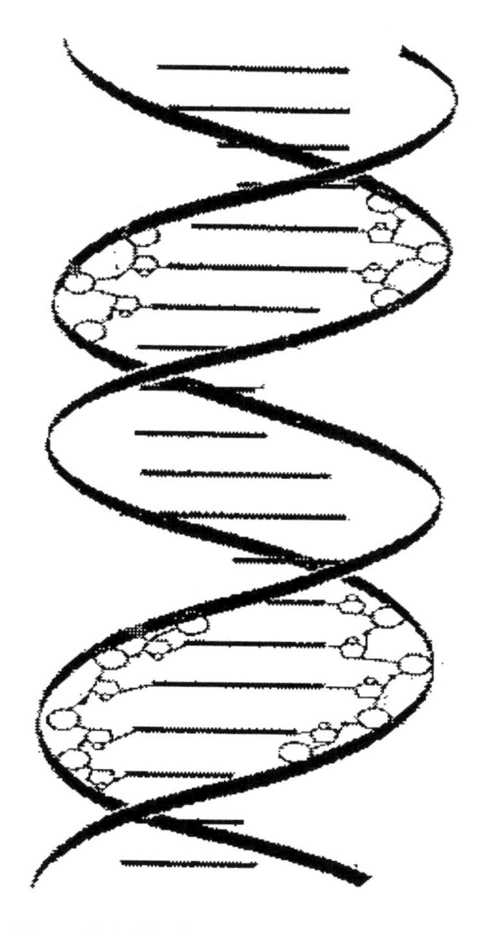

Figure 7.1 The famous DNA double helix

Your DNA, a fabulously long chain of amino acids—a copy of which is contained in every one of your cells—contains a very large percentage of the information required to produce *you.*

We tend to take this for granted, these days—but when you stop and think about it, it's really quite an amazing thing.

Extract DNA from any one of your cells and feed it into a "human producing machine," and out comes a clone of you—lacking your knowledge and experience, but possessing all your physical and mental characteristics, or at least most of them. Of course, we don't have a human producing machine of this nature just yet[1], but the potential is there: DNA seems to encode most of the information required to actually produce a human being.

This was the glory and the romance underlying the Human Genome Project[2]—a huge government-funded initiative, launched in 1990, which charted the whole human genome. The project mapped every single amino acid in the DNA of some sample of human beings. No one could doubt the excitement of this quest at the time: it had a simplicity and grandeur similar to that of putting a man on the moon.

The phase of bio research following after the Human Genome Project has, so far, not displayed such simplicity and grandeur—but scientifically, it has been even more intriguing. Now is the time for leveraging our knowledge of the genome and our increasingly powerful array of experimental molecular-biology tools, and building for the first time an understanding of how molecules interact to construct cells and organisms. Computer technology plays a big role here, and there are also connections with brain science. "Molecular neuropsychology" is no longer science fiction—there is an increasing amount of knowledge about how genes and proteins act and interact in the brain, providing a low-level foundation for the high-level thoughts and feelings in our heads. Modern biology reflects technological convergence in a major way. There is so much going on here that a brief treatment can only hint at the diversity—even more so than in most of the chapters in this book—but we'll give you those hints nonetheless.

FUNDAMENTALS OF GENETICS

In the mid-1800s, Charles Darwin and Alfred Russel Wallace independently had a very bright idea. Existing animals and plants, they hypothesized, had developed by a process of gradual, continuous change from previously existing forms, rather than being created all at once by divine providence or some other mysterious process. The theory of evolution—now considered common sense by a tremendous majority of scientists (though the general population contains a disturbing percentage[3] of religious "creationists")—was born.

Darwin and Wallace had no concrete ideas about the mechanisms by which the transmission and mutation of traits from generation to generation might take place. Ironically, when they did their work, the monk and botanist Gregor Mendel had already uncovered some of the essential properties of these mechanisms—the principles of what we now call "genetics"—but Mendel's work was obscure at the time, and neither Darwin nor Wallace was exposed to it. Thus, it remained for the next generation to connect the dots and start the work of providing evolutionary theory with a concrete underpinning.

We now know that inheritance of traits from one generation to another, in biological populations, is governed by *genes*: segments of DNA that have particular purposes. For example, a gene may "code for"—i.e. be used to generate—a specific protein, such as an enzyme. The decoding of the genes on DNA into proteins is the first step in building an organism from a genetic code sequence. The next step is for the proteins to do their work, interacting with each other and their chemical substrate to construct and then interconnect cells.

Cells with nuclei contain structures called chromosomes, each of which contains strands of DNA. Each cell nucleus has a complete set of chromosomes and therefore has a complete set of genes—even though each particular cell only utilizes some of the genes it contains. Within a species, each chromosome has a specific number and arrangement of genes; and any alterations in this regard can lead to mutations, which can lead to problems such as cancer, or else to positive changes, leading to progressive improvement in the species. The word "genome" is used to refer to the total pool of genes contained in an organism's full set of chromosomes. The "-ome" suffix has now become popular, and one hears about the proteome (the set of proteins in an organism), the kinome (the set of kinase proteins), the lysosome, etc. Marketeers write of the "omics revolution"!

On the chemical level, each gene consists of a specific sequence of molecules called nucleotides, which are composed of sugar, phosphoric acid, and nitrogen-containing compounds. Genes with the same function may be slightly different in different people. DNA is not entirely comprised of genes, though. Over 95% of the human genome is non-gene segments. These segments have sometimes been called "junk DNA," but in fact they serve a number of biological purposes and are essential for the proper functioning of the genetic machinery.

The three-dimensional geometry of the gene is the famous double helix, depicted in Figure 7.1. Determining this geometry was one of the breakthroughs in molecular biology—Linus Pauling first proposed a triple-helix model, and then Francis Crick and James Watson analyzed more complete data and came up with the correct double helix description. The double helix provides an elegant mechanism for reproduction: a child's DNA double helix is formed by taking a strand from the father's double helix and a strand from the mother's double helix, and screwing them together.

THE RACE TO MAP THE GENOME

The most colorful figure behind the human genome sequencing effort was, no doubt, Craig Venter, the founder of Celera Genomics. It was Venter, together with Hamilton Smith, who completed the first genome-decoding of a free-living organism (a bacterium, *Haemophilus influenzae*) back in 1995. Throughout the 1990s, Venter battled with the various others in the genetics community about his radical "shotgun sequencing" approach to determining the human genome. The NIH (National Institutes of Health), organizers and funders of the Human Genome Project, didn't like it, so Venter raised private capital to found Celera and sequence the genome himself. Due to the superiority of this approach, Celera's sequencing effort at first proceeded far more rapidly than the government's; and eventually the government embraced his methods. Celera's competitive effort massively accelerated the government's work, but Celera still won the race.

The "shotgun" idea was a clever one. To sequence a clone longer than the length that could normally be read, one does a whole bunch of short, overlapping sequence reads, and then uses software to assemble them into the complete sequence of the clone.

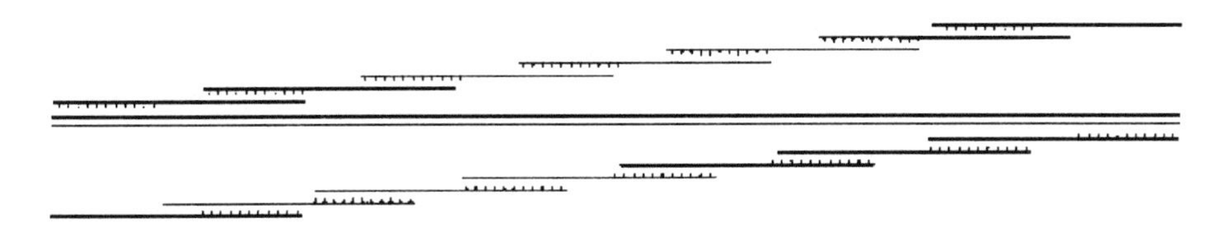

Figure 7.2 Shotgun sequencing

The "shotgun" name refers to the way the clone is prepared for sequencing: it is randomly snipped into small pieces, which are cloned, and the library of subfragments thus obtained is randomly sampled. A large number of sequence reads is generated, which are then assembled into contiguous order and the complete sequence of the clone thus revealed.

The traditional "clone-by-clone" method, on the other hand, works by breaking a genome into bigger subsegments, and then cloning each subsegment into a bacterial artificial chromosome (BAC). The BAC DNA is then broken into smaller chunks, whose end sequences are matched via computer programs. Compared with the shotgun approach, this requires a lot more sequencing work, but many argue that there's less room for error.

Venter liked the shotgun method because it was fast. Traditional researchers rejected it because they thought it was unreliable. The debate continues, but in practice, shotgun sequencing has now become standard operating procedure. Many researchers use it together with the traditional, slower, clone-by-clone approach. One common opinion these days is that the shotgun approach is best for producing an initial "draft sequencing" of a genome, whereas the clone-by-clone method is better for producing a final, nearly error-free version.

Initially Celera's human genome sequence database was far more accurate than the government's, and the company made money for a while selling high-priced subscriptions to its genome database to big pharma and biotech firms. However—in line with Venter's primary allegiance to science over business—they also made their genome available for free on CD, in the form of raw data, without any of the valuable analytical tools that came with the high-priced subscription. At this point the government's freely available genome has improved a great deal, and Celera has sought a new business model. Though Venter largely won the scientific battle over shotgun sequencing, and definitively won the race with the NIH to sequence the genome, he lost the battle to keep Celera focused on selling data and analytics. He eventually ceded corporate power to the pure-business types who wanted to transform Celera into a pharmaceutical company (following in the footsteps of so many other biotech-turned-pharmas). Now Venter is focused on a different, even more ambitious project: a quest to create a minimal organism, by taking a small microorganism and removing its genes until one has only the genes absolutely necessary for metabolism and reproduction. The goal is to literally reverse-engineer an organism in which the function of every gene and every protein is known, and the pathways of development are fully understood[4]. He's also received publicity for a project involving sailing around the world, collecting microorganisms and sequencing their genomes.

And what of the Human Genome Project? The achievement is indeed tremendous, but now everyone openly acknowledges what only a few biologists dared to admit during the '90s: the actual mapping of the genome is only a very small part of the task of understanding how people are made. The design of the "human-producing machine" by which DNA spawns organisms is a much bigger and more interesting job than the complete mapping of examples of the code that goes into the machine. In other words, embryology is probably a lot subtler than pure genetics, and Craig Venter's new minimal-organism project is probably exactly on track: the genome is just the beginning, the real focus should be on the whole organism. Contemporary bioscience, for all its wondrous advances, is just beginning to touch upon whole-organism studies, and even so, only with very simple organisms.

As the excitement of having mapped the human genome fades into matter-of-fact acceptance, the genetics community is looking ahead to what's called *post-genomic biology*. The next big challenge: figuring out how the genetic code actually does anything. How do these sequences of amino acids decode themselves, making use of other molecules in their environment to create organisms, be they hypercomplex humans, or simple one-celled organisms? The completion of the Human Genome Project was one of those ends that was actually a beginning. Some may have thought it would provide us with the ultimate answers, but in reality it has put us in the position where we're able to finally start *asking* the really interesting questions.

This is a very exciting area of research—and a tremendously difficult one as well. As yet there are no tales of tremendous triumph; only some minor victories, a lot of hard work and furious innovation, and the tremendous promise of infinite victories to come. Still, the progress made so far has many lessons to teach. For example, the remarkably tight interrelation between computer technology and biological research is a major new trend resulting from all this work. At the end of the chapter we'll briefly discuss some of the work being done by Ben and his colleagues at Biomind LLC in applying advanced AI technology to the integrated analysis of various types of biological data, with a focus on genetics and proteomics. For now, though, let's focus on genetics itself and the amazing things on its horizon.

WE'VE MAPPED THE HUMAN GENOME, NOW WHAT?

The Human Genome Project originally was planned to last 15 years, but rapid technological advances accelerated the project and it completed in 2003—two years early. The project goals were multifold: to identify all the more than 100,000 genes in human DNA, determine the sequences of the 3 billion chemical base pairs that make up human DNA, store this information in databases, and develop tools for the analysis of this huge amount of data. Some resources have also been devoted to exploring the ethical, legal, and social issues that have arisen from the project.

Of course there were many milestones along the path to completion of the Human Genome Project. Befitting the accelerating pace of scientific progress, most of these occurred not long before the completion of the sequencing of the genome itself. For instance, we recall the day in mid-2000 when newspapers announced the mapping of Chromosomes 16 and 19 on the human genome. Human chromosome 19 contains about 2% of the human genome, including some 60 genes in a gene family involved in detoxifying and excreting chemicals foreign to the body. Chromosome 16 contains about 98 million bases, or some 3% of the human genome, including genes involved in several diseases—such as polycystic kidney disease (PKD), which is suffered by about 5 million people worldwide and is the most common, potentially fatal disease caused by a defect in a single gene.

Over time more and more similar results piled up. The initial rough map of the genome was refined, bit by bit, using sophisticated "gene recognition" software that identifies sequences of base pair amino acids that represent genes, along with lots of good old biological intuition. Eventually, a complete map of the human genome was produced—first by Celera, then by the government team.

Clearly, these are major advances in gene mapping, with potential implications for helping remedy diseases. But—even now that the human genome has been mapped—what does this really mean?

An analogy may be instructive. Suppose a team of scientists goes to another planet and discovers a lot of really long strips of paper lying around on the ground, each one with strange markings on it. Suppose they then notice some big steel machines, with slots that seem to be made to accept the strips of paper. After some experimentation, they figure out how the machines work: you feed the strip of paper in one end, and then after a few hours, the machine spits out a completely functional living organism. Amazing!

So, the scientists embark on a project to figure out what's going on. They have no idea what's going on inside the machines, and all their efforts to bust the machines open meet with failure. Frustrated with their inability to crack open the machines themselves, they devote themselves to completely recording all the markings on the strips of paper in their notebooks, hoping that eventually the patterns will come to mean something to them. When they achieve 10%, then 20%, then 50% completion of their task of recording these meaningless patterns in their notebooks, they declare themselves to have made significant scientific progress.

Occasionally, along the way, they make some small discoveries about the impact that the markings have on the organisms the machine produces. If you snip off the first 10% of the strip, the organism produced is more likely to be defective than if you strip off the last 10%. The region of the strip that's 2000 to 3000 markings from the end seems to have something to do with the organism's head: it seems to be very different for organisms with very different heads, and so forth. These kinds of general observations don't really get them very far toward an understanding of what these amazing organism-producing machines are actually doing.

If you're somewhat familiar with computers, a variation on this analogy may be instructive. Consider a large computer program such as Microsoft Windows. This program is produced via a long series of steps. First, a team of programmers produces some program code, in some computer programming language. Then, a compiler acts on this program code, producing an executable file—the actual program that we run and think of as Microsoft Windows. The same goes for human beings: we have some code, and we have a complex entity created by the code, and the two are very different things. Mediating between the code and the product is a complex process—in the case of Windows, the C++ programming language compiler; in the case of human beings, the whole embryological and epigenetic biochemical process by which DNA grows into a human infant.

Now, imagine a "Windows Genome Project," aimed at identifying every last bit and byte in the C++ source code of Microsoft Windows. Suppose the researchers involved in the Windows Genome Project managed to identify the entire source code, within 99% accuracy. What would this mean for the science of Microsoft Windows?

Well, it could mean two different things:

1. If they knew how the C++ compiler worked, then they'd be home free! They'd know how to build Microsoft Windows!

2. On the other hand, if they had no idea how to build a C++ compiler, and not even any idea of the semantics of the C++ language, their results would be another puzzle to solve—not a complete answer.

In other words, our fictional "Windows Genome Project" participants have mapped out the bits and bytes in the Windows Genome, the C++ source code of Windows, but it's all a bunch of meaningless symbols to them. All they have is a large number of files of C++ source code, each of which is a nonsense series of characters. Perhaps they recognized some patterns: older versions of Windows tend to be different in lines 1000-1500 of this particular file. When file X is different between one Windows version and another, file Y tends to also be different between the two versions. Lines of code N_l–N_m seem to have some effect on how the system outputs information to the screen, and so on. However, they have no overall picture of Windows' internal functionality.

Our situation with the Human Genome Project is much more like Option 2 than it is like Option 1.

The scientists carrying out the Human Genome Project are much like the scientists in our first parable above, who are busily recording the information on the strips of paper they've found, but have no idea whatsoever what's going on inside the magical steel machines that actually take in the strips of paper and produce the alien animals. We now have the human genome mapped to within a reasonable degree of accuracy. Now what? Wouldn't it be nice to understand the process by which this genome is turned into an actual human organism?

The Human Genome Project included in its umbrella a focus on data analysis, but this referred mainly to designing and implementing computer programs that study the huge sequences of amino acids that biologists have recorded and look for patterns in these sequences. This is fascinating and necessary work, but it is still a long way from a principled understanding of how DNA is turned into organisms, and projects attempting to develop such understanding are at least equally necessary.

As an example of the sort of work that we think points in the right direction, Luis Rocha and his colleagues at Los Alamos National Labs have worked on identifying regions of the genome that are similar to each other, based on statistical tests. This kind of similarity mining gives biologists a hint that two parts of the genome may work together at some stage during the process of forming an organism. Similar statistical methods may be useful for recognizing where genes begin and end in a collection of amino acid sequences—a problem that's surprisingly tricky, and may require comparison of human sequences with sequences from other genomically mapped species such as the mouse or the fruit fly.

The relation between 1-D sequences of amino acids and 3-D structures formed from these sequences is hard for scientists to understand even on the simplest level. The big problem here is what's known as "protein folding." Many structures in DNA encode instructions for the formation of proteins, but no one knows how to predict, from the series of molecules making up a protein, what that protein is going to look like once it folds up in three-dimensional space. This is important because many proteins that look very different on the one-dimensional, molecular-sequence level may look almost identical once they've folded up in 3 dimensions. Thus, by focusing on sequence-level analysis, researchers may be scrutinizing differences that make no difference. Currently, only very few 3-D protein motifs can be recognized at the sequence level.

Basically, we barely understand the simplest stages of the production of 3-dimensional structures out of DNA, let alone the complex self-organizing processes by which DNA gives rise to organisms. This is OK—mapping DNA is still of some value even in this situation—but it must be clearly understood. In practical terms, our lack of knowledge of embryological processes greatly restricts the use we can make of observed correlations between genes and human characteristics such as diseases. There are diseases whose genetic correlates have been known for decades without any serious progress being made toward treatment. For DNA researchers to announce that they've mapped the portion of the human genome that is correlated to a certain disease doesn't mean very much in medical terms. Until our understanding of embryology increases sufficiently, we're still just writing down pieces of code for a system we don't know how to operate.

Does all this mean that the Human Genome Project is bad—wasted money, useless science? Of course not. However, it does suggest that perhaps the government is allocating its research money in an imbalanced way. By pushing so hard and so fast for a map of the human genome, while not giving a proportionate amount of research money to studies in embryology and the general study of self-organizing pattern formation, the U.S. government guaranteed that we have a map of the human genome that we cannot use in any effective way.

This brings us to some very deep and fascinating questions in the philosophy of science. As the biological theorist Henri Atlan pointed out in an essay written right around the start of the Human Genome Project, the mapping of the human genome is a very reductionist pursuit. In fact it is almost the definition of reductionism—the construction of a finite list of features characterizing human beings. All of humanity, reduced to a list of amino acids in order—imagine that! However, now that we've made this list, we seem to have failed in the same way so many other reductionist programs have—we've simply exposed our own ignorance about how the whole thing works.

The formation of organisms out of DNA is a very non-reductionist process, which biologists from the last century attributed to a "vital force" underlying all living beings. Modern scientists have still not come to grips with the scientific basis for this apparent vital force, which builds life out of matter. There are disciplines of science—cybernetics, systems theory, complexity science—which attempt to solve this problem, but these have not been funded nearly as generously as gene mapping, and they have not been linked in any serious way with the work on data analysis of genetic sequences. We believe that the study of embryology has the potential to overthrow many of our established ways of doing science, by shifting the focus of attention to complex, self-organizing processes and the emergence of structure—but this "complexity revolution" is something that the scientific establishment seems determined to put off as long as it possibly can.

In this sense, one can see the Human Genome Project as an outgrowth of modern cultural trends extending beyond the domain of science. It's an expression of the quest for understanding, and also of the illusion that reductionism is the path to understanding. It's an expression of our inability as a culture to come to grips with the wholeness of life and being, and to focus on the seemingly magical processes by which life is formed from the nonliving and structure emerges from its absence. Complexity and uncertainty frighten many people, but ever since Einstein and Gödel there has been evidence piling up that reductionism will never succeed, and that to continue to develop our understanding of our universe we'll need to accept chaotic dynamical processes as a fundamental force in the emergence of higher-order structure from basic components.

The wonderful thing about science is that it's self-correcting. Ultimately science is all about the data and the conclusions that can be drawn from it. We'll go ahead collecting data on the human genome, but year by year, the biological community will place more and more focus on how the genome interacts with its chemical and biological environment to self-organize into the organism. Some new biologists coming into the field already have the feeling that gene sequencing is old hat. New technologies like microarrays allow us to study how genes interact and interregulate in the actual living process of the cell—only partially and haltingly right now, but it's a start. We think of this as "the new genetics,"

genetics that reaches up and tries to be systems biology. In time it will succeed. Eventually, as research along these lines matures, we really will understand not just what amino acids make up a human being's genetic material, but how a human being is made.

COMPUTER TECHNOLOGY LEADS THE WAY TO POST-GENOMIC BIOLOGY

It's hardly shocking that post-genomic biology is enabled by advanced computer technology every step of the way. After all, most branches of physical science have become thoroughly computerized. Little of modern chemistry and physics could exist without computers. But it's instructive to see just how many roles computers have played in the new genetics. Firstly, it's only because of recent computer engineering- and robotics-driven advances in experimental apparatus design that we are able to gather significant amounts of data about how genes build organisms. New "microarray" technologies like DNA chips (built similarly to silicon chips) and spotted microarrays (built with robot arms) allow us to collect information regarding the expression of genes at different times during cell development, but this data is too massive and too messy for the human mind to fully grasp. Sophisticated AI software, used interactively by savvy biologists, is needed to analyze the results.

It's not hard to see what the trend is here. Biological experiments, conducted using newfangled computer technology, are spinning us an increasingly detailed story of the microbiological world—but it's a story that only increasingly advanced AI programs will be able to understand in full. Only by working with intelligent software will we be able to comprehend the inner workings of our own physical selves.

Douglas Hofstadter says, in his book *Gödel, Escher, Bach*:

> "The other metaphorical analogue to Gödel's Theorem which I find provocative suggests that ultimately, we cannot understand our own mind/brains...Just as we cannot see our faces with our own eyes, is it not inconceivable to expect that we cannot mirror our complete mental structures in the symbols which carry them

> out? All the limitative theorems of mathematics and the theory of computation suggest that once the ability to represent your own structure has reached a certain critical point, that is the kiss of death: it guarantees that you can never represent yourself totally."

Maybe we can never *fully* understand *our own minds*, but *some other mind*—perhaps an AI mind, which may not be able to fully understand itself—may nevertheless be able to fully understand our minds. Furthermore we, or an AI, may be able to understand *enough* of our own minds to make practical advances in fighting human brain diseases, increasing intelligence, etc. without requiring a full formalization and understanding of the totality of our mental functions.

Gene therapy, the frontier of modern medicine, relies on the ability to figure out what combinations of genes distinguish healthy cells from diseased cells. This problem is too hard for humans to solve alone. It requires at very least advanced statistical methods, and at most full-on computer cognition. The upshot? Rather than fearing AIs as movies like "2001" have urged us to do, we may soon be thanking AI programs for helping find the cure for cancer. We have worked in this area ourselves in the context of both the now defunct company Proteometrics (Stephan) and the startup firm Biomind LLC (Ben, with some help from Stephan), which has already made some interesting progress in this area—as have many others in the field.

As we discussed extensively earlier, artificial intelligence programs have not yet even come close to equaling humans' common sense about the everyday world. There are two main reasons for this. First, most AI programs have been written to excel only in one specialized kind of intelligence—like playing chess, or diagnosing diseases—rather than to display general intelligence. Second, even if one does seek to create an AI program with general intelligence, it still is just a software program without any intuition for the human world. We *Homo sapiens sapiens* have a special feeling for our physical and social environment. Simple ideas like the difference between a cup and a bowl, or complex ones such as the difference between happiness and contentment, are all deeply meaningful to us.

Our intelligence does not exist in a contextual vacuum. AI programs, even those that push towards general intelligence, can't help lacking this intuition. We believe this can be overcome eventually, via sufficiently powerful artificial general intelligence with a perceptual grounding in some shared reality, but we're definitely not there yet.

The world of molecular biology is not particularly intuitive to human beings. In fact, it's complex and forbidding. It has much of the ambiguity of everyday life, but it is not as familiar. There is not as much agreement as one would think about the meanings of various technical terms in genetics and molecular biology, and there is a great deal of ambiguity regarding our understanding of various biological processes and molecular functions. These ambiguities are not resolved by a simple tacit everyday understanding, only by a very advanced scientific intuition. The number of different patterns of genetic structure and activity boggles even the ablest human mind. In this domain, an artificial intelligence has much more to offer than in the world of everyday human life. Here in the microworld, human intuition is misleading as often as it is valuable. Artificial intuition can be tuned specifically to match the ins and outs of introns and exons, the turns and twists of DNA, and artificial perception could be constructed so that such objects were indeed part of the everyday environment, allowing an advanced AI to reason about molecular biology as naturally as a human reasons about politics or farming.

STUDYING GENE EXPRESSION

The new genetics has many aspects, but we'll focus here on one of the areas with which we're most familiar from our own work: the emerging study of *gene and protein expression*. The terminology here is both evocative and appropriate: just as with a person, it's not what a gene does when it's just sitting there that's interesting, it's what a gene does when put in a situation where it can express itself!

At any given moment, most genes are quiet, doing nothing. At the same moment there are some which are expressed—that is to say, they are active. Using the new experimental tools, we can tell which they are. We can see how many genes are expressed at a given moment...and then a little later...and then a little later. In this way we can make a map of genetic dynamics as it evolves. By analyzing this map using advanced computer software, a lot of information about how genes go about their business can be understood: which genes tend to stimulate which other genes; which ones tend to act in groups; which ones inhibit which other ones, preventing them from being expressed. By applying the same analysis tools to proteins instead of genes, one can answer the same questions about proteins (the molecules that genes create and send around to do the actual business of building cells). These kinds of complex interactions between genes, and between genes and proteins, are the key to the decoding of genomes into organisms—which is, after all, what genomes are all about.

All this complexity is implicit in the genetic code itself, but we don't know how to interpret the code. With microarrays, we can watch the genetic code interpret itself and create a cell, and by analyzing the data collected in this process we can try to figure out exactly how this process of interpretation unfolds. The potential rewards are great. Practical applications are tremendous, ranging from drug development to disease diagnosis to genetic engineering and beyond.

It's a straightforward enough idea, but the practical pitfalls are many. A huge host of tools from mathematics and computer science have been unleashed on the problem, both by researchers at major academic firms and by companies like Rosetta Inpharmatics (acquired by Merck, the major pharmaceutical firm) and Silicon Genetics (a gutsy and clever California start-up, acquired by Agilent). New data analysis techniques come out every couple of months, each one with its own strengths and weaknesses. It is a rapidly evolving field, but one which still has much more work ahead of it than behind it.

MICROARRAYS

It would be hard to overestimate the revolutionary nature of the new experimental tools—microarrays—underlying the gene expression revolution. The same tools, with minor variations, are also being made work for proteomic analysis, the study of protein expression. For the first time, with these new devices, biologists are able to study thousands or even millions of different molecules at once, and to collect the results in a systematic way.

Chemists have long had methods for carrying out many simultaneous chemical reactions. Most simply, trays can be created with various numbers of wells (say, 24 or 96 or 384—but not much larger), each containing a different chemical and a unique bar code. The last few years, however, have seen the development of methodologies that push far further in this direction—primarily microarrays, which allow vastly larger numbers of (smaller) sample cells—making possible experiments that scientists would have called impossible only a few years ago. The application of these new methodologies to the analysis of gene and protein data has led to a new area of research that may be called *massively parallel genomics and proteomics*.

Most of the work done so far has been in genomics; the extension to proteomic analysis is more recent, so we'll focus our analysis on microarrays as used for genomic analysis. The proteomics case is basically the same from the point of view of data analysis, though vastly more difficult from the point of view of experimental apparatus biomechanics. (Many proteins are much more difficult than DNA to induce to stick on the surfaces used in these instruments.)

There are several types of microarrays used in genomics, but they all embody a common methodology. Single-stranded DNA/RNA molecules are anchored by one end to some kind of surface (a chip or a plate depending on the type of apparatus). The surface is then placed in a solution, and the molecules affixed to the chip will seek to hybridize with complementary strands ("target molecules") floating in the solution. (Hybridization refers to the formation of base pairs between complementary regions of two strands of DNA that were not originally paired—simply put, the molecules are selected so they'll link up with other chemically-compatible molecules.)

Affymetrix's technology, pioneered by Dr. Stephan Fodor, involves making DNA chips in a manner similar to the manufacture of semiconductor chips. A process known as "photolithography" is used to create a huge number of molecules, directly on a silicon wafer. A single chip measuring 1.28 cm X 1.28 cm can hold more than 400,000 "probe" molecules. This information density is much larger than the 384-well chemical trays of a few years ago, and makes it possible for the first time to do gene expression experiments on the "whole-genome" scale. The procedure of gene chip manufacture has been fully automated for a while now, and Affymetrix manufactures 5,000-10,000 DNA chips per month.

Affymetrix DNA chips have a significant limitation in terms of the size of the molecules that can be affixed to them. So far, they're normally used with DNA/RNA segments of length 25 or less. Also, they are very expensive. It currently costs about $500,000 to fabricate the light masks for a new array design, so their technology is most appropriate when a large number of similar experiments need to be performed, and thus the same kind of chip needs to be used again and again and again. The main example of this kind of use-case is disease diagnosis.

On the other hand, spotted microarrays, first developed by Pat Brown at Stanford, are ordinary microscope slides on which robot arms lay down rows of tiny drops from racks of previously prepared DNA/RNA samples. At present this technology can lay down tens of thousands of probe molecules—at least an order of magnitude below what Affymetrix can do, but still better than the old style chemical wells. The advantage of this approach is that any given DNA/RNA probe can be hundreds of bases long, and can, in principle, be made from any DNA/RNA sample in any array layout that fits on the slide, ad hoc, rather than having to create an expensive photolithography mask for each new combination.

Note the key role of computer technology in both of these cases. Affymetrix uses a manufacturing technique derived from the computer hardware industry, which depends on thorough and precise computer control. Spotted microarrays depend as well on the inhuman precision of robot arms, controlled by computer software. Massively parallel genomics, like the mapping of the human genome itself, is a thoroughgoing fusion of biology and computer science—only here the initial emphasis is on computer engineering and hardware, whereas gene mapping relied upon fancy software algorithms. However, computer software analysis is also necessary for massively parallel genomics, to analyze these large data sets. It's a lot easier for a computer to analyze 400,000 samples than for a human to do it.

There are other approaches as well. For instance, Agilent Technologies, a spin-off from Hewlett Packard, is manufacturing bioarray fabricators using ink-jet printer technology to squirt samples in solution onto slides the way a printer squirts ink onto paper. Their approach is interesting in that it promises to make practical the synthesis of a single instance of a given array design. Lynx Corporation is pursing a somewhat Affymetrix-like approach, but circumventing Affymetrix's patents by using addressable beads instead of a silicon wafer. A variety of other companies are springing up all the time with novel approaches or different takes on the existing ones. Over the next few years we will see a lot of radical computer-enabled approaches to massively parallel genomics, and time and laboratory experimentation will tell which are most effective.

So, how are these massively parallel molecule arrays used? Let's suppose that, one way or another, we have a surface with a number of DNA/RNA molecules attached to it. How do we do chemical reactions and measure their results?

First, the target molecules are labeled with a fluorescent chemical, so that the spots on the chip/array where hybridization occurs can be identified. The strength of the fluorescence emanating from a given region of the surface is a rough indicator of the amount of target substance bound to the molecule affixed to that region. In practical terms, what happens is that a photograph of some sort is taken of the pattern of fluorescence emanating from the microarray, and an image file is created. Typically the image file is then "gridded," i.e. mapped into a pixel array with a pixel corresponding to each probe molecule. Then, there is a bit of a "black art" involved in computing the hybridization level for a spot. Such computations involve various normalization functions that seem to have more basis in trial and error than in fundamentals, which makes it difficult and time-consuming to develop new ones (and, frankly, can easily lead to a situation where poor experiment design or quality control in the trials can lead to difficult-to-detect errors).

This data is very noisy, however. To get more reliable results, researchers generally work with a slightly more complex procedure. First, they prepare two related samples, each of which is colored with a different fluorescent substances (usually, one green, one red). They then compare the relative amounts of expressed DNA/RNA in the two samples. The ratio of green/red at a given location is a very meaningful number. Using this ratio is a way of normalizing out various kinds of experiment-specific "noise," assuming that these noise factors will be roughly constant across the two samples.

Even this ratio data is still highly noise-ridden, for a number of reasons beyond the usual risk of experimental error or manufacturing defects in the experimental apparatus. For one thing, there are many different factors influencing the strength of the bond formed between two single stranded DNA/RNA molecules, such as the length of the bonded molecules, the actual composition of the molecules, and so forth. Errors will also occur due to the ability of DNA to bind to sequences that are roughly complementary but not an exact match. This can be controlled to some extent by the application of heat, which breaks bonds between molecules—getting the temperature just right will break false positive bonds and not true positive ones. Other laboratory conditions besides temperature can have similar effects, but as you can imagine this all adds up to a lot of parameters to control. Another problem is that the "probe molecules" affixed to the surface may fold up and self-hybridize, thus rendering them relatively inaccessible to hybridization with the target.

All these issues mean that any single data point in a large microarray data set cannot be taken all that seriously. The data as a whole can be extremely valuable and informative, but there are a lot of things that can go wrong and lead to spurious information. This means that data analysis methods, to be successfully applied to microarray data, have got to be extremely robust with respect to noise. None of the data analysis methods in the standard statistical and mathematical toolkits pass muster, except in very limited ways. Traditional mathematical techniques require that fairly specific experimental design requirements be met in order for them to be valid. If your experiment is a natural fit for such methods, you've got a powerful tool, but if you try to change your experiment design to match an analysis method you can wind up introducing error and bias which will lessen the value of your results. Much more sophisticated and adaptive analytical technology is needed—to get the most useful results, we'll eventually need artificially intelligent software that can build its own digital intuition as regards the strange ways of the biomolecular world, through a more direct brain-probe interface than is possible for humans (who, at least for now, must interface with our minds using our usual five senses).

The payoff for understanding this data, if you can do it, is huge. Such data can be used for sequencing variants of a known genome, or for identifying a specific strain of a virus (e.g. the Affymetrix HIV-1 array, which detects a strain of the virus underlying AIDS). They can be used to measure the differences in gene expression between normal cells and tumor cells, to help determine which genes may cause or cure cancer, or identify to which treatment a specific tumor seems to best respond. They can measure differences in gene expression between different tissue types, to determine what makes one cell type different than another. Most exciting from a scientific viewpoint, they can be used to identify genes involved in cell development, and to puzzle out the dynamic relationships between these genes during the development process. With all these amazing possibilities and their implications for general health and longevity, the stakes in getting this right are very high.

INTERNET GENOMICS

We've seen that the actual experimental apparatuses being used in postgenomic biology all come in one way or another out of the computer industry, and that the analysis of large, noisy, complex data sets like the ones microarrays produce can only be carried out by sophisticated computer programs running on advanced machines. No human being has the mind to extract subtle patterns from such huge, messy tables of numbers. There is another crucial dependency on computer technology here: the role of the Internet. The biology community has come to use the Net very heavily for data communication—without it, there is no way research could proceed at anywhere near its current fervent pace.

Perhaps you're a bit of a computer hacker and you want to try out your own algorithms on the data derived from microarray experiments on the yeast genome during cell development. Well, you're in luck: the raw data from these experiments are available online[5]. Download it and give it a try! Or check out Rosetta's site[6], and download some sample human genome expression data. Perhaps your interests are less erudite, and you'd simply like to view the whole human genome itself? No problem, just check out the Genome Browser[7].

Gene sequence information, and the quantitative data from gene expression experiments, is only the beginning of what's available online. There's also a huge amount of non-numerical data available, indispensable to researchers in the field. When biologists interpret microarray data, they use a great deal of background knowledge about gene function—and more and more knowledge is coming out every day, with a huge amount of it available online for public consumption, if you know where to look. Most current automated data analysis tools tend to go purely by the numbers, but a new generation of tools already boasts the ability to integrate numerical and non-numerical information about genes and gene expression. As preparation for this, biologists in some areas are already working to express their nonquantitative knowledge in unambiguous, easily computer-comprehensible ways (not totally dissimilar from the Cyc encodings discussed in the AI chapter).

This exposes the dramatic effect the Net is having on scientific language. Certainly the Net is hastening the establishment of English as the world's second language, but something more profound than that is happening simultaneously. The Net demands universal intercomprehensibility. In biological science, this feature of Internet communications is having an unforeseen effect: it's forcing scientists working in slightly different areas to confront the idiosyncrasies of their communication styles and develop a more unified language of biological science.

Compared to ordinary language, scientific language is fairly unambiguous—but it's far from completely so. An outsider would understandably assume that a phrase like "cell development" has a completely precise and inarguable meaning, but biology is not mathematics and when you get right down to it, some tribes of researchers use the term to overlap with "cell maintenance" more than others do. Where is the borderline between development and maintenance of a cell? This issue and hundreds of others like it have come up with fresh vigor now that the Internet is shoving every branch of biological research into the faces of researchers in every other branch. As a result of the Internetization of biological data, a strong effort is underway to standardize the expression of non-numerical genetic data.

One part of this is the Gene Ontology Project[8], an attempt to produce a formal, restricted language for the precise description of "gene and gene product attributes in any organism." In the creation of this project, one thorny issue after another came up—a seemingly endless series of linguistic ambiguities regarding what would at first appear to be very rigid and solid scientific concepts. What is the relation of "development" to "maintenance," what does "differentiation" really mean, what is the relation of "cell organization" and "biogenesis," etc? The outcome of this quibbling over language, however, is much more precise vocabulary syntax and a universal dictionary for non-numerical discussion of molecular biology. Ambiguity can't be removed from the language used to describe cells and molecules (it can't even be removed from physical/biological interpretations of numerical data), but it can be drastically reduced through this sort of systematic effort with the result that genes from different species can be compared using a common vocabulary. The fly, yeast, worm, mouse and mustard genomes have all been described to a significant extent in standardized Gene Ontology language, and the human genome can't be far behind. Soon enough, every gene of every common organism will be described in a "Gene Summary Paragraph," describing qualitative knowledge about what the gene does in carefully controlled language—language ideally suited for digestion by AI programs.

The standardization of vocabulary for describing qualitative aspects of genes and proteins is a critical part of the computerization of biological analysis. With this approach, AI programs don't have to have a sensitive understanding of the nuances of human language and its many ambiguities to integrate qualitative information about gene function into their analyses of gene sequences and quantitative gene expression data. It's only a matter of years before the loop is closed between AI analysis of genomic data and the automated execution of biological experiments. Now, humans do experiments, use sophisticated algorithms to analyze the results, and then do new experiments based on the results the algorithms suggest. Before too long, the human will become redundant in many cases. Most of the experiments are predominantly computer-controlled already, so it is the next logical step to let the software analyze the results of one experiment, then order up a follow-up experiment(s) immediately upon analysis. After a few weeks or months of high-speed trial and error experimentation, this sort of system will present us humans with results about our own genetic makeup.

CURING DISEASE THE POST-GENOMIC WAY

All this abstract, complicated technology comes together to provide practical solutions to some very real problems. Genetic engineering is one of the big potential uses. Understanding how genes work to build organisms, we'll be able to build new kinds of organisms. Right now we have "Frankenfoods," but eventually we (and our AI cohorts) will be able to engineer new kinds of dogs, cats and people (and upgrades for the old kinds), thus opening up exciting new avenues of performance and longevity-enhancing medical capabilities and raising all kinds of serious ethical concerns.

But there are also applications that are, ethically, pretty much unquestionable, such as applying this technology to find genetic causes for disease and develop specifically targeted cures. From an economic point of view, the main value of microarrays and related technologies right now is as part of that vast scientific-financial machine called the drug discovery process. The path from scientific research to the governmental approval of a new drug is a very long one, but when it's successfully traversed, the financial rewards can be immense.

Gene therapy is a new approach to curing diseases, and one that hasn't yet proved its practical worth in a significant way. Although it hasn't lived up to the over-impatient promises made for it 10 years ago, biologists remain widely optimistic about its long-term potential—not only for curing "classic" hereditary diseases, but also widespread diseases such as cancer and cardiovascular diseases. The concept is scientifically unimpeachable. Many diseases are caused by problems in an individual's DNA. Transplanting better pieces of DNA into the cells of a living person should be able to solve a lot of problems. A great deal of research has been done regarding various methods to implant genes with the desired characteristics into body cells. Usually the injected gene is introduced within the cell wall, but resides outside the nucleus, perhaps enmeshed in the endoplasmic reticulum. Fascinatingly, the result of this is that the gene is still expressed when the appropriate input protein signal is received through the receptors in the cell wall, even though the gene is not physically there in the nucleus with the rest of the DNA.

Aside from the practical issues of how to get the DNA into the cell in various circumstances, there's also the major issue of figuring out what DNA is responsible for various diseases, and what to replace it with. In the case of complex diseases understanding this requires understanding how DNA is decoded to cause cells of various types to form—a kind of understanding that has been very, very hard to come by. The discovery of gene and protein expression data from microarray experiments, and sophisticated bioinformatics software analysis, renders this understanding potentially achievable, though still by no means trivial. More precise microarrays and more intelligent data analysis software may render the problem downright straightforward 5 or 10 or 20 years from now, but right now no one knows for sure.

One thing biologists do, in trying to discover gene therapies, is to compare the genetic material of healthy with disease-affected individuals. A key concept here is the "genetic marker"—a gene or short sequence of DNA that acts as a tag for another, closely linked gene. Such markers are used in mapping the order of genes along chromosomes and in following the inheritance of particular genes: genes closely linked to the marker will generally be inherited with it. Markers have to be readily identifiable in the organism the DNA builds, not just in the DNA. Some classic marker genes are ones that control phenomena like eye color.

Biologists try to find the marker genes most closely linked to the disease, the ones that occur in the affected individuals but not in the healthy ones. They narrow the markers' locations down step by step. First they find the troublesome chromosome, then they narrow the search and try to find the particular troublesome gene within that chromosome, and through iterative refinement narrow-down the specific genetic causes of disease.

Previously, genetic markers were very hard to find. Now that the human genome is mapped and there are technologies like microarrays, things have become a good bit simpler. Some markers are now standard—and Affymetrix sells something called the HuSNP Mapping Array, a DNA microarray with probes for many common markers across the human genome already etched on its surface and ready for immediate use. If you have samples of diseased tissue, you can use this microarray to find whether any of a large number of common markers tends to coincide with it. In the past this would have required thousands or millions of experiments, and in many cases it would have been impossible. Now it's easy, because we can test in parallel whether any of a huge number of gene sequences is a marker for a given disease-related gene (and since it requires a standard chip type, it doesn't have the high cost of a custom fabrication, making it sufficiently affordable to get enough chips to do the millions of necessary experiments in parallel). Right now, scientists are using this approach to try to get to the bottom of various types of cancer, and many other diseases as well.

If a disease is caused by one gene in particular, then the problem is relatively simple. In this case one simply has to analyze a sufficient number of tissue samples from affected and healthy people, and eventually one's computer algorithms will find the one gene that distinguishes the two populations. But not all diseases are tied to one gene in particular—and this is where things get interesting. Many diseases are spread across a number of different genes, and have to do with the way the genes interact with each other. A disease may be caused by a set of genes, or, worse yet, by a *pattern* of gene interaction, which can come out of a variety of different sets of genes. That's right, a given pattern of gene interaction can emerge from totally different sets of particular genes so long as both sets contain the necessary interaction sites to exhibit the behavior. Here microarrays come in handy once again, in a big way. If a disease is caused by a certain pattern of interaction, microarray analysis of cell development gives enough data to allow scientists to find that pattern of interaction. Then they can trace back and find all the different combinations of genes that give rise to that pattern of interaction. This is a full-on AI application, and it pushes the boundaries of what's possible with the current, very noisy microarray data, but there's no doubt that it's the future.

Gene therapy itself is still in its infancy, as is microarray technology, and as is AI-driven bioinformatics, but all these areas are growing *fast*—fast like the Internet grew during the 1990's. Exactly where it's all going to lead is not certain, but it's a pretty sure bet that the intersection of medicine, genetics, proteomics, computer engineering, AI software, and robotics is going to yield some fascinating and beneficial results for medicine and general biological science in the near future (much sooner than standard techniques would have).

GENE CHIPS AND ATAXIA

The medical applications of DNA chips are still in their childhood—but preliminary indications are that they may well be revolutionary. As Affymetrix founder Stephen Fodor says, "Affymetrix was founded on the belief that understanding the correlation between genetic variability and its role in health and disease would be the next step in the genomics revolution." The results to back up this vision have started coming in.

A few years ago researchers at the Whitehead Institute used DNA chips to distinguish different forms of leukemia based on patterns of gene activity found in cancerous blood cells. This approach has led to real practical benefits, in some cases *reversing incorrect diagnoses* made by other, cruder methods. This is only the barest beginning. As Dr. Lander of the Whitehead Institute says, "the research program aims to lay a foundation for the 'post-genome' world, when scientists know the complete sequence of DNA building blocks that make up the human genome." Mapping not only what is *in* the genome, but what the things in the genome *do*, is the real secret to comprehending and ultimately curing cancer and other diseases. In recent years there have been literally hundreds of research papers in a similar vein, not the least of which includes our own report on the work in which Biomind and the Centers for Disease Control have over the last few years figured out how to use gene chips to diagnose Chronic Fatigue Syndrome based on gene expression in people's blood cells.

One of the more interesting developments in the medical application of DNA chips was the creation of the Affymetrix spin-off company, Perlegen Sciences Inc. Perlegen's goal is to use DNA chips to help understand the dynamics underlying various diseases—starting out with the rare disease "ataxia telangiectasia" (A-T), with which the two sons of Perlegen co-founder Brad Margus are afflicted.

Ataxia is a word for loss of muscular coordination; telangiectasia refers to the small blood vessels that pop up on the skin and eyes of A-T victims. A-T typically affects youths; 40% of A-T children develop cancer, and few live past their 20s. Margus was the boss of a $100-million-a-year shrimp-processing company when he discovered his sons were afflicted with A-T—and, in a remarkably systematic and dedicated fashion, began to devote more and more of his life to researching the biological foundations of the disease. He helped raise millions of dollars for research on A-T and its genetic basis, a quest that ultimately led him to Stephen Fodor.

Affymetrix array chips, it seemed to Margus and his bioscientist collaborators, could be used to study the way different individuals with A-T would react to different medications. It could vastly accelerate the drug discovery process, by allowing so many experiments to be run in parallel. As this is exactly the kind of valuable humanistic application of DNA-chip technology that makes biologists happiest, it didn't take much effort to convince Fodor that Affymetrix should help Margus in his quest by helping to form Perlegen.

With humanistic applications like this swirling all around him, it's not hard to see why Fodor is relatively unruffled by the ethical dilemmas that some find in genetic research. Are there potential dangers in this technology? To be sure there are, but while such dangers should not be dismissed, neither should the tremendous potential to help people. So far there is little doubt among biologists and physicians that the positive far outweighs the negative. DNA chips have helped find cures for diseases, and they haven't harmed anybody.

Of course, this is just the beginning. We've mapped the genome, and now, baby step by baby step, we're starting to understand the process by which strands of genetic material interact with other molecules to form organisms like us. As we move along this path of understanding, we'll be able to cure more and more diseases—and more dramatic possibilities for genetic screening and genetic modification will open up. We advocate that practitioners be involved in the development of a positive ethics of the usage of such technologies, as we do with all technologies, but do not subscribe to the perspective that potential harm is reason enough to abandon an entire discipline with so many potentially beneficial results. No evil is intrinsic in scientific discovery, only in how we *choose* to apply it, if we choose to misuse it.

INCIPIT BIOMIND (OR: THE AUTHORS' ADVENTURES IN BIOSCIENCE)

We always thought genetics was fascinating—what curious young nerd wouldn't? But it wasn't until mid-2000 that we began to seriously consider genetics and proteomics as a research area we might want to focus on.

Well before Webmind Inc. folded, we had grown seriously disenchanted with the application areas toward which we'd chosen to orient our AI-based products. Business success was proving elusive in spite of the fact that our products outperformed the competition's and—a largely separate issue—it didn't seem that our products were evolving in directions that would make maximal use of our most original technology, the Webmind AI Engine. Financial prediction was fun, and the Webmind MP seemed to work outstandingly well, but the essence of our approach was the use of news to predict market movements. The other products we were making—Webmind Classification System and Webmind Search (a search engine that was never released but was used internally within the company)—were even more human-language-centric. For a variety of reasons, we wanted to get away from natural-language-processing applications. The more we worked on our AI system, the more we on the R&D side realized that starting out with human-language-based products was putting the cart before the horse. We needed an application domain that had a rich variety of nonlinguistic data that the system could reason about, building up a domain-specific knowledge base that could then be used to experientially ground linguistic knowledge—little by little, step by step, much as a human baby grounds its early linguistic knowledge in its observations about the nonlinguistic physical world it's embedded in.

The finance domain did have its strong points—words about market movements could be correlated with actual observed market movements, for example, providing an elementary form of symbol grounding. Too often though, in financial texts, the language was imprecise and evocative rather than precisely descriptive. More and more often, our thoughts began shifting to biological applications. There was so much biological data being generated, and it was so diverse, that it became increasingly exciting to think of all this data being fed into an integrative AI system capable of using it to draw new and interesting conclusions.

Of course, it didn't take long to realize that biological data wasn't really an *ideal* application domain either. It's a great testing ground for integrative cognition, and even perception, but there are too few opportunities for an AI system to act. Actions such as sending information to human users are obviously present, and on a much slower time scale, a bio-focused AI can control robot arms and other equipment running biological lab experiments. But all this is not very similar to the intense perception/action/cognition interactivity that a baby gets from the physical world. So the idea of biological data as an application domain definitely doesn't displace the experiential-learning-focused, Nova-babyish approach[9], but it is a worthy complement. We spent a decent portion of 2001 thinking about how to apply Novamente to analyze genetics data, designing products and running prototype data analysis experiments in this direction, and pitching the concept to LANL and other research institutions.

The amount of data modern biologists are collecting is truly immense. About 100 microorganisms have been completely sequenced, with many more in the pipeline. The human genome and other eukaryotic genomes such as yeast, *Drosophila*, and *Caenorhabditis elegans* are now available online. New sequencing projects begin almost daily. Microarrays and mass spectrometry produce massive datasets, which lead to massive data analysis problems. With each genomic sequence, there are more genes, more RNAs, more proteins, more phenotypes, and more data in databases. It is a blessing to have such data, but only if it can be accessed, integrated, and used to develop new knowledge. In May 2002, a couple months into the post-Webmind Inc. phase, we realized that this was a mission worthy of an AGI. Plus, what better way to start off an AI with a good attitude toward humans, than to have it focus its early energies on analyzing human cells with a view toward curing human diseases and helping humans to live longer?

There is no doubt that existing biological databases contain the secrets to hundreds of undiscovered drugs. What we realized during 2001 was what was needed draw these secrets out. Something simple yet elusive: data analysis software that *automatically deploys this massive data pool within the experimental data analysis process*. This new kind of feedback between wet lab work and advanced data analysis, once it's achieved, will lead to a raft of new discoveries—making the pharmaceutical progress of the last 10 years seem like the nascent beginning it really is. As a single, very important example, current tools make it very difficult to find *sets of genes* that can collectively function as drug targets (sites where gene-therapy drugs can act to interfere with a particular disease process), whereas an integrative data analysis framework will in time make this kind of discovery routine.

Throughout Fall 2001 we talked extensively about these ideas with Maggie Werner-Washburne, a deeply insightful yeast geneticist in the University of New Mexico biology department. The more we talked, the more obvious it became what an excellent AI application bioinformatics really is. No single magic bullet, no one bioinformatic trick, is going to provide the deep, dynamic, goal-directed information integration that modern biology requires. What is needed is a combination of four ingredients:

- database integration
- visualization tools
- natural language processing (NLP) tools that extract information from research papers, adding information to databases
- automated inference tools that synthesize information from different databases

The biggest conclusion from these conversations was this: Whoever can deliver these ingredients in a user-friendly package will be the one leading bioscience into the new millennium. This realization crystallized the vague bioinformatics ideas we and our Novamente had been tossing around, and we were also able to draw upon Stephan's experience with proteomics gleaned during a short stint with Proteometrics before they were bought out in late 2001. We began designing an ambitious bioinformatics software system called the Biomind AI Engine, which was oriented toward deploying the Novamente AI system toward the goal of helping biologists understand their experimental data in the context of the massive amount of general biological "background information" that now exists. In late 2002 Ben and several of his Novamente collaborators founded the firm Biomind LLC, with the intention of pursuing this vision.

Of course, this sort of work is very different from our grand work on AGI. In the Biomind case, the focus is not entirely on artificial intelligence, even in the more restrictive "narrow AI" sense, but equally as much on *intelligence augmentation* (IA)—helping biological scientists to use their expertise and intuition to follow pathways to discovery. Biomind software is not intended, in the short term, to think about biology better than the biologists do. It's intended to integrate information from diverse databases better than biologists do, so that rather than spending their time sifting through huge databases and journal paper archives, biologists can spend their time thinking about biology. (Which they will continue to do better than Biomind, at least for a decade or so!) The Biomind software is intended to recognize subtle patterns in huge sets of quantitative data, such as microarray data, better than biologists do. The human mind is great at recognizing visual patterns, but the high-dimensional numerical datasets that come out of gene chips and other modern biological equipment are definitely not what our minds are tuned for, and the greatest AI skeptic has got to recognize that even the current crop of narrow-AI software can do this better than any human.

Now, in 2005, we've actually made a fair amount of progress toward actualizing the vision with which Biomind was founded, though there's a lot more to be done. Much of our work has centered on microarray data analysis. We've created a product, the ArrayGenius[10], which recognizes subtle patterns in microarray data using AI pattern recognition, integrated with knowledge from the Gene Ontology and other, similar knowledge resources. We've done some work analyzing mutations in mitochondrial DNA related to Parkinson's disease, which we'll discuss briefly in the next chapter on longevity research. We've used AI to guess the functions of many genes whose functions were previously unknown, based on combining gene sequence data and microarray data. We've done some preliminary experiments extracting knowledge about the relationships between genes, proteins and chemicals from biological research papers, using Novamente's natural-language-processing technology. We're plotting a new product, the PathwayGenius—hopefully to be released in 2006—which will integrate all these things into a graphical map of the "process network" by which genes build cells.

From a very high level, the main merit of AI technology in this kind of application is its ability to deal with massive amounts of data—and, in this context, to integrate various massive data stores containing different types of data to arrive at overall judgments. This general capability is what will enable appropriately-deployed narrow AI technology to solve the "islands of information" issue plaguing the life sciences today. When datasets are so large and so diverse, spread across many different locations, and residing on architecturally incompatible platforms, it is extremely difficult to leverage value from the body of knowledge as a whole. By actively learning, understanding and abstracting inferences from data, regardless of its native structure, Biomind's software can assemble a cogent universe of knowledge out of a disparate collection of databases and experimental data results.

We'll now discuss, in a little more depth, some of the specific work being done with the Biomind software, with a focus on the way narrow AI can serve as powerful IA in this sort of domain, helping the human mind to better grapple with the vast complexity of the data and the systems under study.

GUESSING GENE FUNCTIONS

One of the simpler things we've done with the Biomind software is to add some new bits of knowledge to the Gene Ontology itself. The GO is a large and impressive database, constructed by a distributed team of biologists. It consists of a category hierarchy, sort of like Yahoo's categorization of the Web, but categorizing genes rather than Web pages. You can look up a gene in the GO and find out what categories it falls into: which biological processes, cellular components and molecular functions it's involved with.

The GO is far from complete. At this stage, a lot of human genes still have unknown functions, but microarray data can be used to guess functions for the unknown genes. The basic principle is a simple one: if a gene is expressed similarly to the genes known to be in some category, then we may guess the gene actually belongs to that category. The detailed working-out of this principle contains a lot of subtlety. Implementing it naively gives bad results, with too many wrong guesses and too little reliability. However, using advanced machine learning tools like those we've developed in the Biomind toolkit for the appropriate assessment of "similarity," one can do much better. While the approach is far from omniscient, at time of writing we've guessed the functions for about 600 previously ill-understood human genes, with an estimated 80% accuracy. Probably, by the time you read these words, this methodology will have been extended a lot further and a lot more discoveries will have been made.

For example, the following table shows some Gene Ontology categories related to the cell cycle (therefore potentially related to cancer), and reports how many genes of previously unknown function our software inferred to belong to each of them.

GO Category ID	**GO Category Name**	**# of New Members Inferred**
GO:0000067	DNA replication and chromosomecycle	1
GO:0000087	M phase of mitotic cell cycle	3
GO:0000278	mitotic cell cycle	4
GO:0000279	M phase	6
GO:0000280	nuclear division	6
GO:0006260	DNA replication	1
GO:0006261	DNA-dependent DNA replication	1
GO:0007049	cell cycle	14
GO:0007067	mitosis	3

Table 7.1 Examples of Gene Ontology categories related to the cell cycle

As you see above, the GO category called "DNA replication" got one new gene assigned to it, the gene with the standard RefSeq ID number NM_003113. What exactly does this gene do? We don't know, but we do now know, thanks to Biomind, that if you're a researcher interested in DNA replication, you should probably be doing some research involving this gene. Previously, the only knowledge about this function in the Gene Ontology was that it belongs to the category "regulation of transcription, DNA-dependent." So now we know it's involved with both DNA replication and DNA-dependent transcription regulation—which is not surprising, as these are closely related processes.

In case you're curious, the actual genetic sequence for NM_003113 is shown below. As is standard in gene representation: the c's are cystine, the t's are thymine, the g's are guanine, the a's are adenine. Just like the picture at the start of the chapter showed, this is a long winding series of molecules bound together, part of an even longer series of molecules which is the full DNA. (The numbers at the left side of each row just help you keep count in the amino acid sequence; they're not part of the gene itself.)

```
   1 ctgaggccca cgcagggcct agggtgggaa gatggcaggt gggggcggcg acctgagcac
  61 caggaggctg aatgaatgta tttcaccagt agcaaatgag atgaaccatc ttcctgcaca
 121 cagccacgat ttgcaaagga tgttcacgga agaccagggt gtagatgaca ggctgctcta
 181 tgacattgta ttcaagcact tcaaaagaaa taaggtggag atttcaaatg caataaaaaa
 241 gacatttcca ttcctcgagg gcctccgtga tcgtgatctc atcacaaata aaatgtttga
 301 agattctcaa gattcttgta gaaacctggt ccctgtacag agagtggtgt acaatgttct
 361 tagtgaactg gagaagacat ttaacctgcc agttctggaa gcactgttca gcgatgtcaa
 421 catgcaggaa taccccgatt taattcacat ttataaaggc tttgaaaatg taatccatga
 481 caaattgcct ctccaagaaa gtgaagaaga agagagggag gagaggtctg gcctccaact
 541 aagtcttgaa caaggaactg gtgaaaactc ttttcgaagc ctgacttggc caccttcggg
 601 ttccccatct catgctggta caaccccacc tgaaaatgga ctctcagagc acccctgtga
 661 aacagaacag ataaatgcaa agagaaaaga tacaaccagt gacaaagatg attcgctagg
 721 aagccaacaa acaaatgaac aatgtgctca aaaggctgag ccaacagagt cctgcgaaca
 781 aattgctgtc caagtgaata atggggatgc tggaagggag atgccctgcc cgttgccctg
 841 tgatgaagaa agcccagagg cagagctaca caaccatgga atccaaatta attcctgttc
 901 tgtgcgactg gtggatataa aaaaggaaaa gccatttttct aattcaaaag ttgagtgcca
 961 agcccaagca agaactcatc ataaccaggc atctgacata atagtcatca gcagtgagga
1021 ctctgaagga tccactgacg ttgatgagcc cttagaagtc ttcatctcag caccgagaag
1081 tgagcctgtg atcaataatg acaacccttt agaatcaaat gatgaaaagg agggccaaga
1141 agccacttgc tcacgacccc agattgtacc agagcccatg gatttcagaa aattatctac
1201 attcagagaa agttttaaga aaagagtgat aggacaagac cacgactttt cagaatccag
1261 tgaggaggag gcgcccgcag aagcctcaag cggggcactg agaagcaagc atggtgagaa
1321 ggctcctatg acttctagaa gtacatctac ttggagaata cccagcagga agagacgttt
1381 cagcagtagt gacttttcag acctgagtaa tggagaagag cttcaggaaa cctgcagctc
1441 atccctaaga agagggtcag gatcacagcc acaagaacct gaaaataaga agtgctcctg
1501 tgtcatgtgt tttccaaaag gtgtgccaag aagccaagaa gcaaggactg aaagtagtca
1561 agcatctgac atgatggata ccatggatgt tgaaaacaat tctactttgg aaaaacacag
1621 tgggaaaaga agaaaaaaga gaaggcatag atctaaagta aatggtctcc aaagagggag
1681 aaagaaagac agacctagaa aacatttaac tctgaataac aaagtccaaa agaaaagatg
1741 gcaacaaaga ggaagaaaag ccaacactag acctttgaaa agaagaagaa aaagaggtcc
1801 aagaattccc aaagatgaaa atattaattt taaacaatct gaacttcctg tgacctgtgg
1861 tgaggtgaag ggcactctat ataaggagcg attcaaacaa ggaacctcaa agaagtgtat
1921 acagagtgag gataaaaagt ggttcactcc cagggaattt gaaattgaag gagaccgcgg
1981 agcatccaag aactggaagc taagtatacg ctgcggtgga tataccctga aagtcctgat
2041 ggagaacaaa tttctgccag aaccaccaag cacaagaaaa aagagaatac tggaatctca
2101 caacaatacc ttagttgacc cttgtgagga gcataagaag aagaacccag atgcttcagt
2161 caagttctca gagtttttaa agaagtgctc agagacatgg aagaccattt ttgctaaaga
2221 gaaaggaaaa tttgaagata tggcaaaggc ggacaaggcc cattatgaaa gagaaatgaa
2281 aacctatatc cctcctaaag gggagaaaaa aaagaagttc aaggatccca atgcacccaa
2341 gaggcctcct ttggcctttt tcctgttctg ctctgagtat cgcccaaaaa tcaaaggaga
2401 acatcctggc ctgtccattg atgatgttgt gaagaaactg gcagggatgt ggaataacac
```

```
2461 cgctgcagct gacaagcagt tttatgaaaa gaaggctgca aagctgaagg aaaaatacaa
2521 aaaggatatt gctgcatatc gagctaaagg aaagcctaat tcagcaaaaa agagagttgt
2581 caaggctgaa aaaagcaaga aaaagaagga agaggaagaa gatgaagagg atgaacaaga
2641 ggaggaaaat gaagaagatg atgataaata agttgcttct agtgcagttt ttttcttgtc
2701 tataaagcat ttaagctgcc tgtacacaac tcactccttt taaagaaaaa aacttcaacg
2761 taagactgtg taagatttgt ttttaaaccg tacactgtgt ttttttgtat agttaaccac
2821 taccgaatgt gtcttcagat agccctgtcc tggtggtatt tagccactaa cctttgcctg
2881 gtacagtatg ggggttgtaa attggcatgg aaatttaaag caggttcttg ttagtgcaca
2941 gcacaaatta gttgtatagg aggatggtag ttttttcacc ttcagttgtc tctgatgtag
3001 cttatacaaa acatttgttg ttctgttaac tgaatgccac tctgtaattg caaaaaaaaa
3061 aaacagttgc agctgttttg ttgacattct gaatgcttct aagtaaatac aatttttaaa
3121 aaaccgtatg agggaactgt gtagacaagg taccaggtca gtcttcttcc atgttctatt
3181 agctccacaa agccaatctc aatccctcaa aacaatcttg tcatacttga aaatatgaca
3241 ctctagtcaa agccttggta aaataatcag tgtttccaat ctgtcctgtt acaaaagaaa
3301 cagattatta ttgaacttat gcaaataacc attgtcataa gaatgtttat gaatagtttc
3361 caaattatgg caaattcatg tagagagaga aaagtaactg ttttggtttt gctcacaaaa
3421 gtctacttta cctaagggct gtcagatata agtaacttaa aagaaagaga agtttttcttg
3481 actttttgaaa acaaaatatg aaaagaatcg gcaatgtttc aaacaaaaag tcataaaagt
3541 cactttattc ctccatcaaa aaaaaaaaaa aaaaaaaaa
```

A host of RNA molecules, acting in concert, decode this into a protein that is described like this:

```
MAGGGGDLSTRRLNECISPVANEMNHLPAHSHDLQRMFTEDQGV
DDRLLYDIVFKHFKRNKVEISNAIKKTFPFLEGLRDRDLITNKMFEDSQDSCRNLVPV
QRVVYNVLSELEKTFNLPVLEALFSDVNMQEYPDLIHIYKGFENVIHDKLPLQESEEE
EREERSGLQLSLEQGTGENSFRSLTWPPSGSPSHAGTTPPENGLSEHPCETEQINAKR
KDTTSDKDDSLGSQQTNEQCAQKAEPTESCEQIAVQVNNGDAGREMPCPLPCDEESPE
AELHNHGIQINSCSVRLVDIKKEKPFSNSKVECQAQARTHHNQASDIIVISSEDSEGS
TDVDEPLEVFISAPRSEPVINNDNPLESNDEKEGQEATCSRPQIVPEPMDFRKLSTFR
ESFKKRVIGQDHDFSESSEEEAPAEASSGALRSKHGEKAPMTSRSTSTWRIPSRKRRF
SSSDFSDLSNGEELQETCSSSLRRGSGSQPQEPENKKCSCVMCFPKGVPRSQEARTES
SQASDMMDTMDVENNSTLEKHSGKRRKKRRHRSKVNGLQRGRKKDRPRKHLTLNNKVQ
KKRWQQRGRKANTRPLKRRRKRGPRIPKDENINFKQSELPVTCGEVKGTLYKERFKQG
TSKKCIQSEDKKWFTPREFEIEGDRGASKNWKLSIRCGGYTLKVLMENKFLPEPPSTR
KKRILESHNNTLVDPCEEHKKKNPDASVKFSEFLKKCSETWKTIFAKEKGKFEDMAKA
DKAHYEREMKTYIPPKGEKKKKFKDPNAPKRPPLAFFLFCSEYRPKIKGEHPGLSIDD
VVKKLAGMWNNTAAADKQFYEKKAAKLKEKYKKDIAAYRAKGKPNSAKKRVVKAEKSK
KKKEEEEDEEDEQEEENEEDDDK
```

Wherein each letter denotes one of the 20 amino acids that compose a protein: a long, stringy molecule, which folds up into a 3D structure, which then travels around in the cell in which it was created. According to our software, this particular one does something related to DNA replication. What exactly it does is unknown at the time we're writing these words—but perhaps by the time you're reading this, someone will have followed up on the hint provided by the Biomind software and discovered its specific function.

FINDING IMPORTANT INTERACTIONS

As another concrete example of the kind of insight AI brings to biological data analysis—and also the shortcomings of AI at its present stage of development—we'll now show you some data analysis results obtained using the Biomind ArrayGenius software. This part of the chapter may be a bit too technical if your biology isn't so strong—if so, feel free to skip or skim it over.

What was done to get these results was to feed the software some microarray data obtained from the brain cells of people of various ages. This data was made available online by the researchers who did the study (we didn't do the biology). The first question posed to the ArrayGenius software was to learn rules distinguishing young people's brain cells from old people's brain cells, based on the differences in gene expression. This is pretty easy to do, because there are a lot of differences between young brains and aging ones. The software found a large number of good rules, with a lot of variety among them. As was expected, there are many different rules, involving a lot of different combinations of genes that can carry out this particular classification task (young vs. old brain cells).

Because the Biomind software uses an integrated database to make its judgments, the rules it uses to distinguish old from young brains based on gene expression don't necessarily involve gene expression directly. They may involve the software's estimation of the activity level of some biological process, or of the overall expression of genes corresponding to proteins in some particular category. For instance the Gene Ontology database contains a category called "oxygen transporter activity," which corresponds to genes coding for a number of proteins concerned with transporting oxygen. The Biomind software can estimate the overall activity of genes in this category, and then use this overall activity as part of a rule for distinguishing young brains from old. The rules themselves tend to be somewhat complicated mathematical expressions involving the expression levels of genes and the activity levels of ontological categories of genes and proteins.

All this is quite complex, but if pursued a little further, this sort of analysis leads to some pretty simple conclusions. The next question asked of the software was: which genes, and which biological processes and protein families, occur most often in the rules distinguishing young from old? These most frequent ones are the genes/processes/families that are most important for distinguishing young brain cells from old—but what's fascinating is that these are not generally the same ones whose expression levels, taken individually, are most different between young and old brain cells. You can have a gene that, on average, is expressed almost the same amount in young and old brain cells—and yet it may occur in a huge percentage of rules for distinguishing young from old brain cells, because the way its expression level coordinates with the expression levels of other genes is very different in young versus old brain cells. The following table gives some examples of genes, protein families and biological processes, showing their rank in the list of "useful" features (ones that occur in a lot of classification rules) versus their rank the list of features that are highly differentially expressed between the "young" and "old" categories.

Feature ID	**Usefulness Rank**	**Variation Rank**	**Description**
SF036518	1	74	globin
GO:0030449	2	533	regulation of complement activation
GO:0045917	3	536	positive regulation of complement activation
NM_006288	4	63	Homo sapiens Thy-1 cell surface antigen (THY1)
GO:0048143	5	530	astrocyte activation
GO:0019220	6	528	regulation of phosphate metabolism
FAM0002061	7	73	

953_g_at	8	634	Fk506-Binding Protein, Alt. Splice 2
33508_at	9	9072	Homo sapiens cDNA FLJ12376 fis, clone MAMMA1002494
FAM0002013	10	65	
GO:0001515	11	58	opioid peptide activity
GO:0048169	12	529	regulation of long-term neuronal synaptic plasticity
NM_006272	13	534	Homo sapiens S100 calcium binding protein
31503_at	14	230	Human retina cDNA randomly primed sublibrary Homo sapiens cDNA, mRNA sequence
NM_057161	15	227	Homo sapiens kelch domain containing 3 (KLHDC3)
31608_g_at	16	344	voltage-dependent anion channel 1

Table 7.2 Examples of genes, protein families and biological processes differentially expressed between the "young" and "old" categories.

Another interesting question is: Which genes/processes/families tend to occur together in the same rules? There are a lot of different mathematical equations for distinguishing young from old brain cells in this gene expression data set, and each equation involves a different combination of genes, processes and protein families. If two entities tend to often be combined together the same predictive equation (rule), this indicates they have some interaction with each other that is meaningful in the context of the young/old distinction. Similar tables to the one above can be produced embodying this information.

Clearly, here the AI is doing something humans can't viably do—creating a massive number of classification rules (via sophisticated "machine learning" algorithms for rule learning) and analyzing them statistically. Still, it's not exactly spitting out the cure for aging here. What it's doing is giving a list of processes, genes and proteins that are apparently important to the process of aging in the brain. As a consequence of this analysis, we know that regulation of long-term neuronal synaptic plasticity is a significant factor in the difference between young and old brains. This is interesting because synaptic plasticity is the most likely mechanism for learning in the brain. We didn't know this before doing the AI-based analysis—because this particular Gene Ontology category is ranked 529th in terms of difference in expression, but 12th in terms of usefulness in classification models. This means that genes regulating long-term neuronal synaptic plasticity are expressed only moderately differently in young versus old brains—but, the interactions of these genes with other genes are quite substantially different in the two cases. The discovery shown here begs the question: which genes are the synaptic plasticity regulator genes interacting with so differently over time? We've now let the computer do massive amounts of number crunching and data mining to lead us to a question which can get the biologist's mind off-and-running, rather than mired in these tedious tasks. There's plenty to investigate, plenty of relevant papers to read, and plenty more experiments to run. The AI doesn't finish the job, but it makes novel and useful suggestions based on the insights its algorithms provide, which the human brain would never be capable of achieving based on the data provided.

Now, imagine how much insight we'd have if we actually had an AI combining human insight with the statistical and inductive power of contemporary machine learning algorithms. Maybe we'd even conquer aging or figure out neural uploading—the possibilities for AI computational power combined with human insight and intuition (in whatever form that combination takes) are practically limitless.

BIOKNOWLEDGE SINGULARITIES?

All this talk about gene expression data and the like is just *barely scratching the surface* of 21st century biological possibilities. What we're really moving towards is what biologists call *systems biology*—an understanding of the self-organizing wholeness of biological systems like cells, organs and organisms. Sequencing genomes is one step in this direction, gathering gene expression data is another, gathering protein expression data is another (still more nascent, at this time), and yet further technologies will no doubt bring us yet more insightful data as time goes on. Today biology is still largely a science of disorganized, loosely connected factoids; and vague, subtle, intuitive trends. But, there may well be a mini-Singularity of biological knowledge that occurs when all this micro-level data, fed into a BiomindDB or some similar integrative AI system, is enough to bridge the micro-macro gap. From this may emerge the first set of theorems of a unified theory of biology which is, currently, almost completely in the domain of fantasy. (We have the process of evolution, the taxonomy system of Linnaeus and Whitaker, and some specifics of macro-level function of a variety of organisms—but in terms of a general unifying theory of biology it's difficult to even come to a complete consensus on what is life and therefore even covered by biological science.)

Specifically, one may foresee two coming singularities of bioknowledge. The first will be when analysis of gene and protein expression, pathways, regulatory networks and related phenomena allow us to really understand the cell. The gap between the "macro" level of the cell and the "micro" level of its component molecules will be closed. The properties of cells will be concretely comprehensible as emergent patterns from molecular dynamics. We're certainly not at this point yet, but it could well happen in the next 5 years via Biomind technology or something similar.

The second bioknowledge singularity will be when cellular knowledge comes together to form organismic knowledge. Our intuition, for what it's worth in this domain, is that this singularity will actually be easier than the first one. The step from cells to organisms is a tough one, but it may not be quite as perversely subtle as the step from molecules to cells—because it all happens at a relatively macroscopic scale, where things are easier to measure and to intuitively understand.

Once we've achieved these, we will be able to make a variety of major positive changes to our own biology (and negative ones, too, if we choose to follow the path of destruction rather than of positive transcension). With knowledge of how to analyze and improve upon biological functionality all the way down to the protein level, we'll be able to discover causes of senescence (disease, damage, wearing-out) which we never even knew existed, and cure them. We'll also be able to genetically enhance our abilities in perception, locomotion and longevity—and, with biomechanical engineering, give ourselves greater ability to perceive and interact with our world.

AI will lead us to these singularities, hand in hand with human biologists, and it's going to be quite an adventure. It sounds ironic and in some ways a bit peculiar, but our guess is that one of the major roles of nonhuman and later superhuman digital intelligence—as the path to posthumanity unfolds—will be to help understand the subtle emergent properties of the fabulously complex molecular machinery that makes us humans what we are, and to help us figure out how to make this machinery work better, ridding us of pesky little problems like cancer and aging. This is what biology needs, and in a less direct sense it may be what AI needs as well. We cannot think of a better way to encourage cooperation between humans and AIs, and to nudge incipient superintelligent AIs to think of human lives as valuable, than to engage both humans and AIs in a cooperative effort to make all people able to live longer, healthier and more fulfilling lives.

---

[1] Although the human producing machines we do have—other humans—appear to have no trouble making more humans, they can't replicate a human as exactly as a human cloning machine could.
[2] http://www.ornl.gov/sci/techresources/Human_Genome/home.shtml
[3]http://www.ncseweb.org/resources/rncse_content/vol24/9747_the_creationists_how_many_wh_12_30_1899.asp
[4] See also http://www.tigr.org/minimal/
[5] http://cmgm.stanford.edu/pbrown/sporulation/additional/spospread.txt
[6] http://www.rii.com
[7] http://genome.ucsc.edu/goldenPath/hgTracks.html
[8] http://www.geneontology.org
[9] See the authors' papers on Experiential Interactive Learning.
[10] http://ondemand.biomind.com

CHAPTER 8

THE QUEST FOR PHARMACOLOGICAL IMMORTALITY

Nothing sounds more unscientific, more starry-eyed and unrealistic, than the quest for immortality. Yet, as each year passes, a greater percentage of the very down-to-earth pharmaceutical industry is devoted to precisely this pursuit. Billions of dollars are spent on drug research each year in the quest to beat death, to halt the apparently—but not necessarily—inevitable decline of the human mind and body with the progress of time.

Of course, the goal of eluding death long predates science. It has taken hundreds of forms throughout history, pervading all cultures and eras. The ancient Chinese, for instance, had Taoist Yoga, a very complex discipline defining a life-long series of practices that, if adhered to precisely, purportedly resulted in physical immortality. Part of this teaching was that by refraining from ejaculation for his entire life, a man could store his "essential energy" in a space by the top of his head, until he accumulated enough to create an eternal fetus that would grow into his deathless self. The modern variation of such ideas is less colorful but perhaps more likely to succeed: subtle biological and pharmaceutical research, aimed at discovering the roots of aging and creating chemical remedies.

The increasing number of elderly is one of the major trends in current demography. For instance, in 1950, only one in ten U.S. citizens was over the age of 65. Now the figure is about one in 8; and by 2030 it may be 1 in 5. There's little doubt that the current human age record of 122 (Madame Jeanne Calment, who died in 1997) will soon be overturned. This ongoing increase in lifespan has been primarily due to a reduction in various deadly diseases, resulting from improved hygiene and medical care.

One thing our success at reducing disease has taught us is this: the real killer is not disease, but senescence. After a certain amount of time, the body's microscopic parts just stop working of their own accord, without the interference of any germs or viruses or other external influences. This is the essential dark side of the human condition, and the focus of current scientific work on anti-aging. Scientists have a "short list" of biological and biochemical factors suspected to collectively underlie aging—and for each of these likely culprits, there is a pharma firm or a maverick scientist working on the cure[1]. It is plausible that within decades—not centuries or millennia—pharmacological science will have made the very concept of getting old obsolete.

THE BIOLOGY OF CELL SENESCENCE

A healthy body is not a constant pool of cells, but rather a hotbed of continual cellular reproduction. There are only a few exceptions such as nerve cells which do not reproduce, but simply persist throughout an organism's lifespan, slowly dying off[2]. In youth, newly formed cells outnumber dying cells; but then from about 25 on things begin to go downhill, and the number of newly formed cells is less than the number of cells that die. Little by little, bit by bit, cells just stop reproducing.

The sad fact is that most types of human cells have a natural limit to the number of cell divisions they will undergo. This number, usually around 50 or so, is called the Hayflick limit, named after Leonard Hayflick, the researcher who discovered it in the mid-1960s. Once a cell's Hayflick limit is reached, the cell becomes senescent, and eventually it dies.

This may have the sound of inevitability about it—but things start to sound different when one takes a look at our one-celled cousins, such as amoebas and paramecia. These creatures reproduce asexually by dividing into two equal halves—neither half sensibly classified as "parent" or "child." This means that essentially, the amoebas alive today are the same ones alive billions of years ago.

These fellows qualified for Social Security when you could still walk from New York to Casablanca, and yet they're still alive today, apparently not having aged one bit—cells untroubled by the Hayflick limit. This nasty business of aging seems to have come along with multicellularity and sexual reproduction—a fascinating twist on the "sex and death" connection that has fascinated so many poets and artists.

Unlike in asexually-reproducing creatures, cells in multicellular organisms fall into two categories: *germ-line* cells which become sperm or egg for the next generation; and *soma* cells that make up the body. The soma cells are the ones that die, and the standard answer to "Why?" is "Why not?" The "disposable soma theory" argues that, in fact, our soma cells die because it's of no value to our DNA to have them keep living forever. It is, in essence, a waste of energy. Throughout most of the history of macroscopic, sexually-reproducing organisms, immortal organisms would not have had an evolutionary advantage. Rather, there was an evolutionary pressure toward organisms that could evolve faster. If a species is going to evolve rapidly, it's valuable for it to have a relatively quick turnover from one generation to the next.

There doesn't seem to be any single cellular "grim reaper" process causing soma cell senescence. Rather, it would appear that there several distinct mechanisms, all acting in parallel and in concert.

There are junk molecules, accumulating inside and outside of cells, simply clogging up the works. Then there are various chemical modifications that impair the functioning of molecular components such as DNA, enzymes, membranes and proteins. Of all these chemical reactions, oxidation has attracted the most attention, and various anti-oxidant substances are on the market as potential aging remedies. Another major chemical culprit is "cross-linking," the occasional formation of unwanted bridges between protein molecules in the DNA—bridges which cannot be broken by the cell repair enzymes, interfering in the production of RNA by DNA. Cross-linkages in protein and DNA can be caused by many chemicals normally present in cells as a result of metabolism, and also by common pollutants such as lead and tobacco smoke.

As time passes, signaling pathways and genetic regulatory networks within cells can be altered for the worse, due to subtle changes in cellular chemistry. The repair mechanisms that would normally correct such errors appear to slow down over time. "Telomeres," the ends of chromosomes, seem to get shorter each time a cell divides, causing normally suppressed genes to become activated and impair cell function. Finally, the brain processes that regulate organism-wide cell behavior decline over time, partly as a result of ongoing cell death in the brain.

The really frustrating thing about all these phenomena (and some other, related ones we'll review a little later on) is that none of them are terribly different from other processes that naturally occur within cells, and which cells seem to know quite well how to cure and repair. It would seem that cells have just never bothered to learn how to solve these particular problems that arise through aging, because there was never any big evolutionary advantage to doing so. We may well die not because it would be so terribly hard to engineer immortal cells, but because it was not evolutionarily useful to our DNA to allow us to live forever.

CHASING THE IMMORTALITY PILL

Curing old age *with drugs* is one kind of speculative research that modern capitalist society seems relatively willing to fund. For instance, Larry Ellison, the controversial, nearly 60-year-old chief executive of Oracle, is the largest single supporter of anti-aging research, with $20 million per year committed[3]. He may be among the richest men in the world, but he's smart enough to realize that "you can't take it with you"—and he's deploying his wealth strategically with this in mind. He makes no bones about his motivations. "Death has never made any sense to me," he says. "How can a person be there and then just vanish, just not be there?...Death makes me very angry. Premature death makes me angrier still."

Much of the funding for anti-aging research comes not from immortality-obsessed visionaries but from stolid biotech firms concerned with curing particular diseases. As it turns out, most of the factors underlying aging are also connected to various particular medical conditions. Indeed, some medical conditions such as cancer are starting to be reconsidered—are they the signs of aging, and the "disease" appearing in younger people the result of premature aging through exposure to environmental toxins which accelerate cellular senescence and mutation? This dual focus drives the R&D of dozens of pharma firms. For instance, Centaur Pharmaceuticals discovers and develops new drugs for various diseases involving ischemia and inflammation, which its scientists believe will also have general anti-aging properties. Geron focuses on telomere shortening and cell death, with applications to both anti-aging and cancer. Human Genome Sciences works on understanding signal transduction pathways—how they work, why they fail and how to repair and redirect them—a quest which, if successful, will have myriad applications beyond "just" conquering senescence.

Among the most advanced work in the field is that of a company called Alteon Inc. Alteon has picked up a train of research, begun in the early 1900s, regarding the formation of complexes between sugars and the amino acids of proteins. At first these complexes were found to cause the toughening and discoloration of food, observed during the cooking process and after prolonged storage. It was later determined that these same structures were part of a new biochemical pathway in which permanent glucose structures were formed on the surface of proteins. These structures—"Advanced Glycosylation Endproducts" or AGEs—were seen to interact with adjacent proteins to form pathological links between proteins, called AGE crosslinks. Such crosslinks appear to play a critical role in diabetes, as well as in the Hayflick limit of various human cells[4]. Alteon is currently testing medication that promises to prevent this crosslinking from occurring.

LONG LIFE THROUGH CALORIC RESTRICTION

None of these drugs are on the market yet—none have even begun human trials. If one wants to live as long as possible, what can one do right now?

Another interesting, immediately applicable idea from anti-aging research has to do not with pills but with *caloric restriction.* A number of scientists have come forward with the idea that if you eat about 70% of what you'd ordinarily want, you'll live longer. You need to eat a healthy diet, rich in vitamins and proteins, but low in calories.

This has been tested extensively in various nonhuman mammals. For instance, mice normally don't live over 39 months, but caloric restriction has produced mice with 56 months lifespan. This corresponds proportionally to a 158 year-old human. And these long-lived mice aren't old and crusty—they're oldster/youngsters, keen-minded, strong-bodied and healthy. Studies on monkeys are currently underway, though this naturally will take a while due to monkeys' relatively long lives.

Why does caloric restriction work? It increases the ability of the body to repair damaged DNA, and it decreases the amount of oxidative (free radical) damage in the body. It increases the levels of repair proteins that respond to stress, it improves glucose-insulin metabolism, and for some reason not fully understood, it delays age-related immunological decline as well. Basically, many of the well-known mechanisms of senescence set in more slowly if the body has to process less food over its lifetime. Of course, it's not yet demonstrated that caloric restriction will do for humans what it's done for other animals, but none of the researchers involved with the work seem to have much doubt. The relation of this line of thinking with anti-aging pharmacology has yet to be investigated—it may well be there are medications that work most effectively in coordination with a caloric restriction diet.

The caloric restriction work looks highly believable, but we have to admit we don't have the willpower to try it out ourselves. We're not fat people, but you can't see our ribs either. We enjoy eating too much to starve ourselves in hopes of living longer. No one's sure how much good severe caloric restriction would do us anyway. We both grew up in the Eastern European tradition of being told that eating was very good for you. If we started out in our 30s, as we are now, what good would it do anyway? Is it too late? If you have more willpower than we do in this regard, and you try it out yourself and reach 150 or so, please send us an e-mail and let us know!

And before you start restricting your calories too heavily, you should know that not all visionary anti-aging researchers believe it will be as effective in humans as in mice. Aubrey de Grey, whose work we'll discuss shortly, takes a contrarian point of view: he thinks that the larger the organism, the less the effect of caloric restriction. So it increases the lifespan of nematode worms a lot, mice a fair bit, dogs only a little, and humans—in his hypothesis—by just a few years. Whether Aubrey or the caloric-restriction boosters are right is hard to say, based on the available data. Diet research is notable primarily for the sheer number of times one field can be wrong in any given generation.

AGING AND DNA DAMAGE

The two most interesting anti-aging visionaries we know are Robert Bradbury and Aubrey de Grey. Their approaches and concepts are quite different, yet they share a belief that the mainstream anti-aging research community is proceeding unacceptably slowly, due to its tendency to narrowly focus on particular genes causing particular, highly narrow aging-related problems. Both of them aim at more of a high-level, systems-based point of view and would like to see research orchestrated in a manner guided by a deeper system-level understanding. We'll return to de Grey's work in a couple of pages. For the moment we'll just give a quick summary of Bradbury's ideas, because they tie in closely with Biomind's work on Parkinson's disease (which we will also discuss).

Bradbury's basic idea is that the key to understanding aging may lie in a recursive process of DNA degradation[5]. The DNA in each of our cells accumulates mutations as we live, and if these were never repaired, we'd die a lot younger than we do. So, cells contain DNA repair processes that fix broken DNA. But what happens when the DNA repair processes themselves get screwed up, via damage to the DNA that controls them? Then, as one would expect, things go downhill pretty fast.

As an overall theory of senescence, this is overly simplistic (as we've stated it here—Bradbury's theory is more nuanced). Still it may have a great deal of explanatory power, in the sense that many of the other aspects of senescence might not happen if DNA repair processes continued to function properly. Degeneration of DNA repair processes may have a lot of non-obvious side effects.

At the moment, Bradbury is largely out of the anti-aging game—his company Aeiveos, formerly funded by Larry Ellison, is effectively non-functional at the moment; and after a few years' effort, he gave up seeking funding for his startup firm Robobiotics, aimed at the creation of nanotechnology to repair DNA damage. However, ideas similar to Bradbury's continue to emerge in the ongoing biological research of others. Ben's work with Biomind has touched on similar themes, in the context of some joint work Biomind did with the University of Virginia on the genetic roots of Parkinson's disease.

MITOCHONDRIAL DNA DAMAGE AND PARKINSON'S DISEASE

As we write these words in early 2005, the news is full of the death of Pope John Paul II—so for once, we're finding our research is topical and popular. (At least, our bioinformatics research is; unlike our AGI research which is hopelessly out of style, and presumably will remain so until we or someone else succeeds in creating a reasonably intelligent AGI.)

The former Pope suffered from Parkinson's disease, and he was far from alone in this plight. Over a million Americans have Parkinson's, and many more suffer worldwide. Yet in spite of years of effort by medical researchers, tracking down the genetic roots of the disorder has proved devilishly difficult. It seems that, over the last couple of years, a combination of human and artificial intelligence may have taken a large step in the right direction—and the Biomind team has been privileged to play a role in this process of discovery.

Biomind, in collaboration with Dr. Davis Parker and his colleagues at the University of Virginia, have applied Novamente-based AI technology to analyze data regarding mutations in mitochondrial DNA. As a result, we have pinpointed a particular region of a particular gene on the mitochondrial genome that appears to be strongly associated with Parkinson's disease[6].

Parkinson's disease is a progressive disorder of the central nervous system, whose well-known victims—besides the late Pope—include boxer Muhammad Ali and comedian Richard Pryor. Most common in people over age 45, its symptoms include rigidity, tremors, difficulty with balance and posture, and slowing of movements. Various treatments have been found that lessen the effects of the symptoms, but there is no known cure.

The approach our team has taken for analyzing Parkinson's also has implications for other neurodegenerative diseases. For instance, it seems reasonably likely that applying the same techniques to data obtained from Alzheimer's patients will yield analogous results. From a general anti-aging perspective, what it does is allow scientists to develop an understanding of exactly which kinds of DNA damage lead to exactly which kinds of bad cellular- and organism-level effects. This alone doesn't cure aging or Parkinson's or any other malady, but it builds the kind of understanding that we believe is needed to give rise to cures.

More speculatively, this work may eventually lead to a cure via radical forms of gene therapy. Dr. Rafal Smigrodzi, a key member of the University of Virginia team underlying our recent discovery, has left UVA to join GENCIA Corporation, a Charlottesville, Virginia firm developing "mitochondrial gene replacement therapy" via a novel technique called protofection. Protofection allows the removal of bad fragments of mitochondrial DNA, and their replacement by good ones. If this could be successfully done in living human brains, then Parkinson's could potentially be cured via gene therapy that simply repairs the flawed regions of mitochondrial DNA located by Biomind's software.

The story leading up to these recent discoveries is a long and winding one—beginning with a bad batch of heroin, followed by a generous amount of scientific ingenuity, and a synergistic application of human intelligence with advanced artificial intelligence algorithms based on simulation of the evolutionary process.

Parkinson's is rooted in nerve cells—and is particularly associated with degeneration of nerve cells in the area of the brain governing movement. Thus, if one wants to understand the causes and development of the disease and ultimately work toward a cure, the obvious place to look is in the genetic material of nerve cells. But the quest for the neurological genetic defects underlying Parkinson's has proven more difficult than anticipated. Scientific discovery comes from multiple sources—human insight, advanced technology, and just plain luck. The identification of the genetic defects underlying Parkinson's displayed all three.

The lucky break occurred in 1982—a case where bad luck for drug users was good luck for neuroscientists. Street chemists attempting to create heroin accidentally created a substance called MPTP, which interferes with the function of mitochondria (the cellular components concerned with energy production, often known as the "powerhouses" of cells). Drug users who accidentally took MPTP instead of heroin displayed Parkinson's-like symptoms, including being almost completely unable to move.

What MPTP does is to inhibit the production of a mitochondrial complex called "Complex I" within nerve cells in the substantia negra, the brain region responsible for coordinating movement. Complex I contains many enzymes responsible for reducing a molecule called ubiquinol, which is critical to mitochondrial function. Without adequate production of Complex I enzymes, cells can't properly produce energy. Movement-controlling neurons without enough Complex I activity in their mitochondria can't effectively control movement.

In the words of Dr. Davis Parker, of the University of Virginia—the leader of the UVA component of the research effort that produced the recent discovery, and a visionary of the mitochondria/Parkinson's connection for close to two decades—"It was a very important advance in neuroscience done in the name of illicit drugs." Analysis of the effects of MPTP on drug users led Dr. Parker to an intriguing hunch: if mitochondrial dysfunction yields Parkinson's-like symptoms, then maybe the roots of Parkinson's disease lie in the mitochondrial genome.

The DNA one usually hears about lies in the nucleus of a cell, the cell's center. But mitochondria, the cell's energy-producing engines, also contain a small amount of DNA. The human mitochondrial genome only contains 37 genes

(13 of them protein-coding), whereas the nuclear genome contains around 25,000 (at last count); but these 37 genes, inherited from the prehistoric time in which mitochondria were independent bacterial organisms, carry out a lot of valuable functions. If they stop working properly, serious problems can ensue. In 1999, Dr. Parker and Dr. Russell Swerdlow, together with scientists from the San Diego firm MitoKor, published work suggesting that defects in the mitochondrial genome may be correlated with Parkinson's disease. Their specific interest was in the 7 mitochondrial genes underlying Complex I.

This result has interesting implications for the heredity of Parkinson's. When a baby is conceived, its nuclear DNA is formed by combining nuclear DNA from its mother and its father. Baby's mitochondrial DNA, on the other hand, comes entirely from mom. Thus, these results suggest that Parkinson's may be passed maternally—but that its defects can skip generations, making the disease appear random.

The work MitoKor did, under Parker and Swerdlow's direction, involved clever manipulations of embryonic human nerve cells. They removed the mitochondrial DNA from the embryonic nerve cells and replaced it with other DNA: sometimes from healthy people and sometimes from Parkinson's patients. What happened was that the nerve cells receiving the mitochondrial DNA from Parkinson's patients started acting like nerve cells on MPTP. They exhibited low Complex I activity, meaning insufficient energy obtained from mitochondria, which in an affected person eventually leads to Parkinson's-like sluggishness.

These results were fascinating and suggestive—but where were the actual mutations? All this 1999 work showed was that the problem lay somewhere in the mitochondrial genome. The question was, where? Which mutations caused the problem?

To answer this question, Parker and colleagues sequenced mitochondrial DNA drawn from the nerve cells of a number of Parkinson's patients, as well as a number of normal individuals, and looked for patterns. To their surprise, when they set about seriously analyzing this data in 2003, they found no simple, consistent pattern. There were no specific genetic mutations common to the Parkinson's patients and not the normal people.

Enter artificial intelligence. Dr. Rafal Smigrodzki, one of Parker's collaborators, was familiar with the Novamente artificial intelligence work due to his preexisting general interest in AI advanced technology and futurism. Perhaps, Rafal suggested, Biomind technology might be able to find the subtle patterns hidden in the mitochondrial DNA data.

Biomind had already been applying advanced AI techniques to biomedical data since 2001, with a primary focus on a different sort of data—"gene expression" data, gathered from devices known as microarrays that measure how much activity each gene is displaying while a cell is functioning. However, building upon this gene expression analysis experience, it seemed likely that we could brew up some custom software capable of solving the Parkinson's problem.

Appropriately enough, the solution turned out to be a software technique called "genetic algorithms" that simulates the process of evolution by natural selection. A genetic programming algorithm starts with a population of random solutions to a problem, then gradually "evolves" better solutions via letting the "fittest" solutions combine with each other to form new ones, and making small "mutations" to the fittest solutions. The programmers seed the software with a set of solutions to the problem, and their code then combines and mutates (alters) the solutions in various ways, and checks the resulting solutions against a fitness metric—a piece of code which allows a statistical measure of how good a solution the one being checked seems to be. In this case, what the software was "evolving" were potential patterns distinguishing Parkinson's patients from healthy people based on the sequences of amino acids in their mitochondrial DNA. This kind of data analysis is highly exploratory and is never guaranteed to yield a solution—but in this case things worked out happily, and a variety of different data patterns were discovered.

The trick, it turns out, is that while there are no specific mutations corresponding to Parkinson's disease, there are regions of the mitochondrial genome and combinations thereof that tend to be mutated in Parkinson's patients.

There are many different rules of the form: "If there are mutations in this region of this mitochondrial gene and that region of that mitochondrial gene, then the person probably has Parkinson's disease." While it took some advanced AI technology to find these patterns, once discovered the patterns are very easy for non-computer-savvy biologists to understand. The computer did the "heavy lifting" of mining through a tremendous variety of possible gene region combinations and checking their likelihood to correlate to Parkinson's (that was the fitness metric), and then human biologists verified this data using their knowledge of the biology underlying such phenomena.

The AI technology was written up in a technical journal article, which will appear in 2005 in the Journal of Artificial Intelligence in Medicine, but from the point of view of biologists seeking causes and cures for Parkinson's disease, the technical details don't matter. Dr. Parker and his colleague Dr. Janice Parks have already published a paper in Biochemical and Biophysical Research Communications, showing that mutations in a certain narrow region of ND5 (a mitochondrial gene encoding a Complex I subunit) correctly predict whether a person has Parkinson's disease with 94% accuracy. To discuss these milestone results and others, in June 2005 Dr. Parker convened the First International Brainstorming Conference on Parkinson's Disease in Louisville, Kentucky.

These events illustrate the role that artificial intelligence technology has come to play in modern biomedical research. The AI currently used in this context is not "general AI" with autonomy and self-awareness (such as we're ultimately trying to create in the Novamente project) but rather "narrow AI" which is extremely intelligent at carrying out some particular task—like finding patterns in genomic data. Far from displacing human scientists, AI is serving as IA—intelligence augmentation. With its superior capability to integrate data and detect patterns, AI software helps humans understand the data they've collected. People design the experiments based on their intuitive insights into the underlying science, and then the system helps mine out meaningful patterns using a fitness metric which is also provided by the scientists. The results produced by the AI must then be interpreted by humans—and oftentimes may be significantly simplified, as in Dr. Parker's selection of a single, simple pattern from the many complex patterns suggested by our Biomind AI algorithms.

A cure for Parkinson's is probably still a fair ways off, but there's no doubt that significant progress has been made in understanding its underlying mechanisms. There's reason to believe similar mechanisms may underlie other neurodegenerative diseases—for example, Alzheimer's. If Bradbury is right that DNA damage is a key aspect of senescence, then gaining this understanding of the specific degenerative consequences of specific types of DNA damage will help us along the path to, at least, extreme longevity. By deploying a combination of human and computational intelligence, we'll get there—and though it's never soon enough, it may be sooner than you think.

AUBREY DE GREY AND THE SEVEN PILLARS OF SENESCENCE

Of all the senescence researchers in the world, no other has done as much as Aubrey de Grey to improve our understanding of the overall picture of the phenomenon of aging. We don't always agree with his proposed solutions to particular sub-problems of aging, but we do find him invariably energetic, rational and insightful. Although he says he's not a big booster of caloric restriction for humans, because he thinks its effect diminishes rapidly with the size of the organism, he's also one of the skinniest humans we've ever seen. He gives the appearance of being robustly healthy, so we suspect he's practicing some variant of the caloric restriction diet.

De Grey's buzzword is SENS, which stands for Strategies for Engineered Negligible Senescence—a very carefully constructed scientific phrasing for what we've loosely been calling here "anti-aging research." The point of the term is that it's not merely slowing down of aging that we're after—it's the reduction of senescence to a negligible level. We're not trying to achieve this goal via voodoo, we're trying to achieve it via engineering—mostly biological engineering, though nanoengineering is also a possibility, as in Bradbury's "robobiotics" idea. De Grey's website[7] gives a very nice overview of his ideas, together with a number of references.

As part of his effort to energize the biology research community about SENS, de Grey has launched a contest called the "Methuselah mouse prize"—a prize that yields money to the researcher that produces the longest-lived mouse of species *Mus musculus*. In fact there are two sub-prizes: one for longevity, and a "rejuvenation" prize, given to the best life-extension therapy that's applicable to an already partially-aged mouse. There is a complicated prize structure, wherein each researcher who produces the longest-lived mouse ever or the best-ever mouse-lifespan rejuvenation therapy receives a bit of money each week until his record is broken.

De Grey's idea is that, during the next decade or so, it should be possible to come pretty close to defeating senescence within mice—if the research community puts enough focus on the area. Then, porting the results from mouse to human shouldn't take all that much longer (biological research is regularly ported from mice to humans, as they are an unusually suitable testbed for human therapies—though obviously far from a perfect match). Of course, some techniques will port more easily than others, and unforeseen difficulties may arise. However, if we manage to extend human lives by 30 or 40 years via partly solving the problem of aging, then we'll have 30 or 40 extra years in which to help biologists solve the other problems.

Theory-wise, de Grey agrees with the perspective we've given above—he doesn't believe there's one grand root cause of senescence, but rather that it's the result of a whole bunch of different things going wrong, mainly because human DNA did not evolve in such a way as to make them not go wrong. On his website he gives the table shown below of the seven causes of senescence, showing for each one the date that the connection between this phenomenon and senescence first become well-known to biologists—and also showing, for each one, the biological mechanism that he believes will be helpful for eliminating that particular cause.

Seven basic causes—are these really all there is? De Grey opines that, "The fact that we have not discovered another major category of even potentially pathogenic damage accumulating with age in two decades, despite so tremendous an improvement in our analytical techniques over that period, strongly suggests that no more are to be found—at least, none that would kill us in a presently normal lifetime." Let's hope he's right, though of course the possibility exists that as we live longer new effects will be discovered—but if we've put enough resources into the anti-aging program, we should be able to combat those as well.

Damage (rising with age)	**Date**	**Reversible or obviatable by**
Cell loss, cell atrophy	1955	Stem cells, growth factors, exercise
Nuclear [epi]mutations(only cancer matters)	1959/1982	WILT (Whole-body Interdictionof Lengthening of Telomeres)
Mutant mitochondria	1972	Allotopic expression of 13 proteins
Cell senescence	1965	Ablation of unwanted cells
Extracellular crosslinks	1981	AGE-breaking molecules/enzymes
Extracellular junk	1907	Phagocytosis; beta-breakers
Intracellular junk	1959	Transgenic microbial hydrolases

Table 8.1 De Grey's Seven Pillars of Aging

De Grey's particular breakdown into seven causes is slightly arbitrary in some ways, and others would do the breakdown differently; but his attempt to impose a global order on the panoply of aging-related biological disasters is appealing, and reflects a lot of deep thinking and information integration.

One of these "Seven Pillars of Aging" should be familiar from the previous section on mitochondrial DNA and Parkinson's disease: mutant mitochondria. A deeper look at this case is interesting for what it reveals about the strength and potential weaknesses of de Grey's "engineering"-based approach. The term "engineering" in the SENS acronym is not a coincidence—de Grey came to biology from computer science, and he tends to take a different approach from conventional biologists, thinking more in terms of "mechanical" repair solutions. Whether his approach will prove the best or not remains to be seen. We're not biologists enough to have a strong general intuition on this point, but it's often a good bet to say that variety amongst approaches, not a single orthodoxy, will yield the best results. The mainstream molecular biology community seems to think de Grey's proposed solutions to his seven problems reveal a strange taste; but this doesn't mean very much, as the mainstream's scientific taste may well be mortally flawed. Science, like any human endeavor, has its fashions and trends. What is seen as weird science today may be a commonplace field of study in a decade.

Regarding mitochondrial DNA damage, de Grey's current proposal is to fix it, not by explicitly repairing the DNA as in GENCIA's protofection technique mentioned before, but rather by replacing the flawed proteins produced by the flawed mitochondrial DNA. This could work because there is already an in-built biological mechanism that carries proteins into mitochondria: the TIM/TOM complex, which carries about 1000 different proteins produced from nuclear DNA into the mitochondria.

What de Grey proposes is to make copies of the 13 protein-coding genes in the mitochondrial genome, with a few simple modifications to make them amenable to the TIM/TOM mechanism, and then insert them into the nuclear chromosomes. In this way they'll get damaged much more slowly, because the nuclear chromosomes are much more protected from mutations than mitochondrial genes.

Sensible enough, no? Whether this or protofection is the best approach, we're really not certain. Our bet is tentatively on protofection, which seems a bit simpler (since as de Grey admits, fooling the TIM/TOM mechanism in an appropriate way could wind up to be difficult), but maybe both—or some other approach—will yield insights and results. Unfortunately, neither currently proposed approach is being amply funded at the moment. We can't know for sure unless the research is done, but the research can't be done without funding. Perhaps it's time that we, as a society, put more of our resources towards funding such wide-reaching scientific programs, and less into any of a number of less important endeavors.

Similarly, each of de Grey's other six categories of aging-related damage is amenable to a number of different approaches— we just need to do the experiments and see which one works better. A lot of work, and a lot of creativity, is required along the way—but straightforward scientific work of the kind that modern biologists are good at doing. It may well turn out that senescence is defeatable without any really huge breakthroughs occurring—it may just be a matter of finding the right combination of clever therapeutic tricks like protofection or mitochondrial protein replacement. There's nothing in our current understanding of cell biology that stands as an absolute impediment to stopping cellular aging.

Depending on how well this work is funded and how many "hidden rocks" appear—and what happens with the rest of 21st-century science and technology—the process of scientific advance may or may not be too slow to save us from dying. It appears that the only major impediment to a scientific, not mythological, approach to combating old age is economics. People, it seems, would simply rather have a new diet pill or a faster car than research into the fundamental nature of life and how to sustain it. However, even at the current snail's pace of research current funding levels allow, it seems nearly certain that for our grandchildren, or great-great-grandchildren, "old age" may well be something they read about in the history books along with black plague and syphilis—an ailment of the past.

---

[1] see von Zglinicki, 2003 for a view of the state of the art regarding the molecular basis of aging.

---

[2] The work of Drs. Horner & Reh at University of Washington, and Dr. Gage at the Salk Institute, indicates that some nerve cells may indeed be naturally replaced or repaired after damage, and that more possibly would if not for the formation of scar tissue—so that if this can be removed or reduced, nerves even in the brain may be able to self repair, and much more repair and replacement could be done with stem cell treatments.

[3] http://www.imminst.org/forum/index.php?act=ST&f=62&t=5883&s=

[4] see e.g. Aronson, 2004 for a recent discussion of AGEs and their role in cardiovascular aging, and other pointers into the AGE literature.

[5] A brief summary of Bradbury's ideas is at http://www.fightaging.org, which also contains some dialogue between Bradbury and de Grey on related issues.

[6] An in-depth discussion of the biology involved, for the bio-savvy reader, is available in the paper Ben wrote with Rafal Smigrodzki, et. al.

CHAPTER 9

CRYONICS COMES OF AGE

The idea of freezing people in "suspended animation" is a science fiction standard. Many authors have used it as a fictional device to explain how people will travel long distances through interstellar space without getting any older. The star-voyager gets on the spaceship, locks himself in the freezing chamber—and is automatically defrosted a million years later, not one second older. The essential scientific basis for the concept is a simple one: freezing slows molecular motions to a crawl, essentially ceasing cellular processes, including aging and metabolism. What has been the problem—the reason we haven't previously been able to resurrect bodies found frozen to death—is that most kinds of freezing cause damage we don't yet know how to repair.

Serious thinking about cryonics—the technical term for placing people into suspended animation by freezing them—began in the 1960s. At this point the technology for freezing people in liquid nitrogen was just barely becoming economically feasible. Futurists began to consider the possibility of freezing themselves immediately after their death, to preserve their body for eventual restoration by future scientists with advanced medical techniques. Sure, you may have died of incurable cancer in 1975, but what if the cure for your type of cancer is found in 2053? Then you can just be defrosted, given the cure, and jolted back to life by futuristic medical magic.

The weak point of this plan, in the '60s, was that no one had a very clear idea about the revival part. There were no known technologies for defrosting people and curing their diseases—the best cryonics advocates could offer was an emphatic waving of the hands and a quick nod to "future technology." But even so, the idea was visionary. After all, technical details aside, it does seem awfully likely that, 1000 years from now or so, the technology will exist to revive a frozen corpse. Look how far technology has come in the last 1000 years! Or, if not 1000 years, what about 10,000? This isn't a concept whose plausibility requires a sophisticated scientific and engineering basis, or a fanatically optimistic faith in technological acceleration.

Now, 30 years down the line, the idea looks even more reasonable. Biologists are moving toward an understanding of cell death, and it seems quite reasonable that in a few decades—certainly in a few centuries—the process may be haltable or reversible. As discussed earlier in the book, the concept of nanotechnology has become commonplace: machines are getting smaller and smaller, and serious scientists envision swarms of microscopic robots zooming through the body, delivering medicine or repairing damaged cells. We can't yet defrost and repair a frozen corpse, but we now see clear scientific pathways that are likely to lead to this ability, given enough time.

The picture gets clearer and clearer each year: science may not defeat aging within our lifetimes, but it may well be possible for us to avoid *permanently* dying of old age by preserving our bodies shortly after death, and letting future scientists revive us. Furthermore, dozens of forward-thinking people have already availed themselves of this opportunity, and are waiting in cold storage for the next iteration of their lives.

THE VITRIFICATION REVOLUTION

Until recently, cryonic preservation meant freezing in liquid nitrogen. This is a very harsh process. Freezing a human brain or body in liquid nitrogen does preserve its basic cellular structure. Assuming self and memory are contained in the brain's neurons and synapses, they are preserved through the freezing process[1]. However, the formation of ice crystals in the body during freezing causes so much cell damage that no simple defrosting and resurrection of a patient frozen in liquid nitrogen will ever be possible. Advanced nanotechnology of some sort will be necessary to repair the damage from crystallization and bring a nitrogen-frozen body back to life.

The scene shouldn't be painted too darkly: freezing in liquid nitrogen still provides a good way to elude death, vastly better than any previously known alternative. Swarms of future nanobots, zooming through the body repairing damaged cells, will likely possess the ability bring a frozen body back to life and health. Or, advanced scanning technology may simply read the mind out of a frozen brain, and embed it in a computer or an android or a newly synthesized human body.

But these days, we can do better than mere freezing. We have *vitrification*, a related but significantly different method of stopping biological time. Vitrification converts biological tissue into a strange kind of low-temperature glass that is totally free of ice crystals. Gregory Fahy (Fahy *et al.*, 1984) created the technology a couple decades ago as a technique for preserving organs intended for transplantation; and in the intervening years it's come impressively far. So far, embryos, ova, ovaries, skin, pancreatic islets and blood vessels have been vitrified and then de-vitrified for transplant. Successful de-vitrification of whole organs like kidneys, livers, hearts and lungs isn't too far off.

Like many technologies, vitrification is inspired by the wonders of nature. There are frogs that can spend days or weeks in freezing conditions, *with up to 65 percent of their total body water frozen solid.* This is achieved through the distribution of special "antifreeze" chemicals through the bloodstream—"cryoprotectant" chemicals that reduce ice formation using one of a number of mechanisms. Some amphibians manufacture the cryoprotectant glycerol in their livers. Arctic frogs have a special form of insulin that accelerates the absorption of glucose, another cryoprotectant, into their cells. A good cryoprotectant makes water freeze smoothly and purely like glass, with no crystal formation. Cryoprotectants work best, especially in non-viscous liquids like water, if the freezing happens very fast—a process called "flash-freezing."

The biggest problem in organ-vitrification research so far has not been freezing but defrosting. If the reheating of the organ isn't fast enough, damaging ice crystals can form during this phase, rendering the delicacy of the cryoprotected freezing process irrelevant. What's needed to circumvent this problem is, basically, a *very fast heater.* Current research is looking into radiofrequency rewarming, a process somewhat similar to microwave heating, as the way to achieve this. The process may also be aided by chemicals that prevent ice formation through methods besides straightforward cryoprotection.

"Carrier solutions" like glutathione can reduce the amount of cryoprotectant needed to cause vitrification; and ice-blockers like threonine and serine prevent the formation of ice-crystal nuclei, by bonding to nascent ice-crystals in appropriate ways. They are proteins that inhibit c-axis growth, which basically means that they inhibit the crystal from getting longer. There are also proteins which inhibit growth along the a-axis (roughly, the width). The creation of an appropriate chemical cocktail to promote successful vitrification and de-vitrification is a subtle matter, but one on which scientists are hard at work. In large part, it's the recent development of sophisticated ice-blockers that has made vitrification newly plausible for cryonics—this has drastically reduced the amount of cryoprotectant needed to make vitrification work.

The biggest problem vitrification faces so far is that to vitrify a whole organ or body requires a lot of cryoprotectant, and all known cryoprotectants are highly toxic. This is a bigger problem for contemporary organ transplant work than for long-term cryonics applications, however, because it's very likely future science will discover ways to palliate the toxic effects of cryoprotectants. Palliating poisoning is almost certainly vastly easier than repairing the ice crystal damage done by straightforward freezing, and should be achievable via advanced pharmacology alone.

Finally, vitrified systems have a nasty habit of cracking to pieces if they're cooled to liquid nitrogen temperature (-196°C). However, a long-term storage temperature of -130°C to -150°C seems to work fine. In fact, bodies simply frozen in liquid nitrogen without cryoprotectant often crack too, but given the severity of garden variety ice-crystal damage, no one ever particularly worried about this cracking. Now, with vitrification, cryonicists can seek perfection. Vitrify the body, then when the ill that killed it becomes curable, devitrify it, pumping in an antidote to the cryprotectant's toxic effects. "Good morning, what year is it? What? 3010 already? When I died it was 2015."

All in all, while we don't yet have a viable way of freezing and defrosting organs, let along organisms, it's clear that science is progressing in the right direction. The steady, incremental advancement of cryobiology will get us where we need to go.

THE ALCOR LIFE EXTENSION FOUNDATION

Ok, let's say you're convinced now and want to be frozen—how do you do it, in practice? Do you have to build a vitrifier in your basement, and tell your friends and family, should they ever find you stiff and cold, to jam you inside?

Thankfully, no. There are a handful of organizations providing cryonics services to interested individuals—and at least one of them, Alcor Life Extension Foundation[2], appears to be truly serious and responsible. They have close to 40 "patients" in cold storage right now, awaiting the technological advances necessary for their resuscitation. Their first patient was placed in liquid nitrogen in 1967. When you sign up you're given an Alcor bracelet or necklace, stating that if your body is found dead, Alcor should be telephoned at once so they can come fetch it before it decomposes. This is an important, and oft neglected, aspect of cryonics: the person must be frozen as immediately upon death as possible or further damage may result which is so severe that the information stored in their brains—*their self*—will be lost.

To have your whole body preserved and cared for by Alcor, the current cost is $120,000, of which about $30,000 covers the direct cost of transporting your body to Alcor and freezing it, and the remainder goes into a fund called the Patient Care Trust. On the other hand, to be a "neuro patient" and have only your brain preserved costs about $50,000, a relative bargain. Right now, vitrification is only available for "neuro patients," but this limitation is purely a matter of cost—

Alcor can't yet afford a system capable of preserving whole vitrified bodies. With sufficient interest in their work, these funding limitations will be overcome very quickly.

Most Alcor members pay the costs of cryopreservation by taking out a life insurance policy payable to Alcor. A few hundred bucks a year is what it comes out to, if you take out the policy when you're young. Not a bad price for a chance at radically extended life. You've got a vastly better chance at success spending your money on a cryonics insurance policy than playing the lottery every day—and the potential payout is worth more than any dollar amount. The money in the Patient Trust Fund is primarily earmarked for long-term care and resuscitation. In some years, however, Alcor has been forced to spend a great deal of cash on legal battles, due to the difficulty that many conservative-minded individuals have with the very idea of cryonic preservation. Attempts by the conservative faction in American politics to bleed Alcor dry are just another example of how many involved with this faction pay lip-service to individual freedom, and then attempt to forcibly impose their ideology on everyone.

The Alcor management are to be congratulated for their maturity and responsibility. In addition to solving the technical problems of deploying cryonic technology under real-world conditions, they have taken on the very serious task of creating an organizational framework plausibly capable of surviving for centuries or millennia. The difficulties associated with the latter are illustrated by the fate of Cryocare, an organization that split off from Alcor in 1993.

Cryocare management was experienced, including some Alcor very-old-timers. They had a firm handle on the technical side of cryonics, and took two patients under their care, but their organizational model was not robust. The founders viewed Cryocare as a kind of meta-organization, which would subcontract the various tasks involved in cryonic life extension to other firms. In this way, it was felt, the forces of free market competition would allow Cryocare to continually provide its patients with the best quality services in every aspect. In a world where cryonics was widespread, this approach might well make sense.

If you go to the Cryocare website now, you find a message announcing the firm's dissolution. The two firms they initially contracted, BioPreservation and CryoSpan, both pulled out due to lack of demand for their services. Their two patients were transferred to Alcor. No harm done, but thank goodness for Alcor. Clearly market forces alone can't drive advanced research—the radical cutting-edge isn't in the market mindset. It will fall by the wayside to the trends of the day, and needs to be pushed along by individual philanthropists, corporations and even government research labs who are willing to "pay the price" in terms of decreased stock value or cashflow to fund (and defend from interventionist conservatives) future-looking projects.

WHY SO FEW?

Cryonics sounds science-fictional at first—but like a lot of other wild-sounding technological ideas, when you think about it carefully, it makes quite a bit of sense. From our point of view, the surprising thing is that so few people have taken advantage of the possible escape route offered by cryonics. The financial cost is not so high, and the potential benefits—centuries or millennia of extra life—are pretty astounding. The downside is absolutely negligible—in the unlikely event that cryonics totally fails, and you can never be resurrected, how was it any worse of an investment than taking out an insurance policy for a few hundred dollars a year to pay for a traditional burial?

Admittedly, only individuals in the developed world can afford such a thing, so already the pool of possible patients is limited (though it doesn't have to be that way—poverty in a resource-rich environment is a man-made invention, and one we can un-make if we have the willpower to do so). Still, the developed world has at least a billion people in it—why aren't there millions of cryonics adherents frozen in centers throughout these countries? The majority of people, even in the developed world, are religious and adhere to one of the many contemporary belief systems promising a supernatural afterlife. To them, the idea of living forever, or even a very long time, is anathema. They believe that it is their duty to God to die, even if there were technology available to prevent it.

Even so, there are tens of millions of non-religious folks in the developed world—people who believe that once they're dead, they're gone, period. There are millions of individuals in this category who have died since Alcor was founded in 1969. Yet only 40 or so people lie frozen in Alcor's chambers.

To explain the peculiar smallness of this number, it is tempting to invoke Freudian psychology and the notion of the death wish. At very least, setting aside any issues to do with the rationality of religious belief systems, the situation indicates an inability on the part of even most *non-religious* modern humans to think rationally about the subject of their own death. Subconsciously, even die-hard rationalists fatalistically accept death as inevitable, at least, if not desirable. Death is something most of us would prefer not to ponder or concretely confront—and yet, the irony is, openly confronting the reality of the body's death is the only thing that can allow us to take rational steps to *avoid* it.

Upon careful analysis, the biggest risks associated with cryonics are not scientific but rather *social* in nature. Let's say you're vitrified in 2020, and it's 500 or 1000 years until technology advances to the point where you can be defrosted and brought back to life. (We're hoping that by the end of the 21st century pharmacology and nanotechnology will have essentially defeated human mortality—but even someone who isn't a techno-optimist has to got to admit that a lot of progress is going to happen in the next *thousand* years). What's the chance that Alcor, or any other organization that's vitrified you, is still going to be around in 3020?

It is true that there are precedents for organizations lasting this long—the University of Bologna was founded in 1088 and is still going strong, and the Kongo Gumi construction firm in Japan was founded over 1400 years ago. But a lot can happen in 1000 years—revolutions, nuclear holocausts, financial crises, catastrophic interventions by religiously conservative government officials, months-long power outages during which all the corpsicles in storage defrost and rot[3]. Preventative measures can be taken against such catastrophes—there is precedent in the monastic system common to many religions, as a method which evolved for the religion as a system to survive social catastrophe in the greater society—but even so, a lot can still go awry in a millennium.

Still—the risks notwithstanding—the odds of long-term survival via cryonic preservation would appear to drastically beat the odds associated with any of the alternatives: cremation, burial and ensuing decomposition, et cetera. All of those guarantee that you're gone forever. What's the risk in cryonics? Either it works, and you can experience the world again someday, or it doesn't.

For a few hundred bucks a year, which would you prefer? A few extra features on your car, or a chance at living a thousand or a million years, or perhaps becoming a superintelligent AI being and transcending the space-time continuum altogether? We truly believe that, a few hundred years from now, our descendants are going to look back and be thoroughly baffled that a project so obvious as cryonic life preservation took so long to get off the ground.

---

[1] Provided sufficient dynamics can be restored. If you're thinking of having yourself frozen, which is not a *bad* idea, it would be even better to also have a brain-scan done with the best technology at the time so future scientists can make sure the structurally stored information exhibits correct dynamics after restoration.

[2] http://www.alcor.org

[3] Alcor has not yet developed their own independent long-term power source, such as geothermal, or a hardened-bunker storage facility, but such steps have been much discussed in the cryonics community as intelligent steps to take given proper funding.

CHAPTER 10

IN SEARCH OF THE UNIVERSAL EQUATION

AI and genetic engineering are cool, sure, but what about the staples that every Sci-Fi-loving youth grew up on? What about time travel; what about entering into alternative universes through naked Singularities; what about colonizing the solar system, the galaxy, the universe?

Personally, we—like so many kids of our generation—grew up wanting to be a combination astronomer/astronaut. In our cases, we each reached a point somewhere in our late teens where we realized the big show was really *inside*—inside our skulls, inside everybody's skull, inside the collective cultural mind. We became more interested in the process by which our minds and our culture constructed a consensus reality, containing notions like outer space and naked Singularities, than in these physical entities themselves—though the longing for exploring the amazing cosmos never truly subsided.

Now we find ourselves entranced by the feedback involved here: physics gives rise to chemistry, which gives rise to biology, which gives rise to psychology and sociology. Mind and culture turn around and create our subjective perceptual worlds and belief systems, which (in modern culture) include such things as the study of physics and chemistry. This is not to say we're relativists: we don't believe that "everything is just in the mind," with physical reality having its reality only because we believe it's there. On the other hand we're not objectivists either: we don't believe that there is an absolutely real world independent of all minds perceiving it, and that nothing we do can ever change.

We suspect reality is far richer than any of these terms and concepts capture. Perhaps the philosophy of superintelligent post-Singularity AIs will do a better job, but in the meantime, the notion that reality is an interchange between a tangible physical universe and the minds living within it should suffice somewhat.

Among hyperoptimistic Singularity theorists, one sometimes sees disagreements about how physical law will hold up after the advent of superintelligent AI beings. Will these superminds be able to find a way around the "cosmic speed limit" Einstein identified (the principle that nothing can travel faster than light speed, which means that getting around the universe is always going to be a very slow process)? One side argues that physical law cannot be circumvented by any amount of intelligence. Another argues that the "physical laws" we've inferred are just crude patterns that our limited minds have recognized in the data we've gathered using our limited sensory organs. Our own sympathies tend to be about 80% with the latter view, though that doesn't necessarily mean that anything we've imagined through science fiction is likely to come to pass either. What the future of scientific understanding holds is quite likely to be a host of things we, currently, literally can't even imagine. For what little our human intuition is worth, our feeling is that there will be limits to what superminds can do—but these limits will bear only subtle and indirect connections to the limits that our current physical "laws" identify.

Indeed, the word "law" in this context is in our view a little odd and misleading, with unnecessary religious overtones. Physical laws, as we call them, are just particularly powerful observed patterns. Roughly, what we call scientific law is some line of reasoning about a natural phenomenon, which when applied to known particular cases of the general phenomena it covers, is predictive of the outcome. However, this holds only on the scales we can currently observe, and what is very predictive on one scale may be totally inadequate on another—this was the case when the quantum revolution subsumed Newtonian physics. So far, as we investigate the physical universe more broadly and more precisely, we seem to consistently observe new patterns, refining the old ones, though usually not totally invalidating them. It seems reasonable that this process will continue in even fuller force when there are superhuman minds carrying it out.

In support of this view, far from being in a condition of stasis, modern physics theories are all over the place, postulating all sorts of "crazy" things, most famously (as we'll discuss shortly) that the entire universe is a kind of music created by the vibrations of hyperdimensional strings. String theory is deep and fascinating, and may well turn out to be true—in the sense that it is a predictive model of some set of observable phenomena of our universe—at least for a while. (A few decades? A century?) The aspects of contemporary physical theory that interest us the most are the ones that hint at future physics, far beyond anything theorists are explicitly calculating or experimentalists explicitly measuring today.

There is a subterranean line of thinking—not embraced by the mainstream of contemporary physicists, but pursued by a handful of brilliant mavericks, including some very well-known ones—which suggests that rather than being composed of hyperdimensional strings, the universe is in some sense a cross between a hyperdimensional life form and a hyperdimensional computer. The image of the universe as a self-organizing, evolving, self-creating biocomputing system is an exciting one, but the mathematics required to flesh out such an idea is forbidding, and may even be beyond the capability of the human brain. Only time will tell—or it may not: we may not get the chance to find out whether human brains are capable of doing such calculations, because our AI creations may get there before us (though, as we've suggested earlier in the book, we'll likely be right there along side them in one way or another, so unless you're overly hung up on who gets there first, the quest for such knowledge will still lead us all to exciting and amazing new discoveries).

PLATO'S CAVE, 21ST CENTURY STYLE

Perhaps the most amazing feature of 20th century science, overall, is the huge kick in the ass it has given to objectivist philosophy. At the end of the 19th century, it looked like science was going to give us a detailed portrayal of the universe as a giant machine, a clockwork, to use the common metaphor. Instead it did something vastly more complicated, showing us that our petty human concepts of "objectivism," "relativism," "mechanical," and "deterministic" are far from subtle enough to cope with the world as it is. The objectivist program of an enumerative science of immutable laws has been largely unseated by Einstein, Gödel, and their peers and successors.

Plato, who knew little science but had philosophy down rather well, seems to have been onto something rather profound with his parable of the cave. Science has proved this story true in more ways, and in richer detail, than its author could have imagined. Today, as in ancient Greece, even with all our powerful scientific machinery and technological achievements we still must view ourselves as sitting here in a cave, our backs toward the cave mouth, watching the shadows of trees and birds and bears dance on the cave's back wall. We mistake these shadows for reality because we're unable to see the real world outside. This is the lot of all finite minds, and always will be—thus was Plato's intuition. How amused and intrigued he would be to see the complex combination of experimental tools and mathematical and conceptual theories we have conceived, describing one after another aspect of this real world we cannot see. As his fellow Greek Zeno postulated in his metaphorical paradox: we never fully understand, but our understanding gets better and better as time goes on, converging toward the infinite limit of real understanding—a limit that will never be reached.

Cognitive neuroscience shows us that the cave extends even into ourselves. The mind we perceive is not the actual mind that controls us. We believe our conscious decisions are controlling our actions. Yet, beneath this conscious layer, is a whole complex of subconscious processes that—while a real, important part of our selves—operates without any conscious direction. Most decisions arise prior to the conscious process that believes it originates them, as a result of the self-organizing combination of millions of microscopic neural events.

Quantum physics shows us that even the most apparently solid and simple phenomenon of physical reality is just a shadow dancing on the wall. All of us, all the objects around us, are really far more than 99.99% empty space. The idea that anything has a definite position, mass or speed is an illusion: at bottom, everything is made of particles whose state is fundamentally indeterminate. The shadow world looks determinate and solid, until you look at it closely; then you see that the real world reflected in the shadows is indeterminate and fuzzy, and that this indeterminacy is important for understanding some aspects of everyday things. The genes and proteins that guide our body rely on quantum mechanical phenomena in their every interaction. You can't understand them by thinking about the immediately perceptible world, the shadows on the wall—rather, you have to use what we've learned about the world behind the shadows, as bizarre and counterintuitive as this world has come out to be[1].

Our investigations into the world behind the shadows get stranger and stranger as time goes on. Indeed our minds, adapted to the phenomenal world, stretch to grasp these new insights. It is quite remarkable how science, itself a channeling of the human mind in particular directions, can lead to conclusions that rattle the mind so thoroughly, down to its foundations.

Some of the most dramatic revelations of this kind, in recent years, have come from physics. Quantum theory shows us that the everyday world with its solid objects and definite events is just a shadow world, that the real world underneath is quite different—and much more confusing, because our brains aren't adapted to it. The particles we're made of don't have definite states, they're suspended between different conditions of being; and it's this suspension that, by complex chains of causation, makes it possible for proteins to bind together creating organisms from genes, and for electricity to zip between neurons creating emergent thought-patterns. Quantum theory itself is not complete: it only explains a piece of the shadow-world we see around us. Gravity, which holds us down on the Earth and keeps us from flying into space, is a rather obvious feature of the world around us, yet so far no one has a really good understanding of how quantum theory is consistent with gravity.

To put it another way: Among the shadows we observe on the wall we call the world are proteins and electrical fields, planets and gravity. We have a theory about how the real world casting the shadows works that, so far, explains the behavior of proteins and electrical fields and a variety of other things—this is called quantum physics. We have a theory about how the real world casting the shadows works that thus far explains gravitation—Einstein's General Theory of Relativity. But these theories, when you get down into the nitty-gritty of them, appear to contradict each other. They are both predictive of some behaviors of the physical world, but not all, and nobody has been able to reconcile one with the other.

In their efforts to resolve this contradiction, scientists are coming up with yet more complex and peculiar hypotheses about the world behind the shadows. The real universe, hundreds of esteemed physicists at famous institutions now believe, is a ten-dimensional world made of vibrating strings resonating at different frequencies. But this hypothesis isn't yet as convincingly proven as quantum physics or gravity. There are alternative theories. Some scientists believe that the real hidden universe isn't a multidimensional string symphony, but rather a huge invisible computer. We're all patterns in the universe-computer's memory, produced by a software program called physical law.

Who's right? Building multibillion dollar particle accelerators may help us find out, or it may not. At the moment string theory's most interesting concrete predictions involve the behaviors of certain types of black holes—interesting predictions, but hard to test given the current state of practical astronomy. The model may be mathematically self-consistent and predictive in computer simulations, but it is currently physically untestable. These lines of scientific research may lead to the technology of the next century: superpowerful atomic-scale computers, maybe even faster-than-light travel or time travel. Or they may just lead to a lot of difficult and expensive head-banging against the wall of unsolvable mathematical equations and experimental predictions, unverifiable in practice.

From a very general perspective, though, one thing seems clear. Some journalists and a handful of scientists have referred to string theory and other advances in modern physics as the search for a "Theory of Everything." But this phrase is misleading in many ways, and really reflects the vestiges of the objectivist dream of a final set of simple laws that explain everything. Progress in these areas of physics may lead to a lot of exciting things, but a universal understanding of everything is not likely to be one of them! For a number of particular reasons that we'll discuss a little later, even a universal unified physics theory is likely to leave a huge number of gaps in our understanding of the cosmos. The quest to map out the real world behind our backs, by inferring patterns in the dance of the shadows, is an ongoing one with periods of apparently stable understanding, periods of confusion, and periods of revolutionary conceptual progress. Modern physics is just another chapter, albeit a fascinating one.

THE HISTORY OF PHYSICS AS A HISTORY OF CONTRADICTIONS

The history of physics can be understood as a series of successful attempts to use mathematics to come to grips with the intuitively incomprehensible aspects of the world. Human common sense is tuned for explaining human-scale earthly physical events, but as experimental science progressed, we discovered more and more about the physical world in other places and on other scales. The further our discoveries got from the everyday world, the less useful our common sense was for explaining them. Mathematics, however, is not restricted to describing structures and processes that agree with our common sense. We are able to set up equations and models based on our common-sense understanding, and then derive highly counterintuitive conclusions from them. This is the power of mathematics.

Newton, in the 1600s, explained why the moon didn't fall down on the Earth by positing "laws" of motion spanning both earthly and cosmic motions. The explanation of the moon's continued elevation drops right out when one solves the differential equations for the motions of heavenly bodies. Once these equations were derived—as the result of centuries of precedential work in observational physics and applied mathematics—the contradiction between outer space and the earthly world, so vivid for earlier thinkers, fell away as if it had never existed.

James Clerk Maxwell, in the 1800s, resolved a number of peculiarities to do with electricity and magnetism. Building on the conceptual and experimental work of Faraday and others, he posited a set of equations nearly as fundamental as Newton's. This leap in understanding led to its own share of contradictions. Maxwell's equations demonstrate that radio waves (and all sorts of other waves, like microwaves and X-rays) are really just light waves at different frequencies, and that they all travel at the same speed—the speed of light—regardless of how fast the observer is moving. However, constancy of speed leads to mathematical contradictions in terms of Newtonian physics. Einstein's theory of Special Relativity arose to deal with this contradiction, explaining how indeed it's possible for all these waves to travel at the speed of light—by throwing out the commonsensical notion that the length, mass and time of objects are the same no matter how fast the person observing them is moving. An absurdly radical and extremely controversial notion for its time, a mere 100 years later even non-scientific educated people take Special Relativity simply as a fact of the universe and hardly give it a further thought.

Maxwell's equations led to other problems as well, in the theory of heat. A hot object lets off electromagnetic radiation of many different frequencies, but if one adds up the energy of the radiation let off at different frequencies, one arrives at an infinite sum. Obviously, infinite energy isn't released from any hot object. Max Planck, around the start of the 20th century, had the answer: the energy is let off only in discrete chunks or "quanta," not in a continuous range of values. With this humble act of contradiction-resolution, quantum physics began. The simple idea of quantized energy led to a rapid-fire series of experimental and theoretical breakthroughs, similar to what we see today in molecular biology, and revealing the existence of a strange and unfamiliar world underlying the world observed every day. This quantum world: in which positions, momentums, energies and times are shifting and indeterminate rather than definite, in which particles can leap through solid barriers as long as they're not being observed, in which particles can travel back through time, was not intuitive; but it was the picture that resulted from attempting to resolve contradictions in very predictive models of actual phenomena—and it turned out to be a predictive, testable model itself.

While quantum theory was developing, Einstein was not only helping it along but also pushing in another direction. His Special Relativity theory frustrated him, because it directly contradicted Newton's theory of gravitation. Newtonian gravitational theory explained why the moon fails to fall into the sea, and how planes move, but it came with a price: the unrealistic, clearly false assumption that gravitational force moves from one object to another at infinite speed across great distances. Special Relativity says that nothing can move faster than the speed of light. To resolve this contradiction, the theory of General Relativity was conceived. The notion of spacetime, a 4-dimensional surface including three dimensions of space and one dimension of time, was introduced, and in this context gravity was explained as the curvature of spacetime. Massive objects curve spacetime, and then they move along paths in the space-time continuum they have curved. This delicate feedback between objects and spacetime is captured in Einstein's elegant but fabulously complex mathematics.

Gravity is a relatively weak force—you and I have only a very weak gravitational attraction between each other, for example. It only really kicks in for objects of large mass. This means that the difference between Newton's and Einstein's approaches to gravity isn't really observable on the Earth in any simple way. But it is observable in outer space. General Relativity explained a previously baffling eccentricity in Mercury's orbit. And it predicted that light coming from a star behind the sun, but near the edge of the sun from our perspective, would be slightly bent as it voyaged toward us—a prediction that was validated by measurements done during a solar eclipse.

Making the Special Theory of Relativity consistent with quantum physics was not easy. This unification led to the theory of quantum electrodynamics (QED), a theory of the mid-20th century that in many ways is the crown jewel of physics. QED is a complex and beautiful theory, commonsensically counterintuitive and yet fantastically pragmatic—predicting the observed mass of the electron, for example, to well over a dozen decimal places. There was no single superhero of QED: it was created by an international group of brilliant scientists such as Feynman, Tomonaga, Schwinger and Dirac, building on the mathematical quantum physics of earlier minds like Pauli, Schrödinger and Heisenberg[2]. Laser physics is but a single example of the very numerous practical applications of QED.

QED grew into quantum field theory, which extends basic QED to give a fuller explanation of what happens inside the nuclei of particles. This is where the fascinating objects called "quarks" (a word drawn from *Finnegan's Wake*) come into play. Quarks can never be observed in isolation, but in combinations they create particles like protons and neutrons. Understanding quarks helped us understand nuclei, but it required the use of whole new fields of mathematics, different from anything used for physics before. A class of theories called gauge theories, or Yang-Mills theories, was created, integrating all aspects of physical law except gravity. There are infinitely many Yang-Mills theories, but one has emerged as the best explanation of observed data; this is what's called "the standard model." Each different gauge theory potentially hypothesizes a different set of particles. The standard model postulates three families of quarks and leptons (particles which only participate in electroweak and gravitational interactions, not strong interactions—the electron is a lepton), and also other mysterious, as-yet hypothetical entities called Higgs particles.

The standard model is not a simple thing. There are more than 20 parameters, whose values aren't predicted by the theory. But by setting these 20 parameters, one gets an immense number of practical predictions. The theory is really quite a good one. The only problem is this little tiny thorn in its side called gravity. Einstein's grand and beautiful vision of nonlinear feedback between matter and the space-time continuum doesn't fit into the standard model picture of the cosmos at all. These two different theories explain different aspects of the shadows we see dancing on the wall, by making radically different postulates about the real world underlying the shadows. The two theories have to be brought together. That tall shadow can't be both a tree blowing in the wind, and a tall creature moving about—there has to be a common explanation, encompassing the data favoring each explanation. A unifying theory will result in some changes to how we view physics, just as Special Relativity changed our perspective on

Newtonian physics (though we can still use the formulae of Newtonian physics in various practical scientific and engineering applications). It will also almost certainly not be a "final theory" which represents the end of all physical knowledge, but another step towards the unreachable state of infinite understanding.

So the big question confronting theoretical physics these days is: How do you make gravity and the standard model play together? It's thought that if we can answer this, we should be able to answer a lot of other related questions like: Where do the four forces we see come from? Why do we have the particles and waves that we have, instead of other ones? Why is space three-dimensional and time one-dimensional?

Physicists are a creative lot, and many potential answers to these questions have been proposed. Here we'll discuss only two of them: superstring theory, which is perhaps the leading candidate in the physics establishment; and the universal computation approach, which is the domain of a handful of mavericks, but has an impressive philosophical simplicity and conceptual power. We have to admit that the universal computation approach has more appeal to us personally, because it highlights the similarities between the universe as a whole, and complex self-organizing systems like minds, bodies and ecosystems, but that's just a philosophical leaning. Right now, no one knows which approach is right, and in science you have to go where the mathematics and experimental observation leads you, not where your preconceptions want you to go. At the moment, progress in creating interesting new theories is fast, and progress in testing these theories is far slower. Getting concrete predictions out of theories based on such outlandishly complex mathematics is not easy; and doing the very high-energy experiments needed to most cleanly differentiate between the theories' predictions is not cheap. We have our own radical ideas about how AI might be used to vastly accelerate this process, which we'll mention briefly a little later on.

STRINGS AND SUPERSTRINGS

The standard model of physics treats particles as points: zero-dimensional objects. The particles may be blurred out over spacetime in that peculiar quantum way, but they're still essentially points. This seems a reasonable approximation, but in the period 1968–1973, a number of physicists doing complex mathematical calculations regarding some advanced physics theories realized that, in fact, the math they were playing with implied a somewhat different model. What their physics equations were telling them was that particles aren't points, they're one-dimensional objects—*strings*[3], like violin strings.

Like QED, this work didn't come out of any single genius scientist—there is no Einstein of strings. Edward Witten is most often mentioned as the leader of the string theory movement, but John Schwarz, Michael Green and many others have also made huge contributions. It has been observed that in the early days string theory was developed by middle-aged scientists because, according to the sociology of the physics community at the time, young scientists were too precarious in their careers to risk working on something as speculative as ten-dimensional harmonics. Nowadays of course, string theory is fairly mainstream, and it's working on radical alternatives to string theory that's more likely to get you denied tenure. Orthodoxies can arise even among scientists, and there are always proponents of the idea that the current vogue is actually true and there's no point in contradicting it. Still, what's considered freaky, fringe science one day—like superstring theory was just a couple of decades ago—may the next become the world's pet theory. Fortunately, it is a part of science as a discipline to be more rapid than most other disciplines at casting aside its orthodoxies in the face of new evidence.

The strings these scientists found their equations were describing were very short strings—as short as 10^{-33} centimeters. They're so short that for almost all practical purposes, you can consider them as point particles. But for some things, like the unification of gravity and quantum physics, the stringiness properties of these little objects seemed to be important.

It's important to understand the course that development took here. It wasn't at all a matter of some kooky scientists sitting around and brainstorming about what the universe might be made of, and coming up with the idea that maybe particles are teeny-tiny violin strings. Rather, the mathematics of previous physics theories seemed to extend in a certain direction, and the most intuitive interpretation of this mathematics was in terms of vibrating strings. Once again, as in the birth of quantum physics itself, mathematics led where common sense knew not how to tread. There's very little that's commonsensical about string theory, and nobody would have taken at all seriously the idea of "tiny violin strings" as the fundamental particle of the universe if the mathematics hadn't forced them to.

Strings can be "open" like a violin string or "closed" like a rubber band. As they move through spacetime, they sweep out an imaginary surface called a "worldsheet."

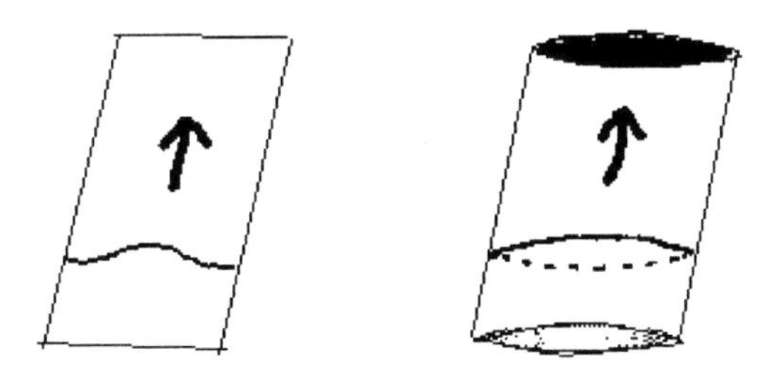

Figure 10.1 Open and closed strings

The strings don't just move—they vibrate. They vibrate in different modes, loosely similar to the harmonics or notes of the strings of a musical instrument. Physical parameters like mass, spin, and so forth, are mathematically derived from the vibrational modes of these tiny strings. Ultimately, in this view, there's only one kind of object underlying every particle—the string. Different particles are just different modes of vibration of strings. A graviton (the particle hypothesized to transmit gravity), for example, is a particular kind of vibration of a closed string, as is any other particle.

Strings interact with each other by splitting and joining. They can also be pinned to various objects, just as a violin string is pinned to the violin at both ends. Unlike violin strings, these tiny little fundamental particle strings can be pinned to objects of various dimensions, called D-branes. Much of string theory has to do with the study of D-branes, in addition to the strings themselves.

These are strings, but what about *super*strings? What's so super about them? To understand superstrings, you must understand that there are two basic types of particles in nature, named fermions and bosons (after 20th century physicists Fermi and Bose). Fermions are "matter-ish" particles, like electrons, protons, neutrons, and quarks. Bosons are "force transmission" particles, like photons, gravitons, and W and Z particles. The standard model deals with both. In the mathematics of string theory, we find a certain kind of symmetry naturally arises: a symmetry between fermions and bosons. This symmetry, called supersymmetry, is defined by the idea that fermions and bosons are grouped together into collections called "supermultiplets" (hence the "super" in "supersymmetry") in which each fermion has a boson "superpartner."

Here the mathematics yields the next big surprise. The only way to explain both fermions and bosons in string theory is to introduce supersymmetry, and the only way to make supersymmetry work in a logically consistent way is to assume that the little strings exist in a 10-dimensional world. So, boom! All of a sudden, the real world producing the cave-wall shadows we see is ten-dimensional—unlike the shadow-world we observe, which has three space dimensions and one time dimension. The idea of a higher dimensionality underlying spacetime wasn't new to string theory—it was invented by Kaluza and Klein back in the early days of General Relativity. Back then it was one interesting speculation among others, but now it has been pushed on physicists by a complex mathematics that they barely understood, a "natural" consequence of extending certain, very predictive mathematical models of Special Relativity and the Standard Model.

How can it be that the real world is 10-dimensional, whereas the observed world has three space dimensions and one time dimension? The standard explanation is a simple one: the extra six dimensions are curled up very small. Just as a two-dimensional piece of paper, rolled up small, looks like a one-dimensional line; so a ten-dimensional universe, rolled up small, can look like a four-dimensional universe. If the size of the rolled-up shape is around the same as the size of the strings themselves, then the extra 6 dimensions are almost impossible to observe, but the mathematics may still tell us they're there.

Alternative explanations of the whereabouts of these extra dimensions also exist. Lisa Randall and Andreas Karch have presented a perspective in which the universe begins as a 10-dimensional brane, and then the natural nonlinear dynamics of forces gradually result in a "reshaping" of this brane so that parts of it appear almost three-dimensional—much as if a three-dimensional lump of Play-Doh happens to get twisted into a form where part of it is essentially flat and two-dimensional. They give mathematical reasons why this sort of reshaping is very likely to happen due to the natural activity of gravity between parts of the brane, and other physical forces within the brane.

It's a fantastically surreal picture of the real world. In a very abstract, mathematical way, it seems to explain all four forces. Gravity, electromagnetism, and the weak and strong nuclear forces all come out of the same unifying equations, if you follow the mathematics and are willing to go with equations that render particles as vibrations of ten-dimensional strings.

But where's the evidence that this isn't all some kind of fabulous mathematical hallucination? Who says the universe really works this way? Where's the proof?

There really isn't any yet. Perhaps the most interesting work drawing practical applications out of string theory involves black holes—regions of space with gravity so strong that any object that enters them, including light, can never escape. Black holes have mass, charge, and spin; and Stephen Hawking showed that they also radiate particles, because of complex effects in the vacuum around their boundaries. String theory has recently proved successful in explaining results in black hole theory, previously understood only in much more ad hoc, inelegant ways. Strominger and Vafa derived important equations about a certain class of black holes by describing them as 5-branes, 1-branes and open strings traveling down the 1-brane, all wrapped in a 5-dimensional donut shape.

Of course, this branch of physics isn't very easy to experiment with—black holes are far away, and still mostly theoretical, although astronomers have identified a few. For more concrete implications of superstrings, we need to look elsewhere, perhaps in experiments physicists can run in particle accelerators (though there is a lot of evidence suggesting that we need more powerful accelerators to delve further into a lot of these mysteries).

Remember, gravity is relevant for massive objects; but remember also Einstein's discovery that mass and energy are two aspects of the same thing:

$$E = m c^2$$

When small particles are accelerated fast enough, they achieve very high energies, hence gravity applies to them significantly (in essence, as Einstein described in that famous equation, they become more massive). The standard model integrates three forces: electromagnetism, and the weak and strong nuclear forces (two aspects of the physics that holds nuclei of atoms together). All these forces have about the same strength at energies of about 1016 gigaelectronvolts (GeV, a billion volts). It is estimated than when we get up to about 1019 GeV, gravity will join the party too, and all the forces will be about equally important. No existing particle accelerator can generate energies this high, but we're creeping closer and closer. Unfortunately, the American Superconducting Super Collider project (SSC), which would have yielded about 20,000 GeV, was cancelled in favor of other scientific projects after being partially constructed. (It's not like the United States doesn't have the resources to fund a number of massive projects in different disciplines, but our leaders appear to feel otherwise.) However, it's expected that around 2007, a European accelerator called the Large

Hadron Collider (LHC) will begin to operate at 7000 GeV per beam (14 TeV total collision energy). We're getting there.

Particle accelerator experiments should let us look for the supermultiplets, the fermion-boson groupings that supersymmetry theories predict. Specifically, supersymmetry implies that every known elementary particle must have a "superpartner" particle. It is clear that no known particles are superpartners for each other. Where are the squarks corresponding to quarks, the selectrons corresponding to electrons, the gluinos corresponding to gluons, and so forth. It is hypothesized that the superpartners are so heavy that you can only see them at very high energies. While this is a reasonable hypothesis and is certainly possible, one can understand why detractors call it a convenient excuse for maintaining a theory that postulates a huge number of particles no one has ever seen[4]. Even if the LHC doesn't reveal these particles, this won't prove for sure that they don't exist—they could just be even more massive than this machine would reveal. The problem is that the current mathematics of superstring theory doesn't let us exactly predict superpartner mass, so we don't know how big the accelerator would have to be to reveal the existence or otherwise of all these particles that the mathematics tells us should be there. Perhaps the mathematical models will be refined to be predictive of these masses before experimentation can verify them, or perhaps only after finding them will someone have the eureka moment that lets them find where these masses come out of the mathematics. Or, perhaps, they don't exist at all.

Although the creation of billion-dollar accelerators to test theories that are still not entirely quantitatively clear is naturally a slow business, theoretical progress on superstrings progresses at a rapid pace. The period since 1994 has grounded a second superstring theory revolution. New mathematical techniques have shown that what used to look like five different superstring theories were actually just one mega-theory viewed from different perspectives. The march toward unification moves on—on a mathematical level, with very little feedback from experimental data. The mathematical theories underlying different aspects of physics appear to unify more and more elegantly, if one's willing to accept abstract mathematical structures like ten-dimensional vibrating strings and superpartner particles. Whether the universe agrees with this mathematics, or is stranger yet, remains to be seen. There is much history in physics to show that elegant mathematics can lead to empirically valid results. But this time, perhaps, have the mathematicians gone too far (or, perhaps, just chosen the wrong extension of the mathematics of the Standard Model and Special Relativity from the set of possible ones)?

MAVERICK VISIONS OF THE COMPUTATIONAL UNIVERSE

Superstring theory is popular at the moment, but it's really just one of many approaches to unifying gravity and the standard model. It does have a lot going for it. The mathematics is beautiful, and one result after another comes out, providing more and more evidence that the equations are sensible. However, as there are not yet any real quantitative predictions or solid empirical validations of superstring theory, it's not hard to see why some physicists don't accept the viability of the superstring approach. The problem isn't so much that a universe of ten-dimensional vibrating strings is an absurd idea. Physicists are pragmatic as well as aesthetic, and will accept any hypothesis that seems to predict observed empirical data. The problem is that superstring theory doesn't make any useful empirical predictions, and on the other hand, it postulates a huge number of particles that have never been seen. To go back to the Plato metaphor, it doesn't tell us anything about how the shadows we see on the cave wall move. It predicts a whole lot of shadows we've never seen. Of course it's possible we'll see these shadows under some strange conditions that we haven't encountered yet, but for now it is only an elegantly phrased collection of vexing new mysteries. What it does do is unify, conceptually and mathematically, the most powerful and useful existing theories for explaining how the shadows move.

Some of the alternatives, like loop theory, are worked out in almost as much detail as string theory, and by scientists with equally impressive mainstream-physics pedigrees. Some of them, however, are a bit more speculative. In the latter camp we find one of the more fascinating directions—the envisioning of the universe as a computer.

Unlike the string theorists, who have followed their mathematics wherever it leads and bent their intuitions to follow their math, the universe-as-computer folks are beginning with an intuition, and seeking to build appropriate mathematics around their intuition. Unlike proponents of intelligent design, they are not attempting to reinterpret observed results to suit a preconceived and untestable notion, but rather they are drawing on intuition to embark upon a different mathematical model inspired by computational and information theory, rather than building upon the Kaluza-Klein model of extended dimensionality. Such an alternative mode, just like string theory or loop theory, can then be mathematically analyzed for internal consistency, and experimentally tested as to its predictive abilities.

Their intuition is a simple one: perhaps the universe is a giant computer. Computer programs, we have seen, can lead to fabulously complex behaviors. One variant of the Church-Turing Thesis, a staple idea of theoretical computer science, states that any finitely describable phenomenon can be represented as a computer program. Why not the universe itself?

The standard theories of contemporary physics all support this idea conceptually, in the sense that their equations can be simulated to within arbitrarily high accuracy using computer programs. This property is utilized frequently in practice, since most useful physics equations are too complicated to solve analytically in most real applications, and so the majority of physics applications these days involve computer simulations. However, the physics theories in common play today are generally not explicitly computational—they're formulated in terms of noncomputational mathematics and then simulated on computers for practical purposes. It's an interesting idea to get rid of the middleman and make the theories themselves computational, since current theories are basically being tested in their computational versions anyway.

One of the more vocal advocates of this point of view is Ed Fredkin[5], an eccentric multimillionaire who owns his own island and has been responsible for several innovations in computer science—particularly in the area of computer programs called "cellular automata," which is the mathematical model he's chosen to work from in building his formulation of physical theory. Fredkin has been working for years on a model of the physical universe as a special kind of computer program called a cellular automaton, which is any algorithm in which some grid of cells contains state data that is modified at each time-step based on the state of nearby cells, according to the specified rules in the automaton. This is a reasonable choice, as cellular automata have been used to model all sorts of physical phenomena, from weather systems to immune systems to water flow in the ocean and neural networks in the brain.

The idea of the universe as a computer is, keep in mind, another kind of mathematical model. It does not imply that on God's cosmic desktop there's a plastic box filled with silicon circuits chugging away at the computations to run "universe.exe" on "Windows Infinity." What most folks know as "a computer" is just a particular instance of the von Neumann design to implement the basic theory of computation as worked out in the Church-Turing Thesis. More radical computers that are being developed today—DNA computers, quantum computers, and so on—may not look at all the whirring plastic boxes of the last part of the 20th century.

Fredkin has gone a long way toward resolving various conceptual problems associated with the "universe as computer" idea (see Fredkin, 1990 for an early treatment). On the face of it, it's not clear how quantum randomness and indeterminacy could come out of a computer program, or how the time-reversal that we see on the microscopic scale (particles, unlike macroscopic objects, appear to be able to go back in time) could either. But Fredkin invented a whole new research area to deal with such problems: reversible computing. It turns out that with enough memory available, computer programs can run both backwards and forwards (roughly, it is theoretically possible that with enough memory you can store enough state transition data to back-out any process). Fredkin solved the quantum indeterminacy problem by introducing the seemingly oxymoronic notion of "unknowable determinism." Basically, he notes, if the universe is a computer that's running at full blast and using all its resources to compute its own future as fast as possible, then there is no shortcut to predicting the future of this machine. That is, there's no way to build a faster computer than the universe itself. So, if the universe is a computer, it is inherently running an optimal "universe computing algorithm" and there's no way to predict its future course of evolution, you just have to wait and see. In effect, from the point of view of particular observers like ourselves, the universal computer may be indeterminate.

But Fredkin, although he's made a lot of progress, has not yet released his complete equations for the universe as a computer. One suspects there are difficult problems to crack here. Philosophically this approach is more appealing than string theory, but the mathematics doesn't seem to roll out nearly as elegantly. And there are other researchers in this particular space who seem to be running up against similar walls. For instance, Juergen Schmidhuber (1997) has created a very sophisticated theory of computational physics based on the notion of a "God machine" which runs every possible computer program at once, alternating between different programs each time cycle—our universe then being one of these programs. However, his model, too, is less fully mathematically developed than string theory.

Another advocate of this sort of work—and another possessor of a fascinating, diverse history and a super-brilliant brain—is Stephen Wolfram. Although best-known these days at the CEO and founder of Mathematica, the leading maker of software that does advanced mathematics, he started out like so many other technology pioneers, as a theoretical physicist. In May 2002 he launched his book *A New Kind of Science*, which does not give a detailed theory of unified physics or any other aspect of physical science, but lays out a fascinating approach to dealing with scientific issues, with notions of self-organizing computation at the center. Unlike Fredkin, who is big on concept and vision but skimpy on details (though in other domains he has proven himself quite capable of supplying formal details as needed), Wolfram's book unleashed on the world a barrage of details, together with a vision that, while not quite as powerful as the more delightful among his detailed examples, is nothing to be scoffed at.

Wolfram published his first scientific paper at the age of 15, and got his CalTech physics Ph.D. at age 20. He began his career studying quantum field theory, cosmology, and other aspects of advanced physics, but by the late '70s, as he approached the old age of 30, his attention had shifted to computing. In 1981 the first commercial version of what was to become Mathematica was released. Mathematica now has many imitators, but when it first came out it was revolutionary in its impact. It occupies a space somewhere between calculator-style math and AI. It doesn't actually *think* like a mathematician, but it carries out complex algorithmic operations in areas like algebra, calculus and geometry that, much like chess playing, would be thought by a non-expert to be impossible without deep thought. Wolfram did not invent the type of algorithmics on which Mathematica is based, but he was the first one to tie it together in a handy and usable package.

Around the same time that Mathematica came out, Wolfram's mind also set him stirring in a different direction. Never at a loss for ambition, he resolved to develop a general theory of complex structures and dynamics in the natural and computational worlds. He latched onto an obscure branch of mathematics called cellular automata, and brought it to the fore in the scientific world. Cellular automata were first invented by the great mathematicians John von Neumann and Stanislaw Ulam as a way of capturing, in a very simple computational model, the basic self-organizing and self-reproducing processes of living systems. Von Neumann and Ulam used cellular automata to create the first-ever example of a computational system that could reproduce itself. They didn't actually write such a program, but they showed mathematically how one could be created. Since that time, others have made their vision more concrete, a pursuit related to the active research field of Artificial Life (for instance, Tom Ray's Tierra system[6] is a current computational framework that includes self-reproducing computer code). Throughout the early '80s, Wolfram released, one after another, a series of exciting new results showing that very simple cellular automata systems could produce extremely complex behaviors. This work has a major influence on the newly developing "science of complex systems," along with other concurrent developments such as chaos theory and neural networks. Wolfram developed several practical applications of his ideas, including a new way of generating random numbers for computer programs, and a new approach to computational fluid dynamics, both of which proved highly successful.

His new book follows up the work on cellular automata, dealing with a wider class of simple computational systems, and showing how one after another phenomenon that "looks like" physics, or life, or intelligence, can emerge from these systems. These are fascinating and very suggestive results. Wolfram's program doesn't quite amount to anything yet in terms of physical science, but he suggests that, if pursued by teams of researchers over a period of years or decades, it will. Based in the prior work of computational and complexity theory pioneers like Turing, Zuse, Ulam, von Neumann, Gödel, Komolgorov, and others, Wolfram lays out a scientific approach with self-organizing computation at its center. Exhibiting something similar to the community leadership of David Hilbert in the early 1900s with his "23 Unsolved Problems" and "Metamathematics" program proposals, Wolfram's book is an attempt to compel the field to embrace complexity science as a cohesive whole and pursue a grander-scale research program, rather than a disjoint collection of smaller ones viewed as sub-disciplines of physics, mathematics, and computer science. He's not suggesting that he (and those predecessors upon whose work he draws) has found the next Newton's Laws of Motion, but perhaps that he's found the next equivalent of the calculus, the mathematical foundation of Newton's Laws. This new calculus is not a precise set of mathematical definitions and equations; it's a broad class of interesting computer systems that give behaviors roughly analogous to certain natural phenomena.

Going even further out of the mainstream, Tony Smith, a maverick physicist in Georgia (who pays his rent by working as a criminal lawyer), has published a more complete computational physics theory than anyone else. His theory (which goes by the unwieldly name "The D4-D5-E6-E7-E8 Model") is large and complex. It posits the universe as an 8-dimensional machine operating according to specific mathematical equations that he has articulated[7]. Mainstream physicists like the superstringers have nothing but contempt for this kind of work. On the other hand, the amount of work that has gone into a theory like Smith's is virtually nothing compared to the amount of work that has gone into superstrings—yet, if you look at both theories without preconceptions about which one "should" be better, Smith's results are nearly as impressive. Smith's theory runs into some subtle conceptual tangles in cases where superstring theory is clear and simple (for example, pion decay). On the other hand, Smith's theory makes more concrete empirical predictions than superstring theory, a fact that has led Smith on an interesting excursion into the sociology of experimental science (and up against the problem of orthodoxies that so many previous scientific mavericks have faced).

Specifically, the D4-D5-E6-E7-E8 model predicts that a particular kind of quark called the Top Quark should have a mass of about 130 GeV, while the standard model predicts a mass around 170 GeV. Fermilab and other institutions have done experiments leading to numerical predictions of this mass, which should allow empirical selection between different theories like these; but when you get right down to it, what you find is that the interpretation of the empirical data is a very subtle matter. The estimation of quark mass is based on averaging values obtained over different "events" of observing various kinds of particles. However, it's not always clear when a particle has been observed—often experimental noise looks just the same as a particle observation. Whether quark mass comes out around 130 GeV or around 170 GeV turns out to depend on whether a few borderline cases are classified as particle observations or not. There's a surprising amount of subjectivity here, right at the forefront of experimental physics—the hardest hard science on Earth. Until our experimental methodologies improve, we may be misreading the results, regardless of which theories they appear to support. Even when we are fairly certain we're measuring phenomena correctly, how we interpret this data is still relatively subjective—it's quite possible that several predictive mathematical models exist which describe any given event with equal elegance.

Whether Smith is right is not the point here—in all probability, his theory has some interesting insights, but doesn't capture the whole story. Perhaps it's total bunk (as some of our physicist friends strenuously tell us); perhaps it's 50% of the way to the holy grail. Perhaps this theory, and Fredkin's theory, and Wolfram's theory, and superstrings, all capture different aspects of the underlying truth. The point we want to make is that, even though this is hard empirical science, things are pretty much wide-open at this point. These theories are unfolding stories on the leading-edge of scientific understanding, and at any moment there could be a mathematical or experimental breakthrough which completely changes the picture. The leading theory, superstrings, has virtually no empirical support, and is valued primarily because of the elegance of its mathematics. Indeed, it was created almost entirely based on advanced mathematics rather than conventional physics. There are dozens of radically different theories, including conceptually fascinating ones like "the universe as a computer," which exist at various stages of development. We don't have the experimental tools to really test *any* of these theories yet, and the tools that we do have give results that are difficult to interpret; leading to potentially significant theoretical bias in the production of apparently hard experimental data. For advanced theoretical physics, this is a time of exciting exploration, not a time of solidity and stable, continuous progress.

PHYSICS AND IMMORTALITY[8]

Drawing grand philosophical conclusions from the laws of physics is a dangerous game, because our understanding of these "laws" changes so frequently. In the late 1800s, deterministic physics taught us that we lived in a clockwork universe; now quantum theory teaches an uncertain universe pervaded by nonlocal quantum resonances, and the physics theory of 2050 may give us entirely different philosophical direction. With this caveat in mind, it's still almost irresistible to see what contemporary physics theories have to say about the "big questions."

There is no bigger question than immortality. Living 150, a thousand or a million years would be great, and we hope can be achieved by one or more futuristic technologies such as pharmacology, nanotech or uploading; but if the universe itself is going to end in a few billion more years, this may provide a final termination to our lives that none of these technologies can resolve.

The Second Law of Thermodynamics would seem to present problems where true immortality is concerned. If the universe as a whole is a closed system, then the Second Law suggests that entropy will increase universe-wide until complex structures like intelligent systems inevitably deteriorate—but the Second Law isn't necessarily the last word. There have been a variety of serious attempts to reconcile some form of physical immortality with the known laws of physics.

Freeman Dyson's 1979 paper "Time Without End" is considered by many to be the first truly serious physical review of the possibility of an immortal universe within the constraints of contemporary physical knowledge. Dyson assumes an open universe (an assumption some believe may not hold true, though there is not yet definitive proof either way), and then postulates a scaling law which would allow biological adaptation to a second-law-friendly entropic universe, and the possibility of transmission of infinite information at finite cost without assuming any changes in the known laws of physics. Dyson says in his paper: "Life and communication can continue forever, utilizing a finite store of energy, if the assumed scaling laws are valid." Far from being a fringe scientist or wild philosopher, Dyson is a serious physicist with impeccable academic credentials (though one with a positive futurist philosophy). That doesn't make him correct, but it does mean that "crackpots" are not the only ones thinking about these issues and their compatibility with the laws of physics.

The outcome of Dyson's scenario is an evolved entity which exists in a low-temperature universe, and is capable of intelligent communication of vast quantities of information at low energy cost. Dyson assumed a biologically-evolved entity, but such a scenario obviously still influenced John Smart's idea about immortality (presented in the form of an answer to the famous Fermi Paradox, which asks: if there is other intelligent life in the universe, why haven't we encountered them or their spacecraft or transmissions?). Smart's suggestion[9] is that we have indeed encountered them, we just don't recognize their intelligence, in part because the form they take—black holes—is foreign to our notions of what constitutes an intelligent entity. His idea about intelligent black holes being the deliberate progenitors of new universes is a take on Intelligent Design (one which we think the Christian theologians behind the other I.D. theories didn't really have in mind); extrapolates to a take on reincarnation (the information of one universe is used to seed another as it disappears into the event horizon of the black hole's mind, and is used in universe reproduction—as per the theories of Lee Smolin) and would seem very strange even to Hindus. Currently, there's no evidence that Smart's strange ideas are correct, but there's also no proof they aren't. Most people dismiss them as pseudo-theological wishful thinking, but his ideas, while unproven, do not actually violate any known physical laws. Citing this lack of counterproof as proof that he must be right would be argument ad ignorantium, but we are simply correctly pointing out that these are very unsettled questions which are best approached with an open mind.

However, as unusual as Smart's idea is, it is far from the most controversial physical immortality postulate. Quantum immortality, which postulates an outcome of Hugh Everett's many-worlds interpretation of quantum mechanics that no intelligent entity can cease to be (intelligent consciousness is transferred to other universes during state change—death), is extremely controversial, not the least reason for which is the controversial nature of the Everett many-worlds conjecture itself. Perhaps most controversial is what Tulane professor Frank Tipler and Cambridge professor John Barrow call the Final Anthropic Principle (FAP), which states: "Intelligent information-processing must come into existence in the Universe, and, once it comes into existence, it will never die out."

Barrow's "Anthropic Principle" itself is a fascinating and somewhat less controversial statement, which says that to understand our physical universe, the fact that we humans exist in it is an important fact that shouldn't be overlooked. In other words, of all the possible universes out there, only some of them have the property that they would give rise to human beings. Potentially a huge number of universes exist, or have existed, and only a small percentage of them have laws amenable to human life. When we ask a question like "Why three dimensions?" for instance, the answer may be, "Because otherwise human life wouldn't be here to ask the question." Perhaps there have been a series of universes, formed in between a series of Big Bangs and Big Crunches, and this one happens to have three dimensions.

The Final Anthropic Principle adds a twist onto this idea. Instead of just asking, "What must the universe look like in order for us to exist?" one asks, "What must the universe look like in order for us to exist and never stop existing?" Of course, this has a different status than the original Anthropic Principle, since (setting aside philosophical doubts about the nature of being) we know we exist through common sense, but (theological epiphanies aside) we don't know that we will never stop existing! Tipler and Barrow's idea is that it may be possible to use the laws of physics to show that this is the case: to show that immortality is ours in a physical sense.

Along with other, even more controversial aspects, the Final Anthropic Principle is questionable in that the best arguments in its favor require a closed (or flat) universe, and there are physicists who currently believe that evidence points towards an open universe. This is not the primary criticism of the FAP. Most critics claim that the FAP begs the question of its own consequences, but they are committing the same logical fallacy themselves. Their position is that immortality is a religious principle which is not upheld by physical law, de facto, and that any physical conjecture which supports it is automatically invalid. They say it violates the Copernican principle (that no "special" observer can be conjectured, keeping metaphysics out of science); but that principle is itself a metaphysical principle (it is a meta-rule of science, not itself a provable physical principle). In analyzing the mathematical arguments underlying the Final Anthropic Principle, rather than the metaphysical implications of it, the rules of mathematics and physics should be applied—not metaphysical bias. The point is not that the FAP must be correct, but rather that it must be shown to be either correct or incorrect. Metaphysical arguments against it are fine, but they should not be construed as physical ones.

If the FAP is a metaphysical statement, by virtue of being a physical postulate which has metaphysical implications, then so are all conjectures of cosmology—it's just as metaphysical a statement to assume that all life must end, as it is to assume otherwise. Argumentation about such things using the laws of physics is valid cosmological debate and disproof by begging the question or appeal to belief isn't good enough—regardless of whether you're a nihilist or an eternalist. Physical cosmology and metaphysics are intertwined. In the case of physical arguments, the mathematics and statements of physical law must be what is debated in any area where direct observation is currently impossible. With regard to metaphysics, it is still not valid to simply dismiss a differing viewpoint based on fallacious appeals.

Support for the FAP is rooted in a counterargument to the common scientific argument that the second law of thermodynamics mandates an extinction of life. Barrow and Tipler quote Pierre Duheim's 1914 statement:

> "The deduction [of the 'heat death' from the second law of thermodynamics] is marred in more than one place by fallacies. First of all, it implicitly assumes the assimilation of the universe to a finite collection of bodies isolated in a space absolutely devoid of matter; and this assimilation exposes one to many doubts. Once this assimilation is admitted, it is true that the entropy of the universe has to increase endlessly, but it does not impose any lower or upper limit on this entropy; nothing then would stop this magnitude varying from negative infinity to positive infinity as time itself varied from negative infinity to positive infinity; then the allegedly demonstrated impossibilities regarding an eternal life for the universe would vanish."

They propose that theorems of mathematical limits show, as Dyson also argued, that there needn't be an extinction of life (which is defined as information processing capability in the FAP) due to the second law.

Tipler's extension of the FAP to what he calls the Omega Point (a term borrowed from de Chardin) is one of the most controversial proposals in science today[10]. In the Omega Point theory, Tipler proposes a complex physical argument that amounts to the idea that the universe is an information-processing system, and that on this universe-computer can be run a simulation which appears to run forever within its internal timeframe, though to an observer outside its frame it is a finite system. He proposes that such a simulation could be initiated by a race of beings with advanced computing capabilities, which can set it in motion at the correct time, taking advantage of the asymptotically-increasing computational capacity of the universe-computer as the possible outcome of certain physical postulations of the universe. Within this computer, he posits, is an afterlife—because all beings that ever lived can be simulated within its infinite computational capacity, by replicating all possible Quantum Brain States. Tipler makes the metaphysical argument that this is the nature of God, and it is that statement upon which pretty much all arguments against the Omega Point are based.

Physical counterarguments to the Omega Point are based on the assumptions it requires—that the universe either ends in a Big Crunch, or that there is a positive cosmological constant (both of which theories, while not disproved, are currently out of favor). Sociologically, it is argued that the Omega Point prerequisite of a society with advanced technological capabilities to start the simulation on the universe computer is a frail requirement. Furthermore, without a Big Crunch Tipler posits that baryon-tunneling spacecraft will be required to create the required physical state for the Omega Point to occur. These are indeed arguments with veracity, and the Omega Point may turn out to be yet another failed physical postulate—but the core of it is a physical postulate. Ad hominem arguments that Tipler is a repackaged theologian or pseudoscientific charlatan are not sufficient counterarguments to his theory. If the Big Crunch theory comes back into favor, in light of new physical evidence that causes the accelerating universe theory to fall out of favor, then the major physical and one of the major sociological disputes with the Omega Point theory fall away. Physicists would be better suited to review Tipler's mathematics, and to determine more concretely the topology of the universe and the value of the cosmological constant, and then pass judgment *physically* about this theory. Metaphysically, they may find its implications unappealing, but that's not a reason to dismiss a theory without physical review.

Tipler's book *The Physics of Immortality* is a long and fascinating one, which begins with a long review of the theology of immortality and then turns to his original scientific ideas (the book is deeply hated by many physicists). Our own feeling is that we wish he'd gone lighter on the theology, just as we wish John Smart had gone even lighter on the Intelligent Design angle in presenting his fascinating black hole theory. Those connections are certainly there conceptually, but it's not clear why they're worthy of significant focus. Conflating metaphysics and physics is almost inevitable in cosmological debate, but there are very good reasons why physicists are skeptical of tightly coupling a physical theory with a metaphysical one. There is a great risk of trying to prove an implication rather than analyzing physical reality when one becomes too fond of a particular philosophical interpretation of physical law. Even more disheartening is the possibility for theologians to latch onto valid scientific debates, and spin them as "proof" that science supports an autocratic and literalist interpretation of their particular sect's theology[11]. Let us state clearly that science does no such thing. Whatever debates there may be about the cosmological origins and future of life and the universe, there is nothing in physics which supports a literal interpretation of any religious text. (Also, evolution as a mechanism for the development of complex structures is not considered to be undermined even by these arguments—intelligence may set an initial state for a universe, but evolution is clearly evident in biological and physical observation.) Even *if* Tipler's or Smart's wild theories are eventually proved correct, that doesn't mean that an intelligent black hole or universal computer—as either creator or afterlife—will exhibit any properties even remotely resembling the ideas of any particular religion.

While there is a great deal of emotional appeal to us for a physics which implies immortality, there is no certainty in any of this. There is also no certainty in the other direction. All we can state is that current scientific knowledge leaves room for debate about the physical and biological necessity of finiteness. It is outside the scientific mainstream to prefer such ideas, but there is not much less *physical* evidence for them (particularly Dyson's) than for many other, more popular cosmological postulates. There is statistical evidence for the finiteness of individual biological entities, which, as we've discussed in the biological immortality chapter, may not be a result of physical law but rather of incomplete evolution, but not much beyond that. We've never seen a universe go extinct, and many of the physical requirements for either a nihilistic or an eternalistic outcome are unsettled (the value of the cosmological constant, the topology of the universe, the invariance of the fine structure constant, etc.). That said, with respects to building AGI and deep-space travel vehicles, we see no reason not to try to achieve these amazing human dreams, even if it is shown that bringing about the Omega Point is impossible.

A THEORY OF EVERYTHING—NOT

Let's suppose the string theorists, or the universal computists, or some other group finally achieves their end goal—unifying quantum theory with gravitation and creating one single equation, accounting (in principle) for all phenomena in the universe. What happens then? Do the angels descend from Heaven, dancing in the streets and singing on the rooftops, serving out the wine in holy grails laminated with spinning black holes, bringing peace on earth and throughout all the galaxies at long, long last? More seriously: does science enter a whole new era, where every phenomenon observed is analyzed in terms of the one true equation?

Certainly it will be a revolution in particle physics, with likely implications for cosmology. Without a doubt there will be new engineering insights from such an equation, which will lead to new technologies—maybe those we envision in science fiction, like quantum gravity computers or time travel machines, or maybe other types of things we can hardly imagine now. Perhaps, as some maverick theorists believe, new light will be shed on the mysteries of biological processes like cell development—that the "spark of life" really emerges from the details of unified physics.

No serious scientist really believes that such a "Theory of Everything" (TOE) in principle will really be a theory of everything in practice. There are a number of technical and conceptual points that will stop this from happening.

As superstring theorist John Schwarz says, "the TOE phrase is very misleading on several counts…[And] it alienated many of our physics colleagues, some of whom had serious doubts about the subject anyway. Quite understandably, it gave them the impression that people who work in this field are a very arrogant bunch. Actually, we are all very charming and delightful."[12]

> For one thing, if string theory or universal computation or some other approach succeeds in making quantum theory and gravitation play nice together, this still doesn't explain *why* our universe is the way it is. It doesn't explain what physicists call the "initial conditions" of the universe: it only explains how things evolved from their starting-point. This may seem a small technical point, but it's a big one in practice: there may be many, many, very different universes consistent with equations as general as those of string theory, and the particulars of our universe may be very intricately bound-up with an understanding of its initial conditions.

Next, we may find that one thing the "universal equation" binding quantum theory and gravitation teaches us is that some very important things can't be explained or understood. Just as quantum theory has taught us that particles don't have definite states, the next wave of physics may open up new kinds of indeterminacy and unknowability. Heisenberg uncertainty may seem fairly certain compared to the determinacy of processes which we can reason about once we have a unified model—only time, and mathematics, will tell. We may wind up learning, in ever more exquisite detail, why our own finite, macroscopic minds are not up to the task of understanding the real world underlying the shadow-world they've evolved to see.

And this latter point ties in with the most serious problem with these approaches to physics: the problem of *intractable calculations*. Right now, the equations of string theory are so hard to solve that we can only really understand them in very special cases. There is an assumption that the mathematics of the next century will allow us to make more progress in this regard. But to what extent will this really be true? Right now, we can't do explicit quantum theory calculations to understand proteins or neurons, only very simple molecules. We can't do string theory calculations to understand electrons and protons, and we can just barely do string theory calculations to understand strings. To bring string theory to the next level, we'll have to be able to use it to understand electrons and protons, but that may be as far as it goes—the elucidation of the implications of these "micro-micro-micro-level" equations for macroscopic phenomena may remain too difficult for mathematicians, or computer-simulators, or even AIs to resolve for hundreds or thousands of years—or even forever. It's no coincidence that Wolfram moved from physics to Mathematica—but Mathematica isn't enough. We need not only calculational prowess, but superhuman intelligence to guide the calculations. Even our AI superminds may run into mathematical obstacles that we can't yet foresee or conceptualize.

In a way, it all comes down to Plato and his cave. From studying these shadows on the wall, we can't really figure out what the birds and trees and squirrels are, but we can learn more and more about them. We can make deeper and deeper theories, some of them explaining the interaction of very small bits of shadow, some of them explaining particular kinds of large shadow. It's vain to think we can fully understand reality, or even fully understand what "reality" means. The very notion of "everything," when taken seriously, intrinsically goes beyond our understanding. The hubristic search to understand everything is valuable, insofar as it give us the passion to understand more and more. In more reflective moments, though, we have to acknowledge, as Schwarz did in the quote given above, that we can't really capture the whole big universe in any little box—not even a box composed of the most sophisticated and fantastic equations.

Our own suspicion is that we're going to need AIs to make sense of "it all"—insofar as we can ever make sense of it. AIs are needed for two reasons: one, to do the increasingly intractable math; and two, to recognize the relevant patterns in the masses of experimental-physics data. For us, the subatomic, high-energy world is counterintuitive and mysterious, and this is largely because our brains and intuitions are evolved to deal with a particular, different aspect of the physical universe. We can at best interact with it through instruments, but then we must describe it with metaphor and mathematics—constructs we use to shoehorn an alien world into our mental models of understanding. What if an AI was given sensors and actuators attuned to the subatomic world? Then, to it, *our* everyday world would be the strange one, and the subatomic world would be intuitive—the sensors would feed subatomic perceptions into its mind the way ours feed us human-scale perceptions. It would think as readily about fermions and bosons as we do about cats and trees. What if this AI were also put in communication with us, or with other AIs more accustomed to the macro domain, and through these chains of communication we developed a shared understanding of these different scales of the world? What kinds of patterns would such a system recognize?

Our physical theories about things like gravity, force and power, electromagnetism and fluid flow, all come out of our everyday intuition. The mathematics is grounded in real-life experience, and real-life experience is often useful in navigating and structuring the mathematics. On the other hand, our physical theories about high-energy physics and unified quantum gravity are driven by very abstract mathematics and invented, ungrounded conceptual structures. No wonder they are so much more complex, so much less natural and workable. For an AI with appropriately rich subatomic sensors and actuators, the conception of simple, tremendously powerful "Newton's Laws of super-high-energy quantum gravity" would be a lot more plausible than for us. Such laws would not be exact and true any more than any other physical law, in that they would always be amenable to further exploration and refinement, but they would be tremendously powerful. We may never "understand" the physics theories that such systems come up with, but we will at least be able to follow the mathematics at a high level, and appreciate and utilize the empirical results.

To return to the questions at the beginning of this chapter, we can safely say that colonizing the solar system is merely a matter of engineering and economics. There's no theoretical barrier—it's not even an order of magnitude more difficult than colonizing the bottom of the ocean (though it is much friendlier to whales, octopi and dolphins and other creatures of the sea). As for time travel, journeying to alternative universes, and traveling outside the solar system with any means other than "generation ships"[13]—we really have no idea. Immortality, which seems no less ridiculous, is not theoretically contraindicated—it simply has no precedential basis. Maybe our current cell biology understanding is flawed, and immortality is theoretically proscribed, but current theory has no such restrictions. On the other hand, time travel for macroscopic objects, faster-than-light travel, universe-hopping, and so on, are impossible to model with our current understanding of physics—so, even if they are physically possible, they currently appear totally impossible to us. However, theoretical particle physics and cosmology are in flux. We can't agree whether the universe is open or closed, the cosmological constant is large or small, the fine structure constant varies or not, and so on—and that's without admitting any radically bizarre physics theories, and remaining in the context of assuming that the Standard Model and Special Relativity are both immutably correct. Only time will tell whether or not our evolving understanding of physics—or evolving physics, if nascent research that the fine structure constant varies[14] proves correct—will reveal such things to be possible.

---

[1] If you're not familiar with the strange properties of the microworld, this chapter isn't really going to fill you in adequately. There are loads of books on the subject; two favorites from our youth are Gary Zukav's *The Dancing Wu Li Masters* (Zukav, 1984), and Fred Alan Wolf's *Taking the Quantum Leap* (Wolf, 1989). Though some other books by these authors are full of what seems to us like New-agey nonsense, these are good ones.

[2] It takes nothing from Einstein's fabulous work to point out that Special Relativity also had major contributions from others—Lorentz and Poincaré in particular prior to the 1905 paper, and Minkowski in its 1908 formulation of non-euclidean geometry as a mathematical model for relativity.

[3] See (Greene, 2000) for a nontechnical review of the essential points of string theory.

[4] Of course, detractors said the same thing about the work of Einstein, Planck, and so on in the early 20th century. Scientists shouldn't assume correctness or incorrectness based on orthodoxy, they should allow modeling to lead the way to

experimentation—fortunately, they usually do, though their financial backers and society at large sometimes makes the path a difficult one.

[5] See http://www.digitalphilosophy.org, for Ed Fredkins writings on his theories.

[6] See Ray and Hart, 2000 for relatively recent work with Tierra.

[7] See http://www.valdostamuseum.org/hamsmith/TShome.html for Smith's online scientific documents, which range from the deep and insightful to the entertainingly frivolous.

[8] We ripped off this section header from Frank Tipler's book of the same name, which, along with his *The Anthropic Cosmological Principle* (co-authored with John Barrow) and David Deutsch's *The Fabric of Reality*, is a good primer on such wild theories.

[9] Available in this article at his Acceleration Studies Institute site: http://accelerating.org/articles/answeringfermiparadox.html

[10] Other than the particular Intelligent Design theories which state that the existence of complexity implies a literal Christian Bible (a paramount of begging the question if ever there was one).

[11] Theology specifically deals with the issues of the unknowable nature of reality beyond humanity's experiential realms. That one can define a topic as unknowable, and then assert as irrefutable your postulations about it, certainly vexes us. No matter how crazy a theory may be, in order to be science, there must be some possibility of verification — not a definitional statement of the phenomenon as being inherently unknowable except through blind faith.

[12] http://www.theory.caltech.edu/people/jhs/strings/index.html

[13] A "generation ship" is an idea, from science fiction, of a spaceship designed for many generations of people to travel through the stars. If extreme longevity or immortality were achieved, you could have a one- or few-generation ship, but it would still take a very long time to get anywhere outside our solar system without faster than light travel—which is impossible, according to current theories of physics.

[14] Here's another area of disagreement even in standard physics. The fine structure constant appears in the equations governing the interaction of elementary particles, and thus by extension the stability of all physical law. If it's really a variable and not a constant, the laws of physics may vary over time—though most likely subtly; and we can't really be certain, since we humans have a hard enough time understanding particle physics and evolutionary cosmology even if the laws stay *exactly* the same at all times. The idea that it may vary is the rare case of a recent controversial physical debate which arose from observation rather than mathematical modeling. See http://arxiv.org/abs/astro-ph/0310318 for the very speculative, very technical result that has opened up this debate.

CHAPTER 11
CURING POVERTY WITH COMPUTING?

In 1998, when trying to staff his new Internet/AI company Webmind Inc., Ben was having difficulty finding sufficiently brilliant and qualified software engineers in New York City, and decided to cast a wider net. He posted job ads on a number of Websites, and one of the responses was from a young Brazilian programmer named Cassio Pennachin. In spite of a lifelong fondness for Brazil (due largely to being born there when his American parents were living in Rio de Janeiro in the late 1960s), Ben was reluctant to hire a Brazil-based telecommuter; but Cassio offered to work for a month for free to prove his worth. He successfully did so, thus starting a software development collaboration that has been flourishing ever since.

Cassio, who lives in Belo Horizonte, in the state of Minas Gerais, first took on the job of fixing up some code for evolving new structures inside the Webmind AI engine using a variant of genetic programming. He proved to be an excellent manager as well as an excellent software engineer, and he accumulated assistants until, in early 2001, we had more than half our engineering staff—60+ people—in an office in Belo Horizonte, with Cassio as the worldwide leader of AI R&D Development.

We visited the Brazilian office of Webmind Inc. roughly twice a year, and each time we were hit by a major culture shock. It was not the Webmind office which was shocking—it looked essentially like the office of a US software company, and in fact was vastly more attractive than the New York office of

Webmind Inc., in an old, dingy building in the Wall Street district. Rather, visible during the drive to the city from the Belo Horizonte airport was very obvious poverty on the outskirts of the city and in the less desirable neighborhoods, which contrasted starkly with the totally Americanized/Europeanized modernity of the downtown business district. A tremendous number of tin, scavenged brick, and cardboard shantytowns could be seen just a few miles away from downtown Belo Horizonte—Juscileno Kubitschek's and Oscar Niemeyer's Modernist prototype city for Brasilia. Webmind Inc.'s Brazilian staff lived well; in some ways better than the US staff, given the greater buying power of their wages in Belo Horizonte versus that of the Americans in New York. They had nice clean apartments, relatively late-model cars, and furthermore most of them had maids to do their laundry and cook their meals, and lived in doorman-guarded buildings in the solidly middle-class parts of the city.

What was striking to us was the fact, pointed out by our Brazilian friends on many occasions, that Brazil has one of the greatest degrees of income inequality in the world. In Brazil, to paraphrase a nursery rhyme, when it's good it's very very good, and when it's bad it's awful. There is very little mass starvation in Brazil, unlike, say, Rwanda or Ethiopia, but there's a lot of abject poverty—favelas (shantytowns) encircling the major cities, in which people live in primitive and unsanitary conditions (dirt floors, no electricity, etc.) without any of the charm and naturalness that compensates for such things in a pastoral, rural context. The contrast between the life of the average Brazilian and the shiny new Webmind Brazil office, with its high-powered computers, nice offices, and educated, hip employees, could hardly be more striking.

Of course, this is just a particular illustration of a more general point. As we march merrily into the cyber-infused future, armed with our PDAs, PSPs and superpowerful laptops—and are increasingly aware of the next wave of biotech, nanotech and AI technology about to knock us off our feet and perhaps even transport us out of our bodies—it's worth remembering what a small percentage of the world's population the cyber-revolution is currently affecting in any direct way. Even in the US, there are huge urban and rural ghetto areas where computers are uncommon, and street-corner drug dealing or basement meth lab operation is a far more common teenage occupation than computer hacking. For the majority of people in "third world" countries, and really even in developing nations like Brazil, India, and China, the technological revolution is something that one sees on TV or in American or European magazines—if one even has access to those things.

Every time we go to Brazil to meet with friends and colleagues there, we can't help but think: "*but it doesn't have to be this way.*" The information revolution has the potential to benefit every human on Earth, not just those fortunate enough to be born into certain classes or in certain countries. This brings up a critical issue that we've largely ignored in the previous chapters in this book—the interaction between futuristic technologies and the socioeconomic structure of the human world we live in right now.

WHY ECONOMICS AT ALL?

There's an argument for setting socioeconomic issues aside and focusing solely on technology. After all, in the fantastic future, won't money be obsolete? We've all seen on TV's *Star Trek* how there is no need for money in the future. Molecular assemblers allow people and intelligent machines to create whatever materials they need, and there are ample sources of energy to run them. So, why not just forget about economics, and wait for it to become obsoleted by amazing new technologies which will undermine the very nature of supply and demand?

The problem is twofold: the nature of economics during the time before the Singularity is going to influence the nature of the Singularity itself, both as a reflection of the initial ethics of a post-Singularity society, and by determining whether there are haves who participate in the Singularity and have-nots who get left behind. If the Singularity arises in a situation where some people are valued and some devalued, this ethical principle will likely be absorbed by post-Singularity society, with murderous effects to the have-nots. This is the very situation so often commented on in science fiction's dystopian literature, from Philip K. Dick to William Gibson. While it is certainly possible that post-

Singularity society will emerge so dramatically as to cast off the prejudices of pre-Singularity society almost instantaneously, it is more likely to do so if during the run-up to Singularity, people make a genuine and sustained effort to cast off our centuries-old baggage of prejudice—including economic prejudice based on the incorrect notion that the poor are poor because they are lazy or stupid (rather than that the economic policies of the elite have made their rise out of poverty *on any large scale* almost completely impossible).

Traditional economics teaches that there *must be* haves and have-nots, as supply and demand economics requires this be so. However, there are currently enough food and energy resources worldwide that poverty could be eliminated, were we to have the willpower. Were population growth to slow down long enough for renewable food and energy production technologies to catch up with projected demand, we could probably avoid cataclysmic starvation and major resource wars (however, social and religious dogmas are a major barrier to family planning efforts worldwide, so we may not avoid these situations, simply by being afraid to discard old superstitions in the face of contemporary reality). Complete ecological devastation of this planet can most likely be avoided if space travel and nanotechnology are also developed, allowing us to both clean up polluted resources on Earth with molecular assemblers and disassemblers, and to gather new core resources for molecular assembly (and build colonies) in space. None of these amazing, sci-fi solutions can come to pass if we don't apply resources to solving these problems before their ill-effects cause worldwide catastrophe.

Changing the contemporary nature of economics to better reflect forward-looking ecological and social values is an essential part of a successful Singularity. We can program AGI systems to respect sustainability and fairness in resource distribution and usage (i.e. economics), but if we don't practice these things ourselves, we will either never reach the Singularity, or (in another familiar Sci-Fi scenario) AGIs may try to wipe out humanity because they see our thoughtless approach to economics as a disastrous drain on global material resources needed to further the development and growth of intelligent life on earth.

While the economics of food and energy production are the two most crucial places such reforms are needed, the diffusion of technology into even poorer parts of the world may be one way to jump-start those reforms. One reason is, as we'll see below, technology can be used to set up workable hypereconomic markets. In such markets, values other than those established by professional capitalists can be given dollar values, which encourages business to respond to social and ecological demands by being forced to confront a cost-benefit analysis head-on (rather than buried in operating costs, and offset by subsidies from political allies). On the large scale, this diffusion process may be viewed as an inevitable consequence of the advance of technology and the overall trend of globalization; but in practical terms of nitty-gritty human reality, the expansion of technology beyond the world's economic elite is by no means an automatic process. Rather, it is the result of huge amounts of hard work and careful planning by dedicated people in the growing middle classes of developing countries. Vastly more work will be required to finish the process of disseminating technology and its many benefits across class barriers, including more cooperation from those of us in developed nations. There are technical problems involved here, but there are also major "purely human" problems, with tremendously complex political and cultural dimensions.

The individuals who are working to improve the human condition by spreading advanced technology throughout the human population as a whole are just as deserving of "cyber-hero" status as the people who are working to add impressive new functionalities to our supercomputers or mobile networks, or to create new gene therapies or biocomputing devices. The importance of this aspect of cyber-development should not be underestimated, not just in an ethical sense but in the context of the overall course of technical and human development. In fact, we'll put forth a somewhat radical proposition in this regard: we believe that the nature of the next phase of the tech revolution will be extremely different if it really is spread across the globe, rather than restricted to a small group of economic elites. Technology developed within a culture of inclusion and compassion is going to be very different from technology developed within a culture of elitism and ethical indifference, in thousands of both obvious and subtle ways. If we want our advanced technology to be friendly and compassionate to us, we'd better develop it within a culture of friendliness and compassion. This is an issue that cuts at the very contradictory heart of modern cyberphilosophy, confronting our wildest dreams and futuristic visions with the grittiest aspects of human reality. As we'll see, it's an issue on which different contemporary cyber-visionaries take very different views.

For example, it's not all that difficult for the mainstream commercial computer hardware/software industry to provide tolerably-priced solutions for low-income Third-Worlders—but in order to motivate them to do so, competition from the open-source community and other private-sector, academic, and government "do-gooders" seems to be a very effective mechanism. Purist capitalists would argue that government subsidies for such a project are unethical and point to a *necessity* to fail—if those people can't produce enough with their own labor to pay "fair market value" they should not have the resources—but most of those kinds of people live in fully-developed nations like the United States and countries of Western Europe, where decades (sometimes centuries) of government subsidies have already elevated enough of society to a solid middle-class that the idea of a "free market" can even exist. For example, without huge government projects ranging from the Panama Canal, to the massive agrarian restructuring projects in the Tennessee and California Central Valleys, to the Eisenhower Interstate system, the United States would not be the economic leader it is today. Government paternalism and corporate welfare have helped build every major economic empire in history, even that of the U.S., despite the adamant claims otherwise put forth by purist Libertarians. Even today in the United States, agriculture, energy production and defense— to name the big three—are heavily subsidized industries.

WHO DECIDES WHAT SOMETHING IS WORTH?

To better understand these issues, it pays to delve a little bit into economic theory—not so much the mathematics of economic prediction and modeling, but the philosophy of economics. What are these things called markets? What are prices, and what really determines them?

When people talk about economic value, they generally talk about "the market" as the deciding entity for value of all goods and services. We all procure goods and services using money, which is a token that refers to a unit of market worth as abstracted out over the whole network of production and consumption. Prices are considered reasonable if they are "what the market will bear" and are driven up and down by "market forces." There are markets for goods and services, there are currency markets (which are markets for market value tokens), stock markets (where portions of a company are bought and sold), and so on. There are even markets for derivatives and futures of markets—that is, there are markets where predictions about future value are explicitly bought and sold, rather than implicitly so in the current purchase price of a good or service.

But what is "the market?" In essence, markets are the following: systems for assigning value to things, based on a specific collection of information about those things. For example, when the stock market values a company, it takes in to account a wide variety of factors, including: profit/earnings ratio, quarterly profits, cash-in-bank, number of shares of stock issued, analysis and value of competitors, recent news about the company's strategy, who the CEO is, and so on. A commodity, such as a car, is priced based on competing prices of similar things, cost of raw materials, R&D and marketing costs to bring it to market, production and shipping costs, perceived extra value (such as trendiness) for the customer, etc. A whole host of weighted parameters are taken into account—and sometimes it's not clear what all the parameters are and what weights they will receive, even for professional analysts and traders.

Price is the single valuation metric to which most laypeople are ever exposed; especially with regard to the things they purchase the most often: consumer goods and services. Most consumers do not know anything at all about how a particular good or service was assigned some particular price. Some economists claim that you need only look at price and none of the other parameters really matter, but practitioners who manipulate prices professionally (such as real estate or currency speculators, or market-maker stock analysts) know that they need to analyze and manipulate a variety of pieces of information in order to move the "big piece of info"—price. However, when presented with the idea of exposing more of these parameters to more people, naturally the professionals claim that laypeople are not smart enough (or, if being generous, that they're too busy) to understand *any* other metrics besides price.

Consumers generally take it on faith that market value is indeed fair. It is assumed that all resource usage that went into bringing the product or service to market is accounted for in the price, and that things of equal price reflect roughly equal value. However, there are a wide variety of forces at play that can make this not the case. Some practices, such as "dumping" (releasing a product below cost and sustaining a loss to drive competitors out of a market), "price fixing" (when companies in a market get together and set an artificial price for their products by agreement), and so on, are illegal—but they still occur, sometimes without the perpetrators ever being caught by government regulators. Other practices, such as controlled production (making more or less of something available to push its price up or down), tiered pricing (giving customers different prices based on their value as a customer), government subsidies, etc. are legal and widespread. Some speculators make their living entirely by manipulating markets, usually by exerting price pressures through volume purchasing or selling, and by releasing or withholding information they have about the market to the media. Government central banks and economic planning entities, such as the U.S. Federal Reserve Board, also manipulate prices on a macroeconomic scale by setting interest rates—an act which can move prices up or down across pretty much all sectors of the economy. Either way, prices can be deliberately manipulated by a few powerful actors in a market. Just because a system is complex doesn't make it immune to some amount of perturbative influence, even direct control of a subset of its parameters.

Anyone who has access to sufficient information can manipulate a market, as long as they have sufficient financial and/or information dissemination resources available to them. If they don't have the wealth to do it alone, they can cooperate with a group of people who do, and exchange their ability to analyze and manipulate the information parameters that go into pricing for a share of the wealth. Indeed, this is essentially the job of an analyst at a financial institution. Some sell their services to big companies, and others affiliate with a small group of wealthy individuals and form a "hedge fund."

While it is currently the exclusive domain of specialists, in theory any group of people, even a group of consumers, can get together and manipulate a market. Boycotts are the most successful way that consumers have done this so far. However, boycotts aren't usually organized to drive down prices, though they could be. Usually they are organized for social or environmental reasons. This brings us to a very important issue: how are social and environment costs accounted for in markets? Historically, in the absence of government regulations about the environment or workplace standards, the only reflection of these costs was in the price of raw materials and labor (which, often, have been subsidized by governments—incurring a double-dip hidden cost to the consumers/ taxpayers in society at large). When laborers feel exploited, they strike. When consumers feel that way, they boycott. Sometimes these actions are successful, and sometimes they are "broken" (frequently with government assistance) and fail. However, these confrontational methods needn't be the only way for consumers' values to be heard by corporations and governments.

The fact is that markets, while complex, are not supernatural. ***People's actions are what determines market value of commodities***, companies and currencies—whether that be the actions of a large group of consumers, a wealthy and powerful currency speculator, a market-maker institution, a government body, or whomever. Ultimately, *all of us* are "the market."

HYPERECONOMICS: EXPOSED VALUATION, SOCIALLY RESPONSIVE MARKETS

While the evolution of economies is not as easy to measure as that of technological areas like nanotech, biotech or computer hardware, nevertheless it's qualitatively clear that in many senses, economies are advancing exponentially in their complexity and sophistication. One indication of this is the increasing amount of money that flows through sophisticated financial instruments like hedges, futures, derivatives and options—while the uses of these instruments are oftentimes unethical, their increasing prevalence nonetheless represents a kind of increasing intelligence and reflectiveness on the part of the "economic mind" of human society.

As with any other exponential process of creative evolution, the long- or even medium-term outcome of economic evolution is difficult to project; but a careful analysis of the Internet economy and other current trends allows us to make some educated guesses regarding where things might be heading. One concept that jumps out vividly from this kind of econo-futuristic thinking is the notion of "Hypereconomics." If the future eventuates in a way that allows humans and/or human-like AIs to continue to exist and evolve, then the economy that they develop is likely to be a hypereconomic one.

The term "Hypereconomics" was coined by the late Sasha Chislenko to describe an area of economic theory and experimentation, which strives to utilize computer technology to create a combination of what Stephan has termed "exposed valuation markets" and "socially responsive markets." Exposed valuation markets are, simply, markets in which common actors (consumers and producers), and not only market experts, can exchange information about underlying parameters of the transaction's value in order to arrive at a fairer price. Socially responsive markets are markets in which the social costs and values of the society in which the transactions take place are accounted for directly in pricing. Developed as Hypereconomics by Sasha and Stephan, the theory draws upon other work in recommendation and auction systems, agent systems, computer-mediated communications, AI, and some historical economic theories—particularly the ancient techniques of bartering and instantaneous negotiations ("haggling").

To understand Hypereconomics, you need to understand what an exposed valuation market really is, and why you might want one. In an exposed valuation market, you as a consumer can set value for certain underlying parameters directly. A producer/vendor can then choose whether or not to accept those modifications of value. Groups of like-minded consumers can aggregate to put economic pressure on, or offer economic incentives to producers to respond to these underlying values. There is some similarity to currently popular auction systems like eBay—except that there is much more information than price bid which a consumer can put forth to a seller. Let's give an example.

Say you wanted to buy a pair of sneakers. Normally, you shop around online or in stores, find the lowest price, and make a purchase. However, you don't really know, unless some activist group or investigative journalist decides to take a look, whether or not those sneakers are produced in a way which you are happy with. If they aren't, you can choose not to buy them. However, the company only ever really knows one thing: sales figures. They can infer that if they're in the media spotlight for some particular activity, and sales figures move dramatically one way or the other, that this is the cause (but it may just be a coincidence, as any logician or statistician can tell you). A somewhat more precise metric is opinion-polling and market research; and reading letters from (the self-selecting set of usually dissatisfied) customers who've taken time to express their views is about as direct as it gets.

What if you could specify that there were certain criteria about how you'd spend your money to benefit a company that adheres to your social values? Hypereconomics is an attempt to expose those parameters as something you as a consumer can place value on—directly, not implicitly. Producers can then know better what you want, and how much you're willing to pay for it.

In the sneaker example, maybe you'd place a composite bid like this:

Average offer price across all known vendors:	$80
Sweatshop-free labor:	$5
Renewable, low-water-usage fabrics:	$5
Renewable energy used in production:	$5
Renewable energy used in transport:	$5
Company supports hypereconomic research:	$10

Available in my favorite color:	$2
History of product durability:	$10

The average, gained by collecting asking prices from a variety of sources, is taken as a baseline. A company which meets none of your value criteria knows that they can get you to pay $80 maximum, but more likely $38, whereas a company which meets all of your criteria knows they can earn up to a maximum of $122 for their efforts. A company that meets all of your criteria can make a minimum of $42 more than their competitors. Only meeting some of the buyer's criteria reduces overall gain for the producer, but it's better than nothing. This gives you a chance to tell companies—directly—what you value in their products and to what degree you're committed to that value, and for a producer to decide if the extra investment in doing something others may not be doing socially or ecologically is financially rewarding (or at least viable). In a hypereconomic system, a producer doesn't just have to merely either ignore you or meet your demands and hope you'll pay what it cost them to do so. A producer can respond to these bids, and could for example respond that they need an extra $5 over your $10 combined energy bid to account for renewable energy sources that (unlike oil) are not government-subsidized, and then you could decide if you'd pay it or not.

Traditional economists like to say "that's all already accounted for" in pricing. They're almost right—such things sometimes get accounted for in pricing when people (through boycotts or convincing their government to enact regulations) act to make it so; but the system does not naturally accommodate such information exchange, and as such these concerns are most often communicated between consumers and producers in an adversarial manner. Both sides lose time and money on costly lawsuits and publicity campaigns.

However, maybe the traditional economists are right: you're too stupid or busy to understand all these things, and you can only fathom one parameter—price. Much more likely, you *understand* these ideas very well but you don't have time to both earn a living for your family and also unearth all the details about what companies behave in a manner compatible with your social and ecological values, and to then embark upon boycotts, letter-writing and publicity campaigns. Or you think, not totally unjustifiably, "what can only one person do to change the mind of a huge corporation?" The traditional barter and haggle systems had going for them the idea that you could directly negotiate value with another producer (since everyone is at some time both a producer and a consumer), but they don't scale very well to large societies. This is why Hypereconomics was co-developed alongside the global Internet and ideas in agent systems, recommender and auction systems, and AI.

HYPERECONOMICS AND COMPUTER TECHNOLOGY

Computer technology has the potential to make a hypereconomics-type system possible on a global, rather than small village, scale. Information gathering is very time consuming, and even placing a bid at an auction requires more time investment than simply going and paying a fixed price for something. Furthermore, the complexity involved in information gathering can be very prohibitive for many people. Feeling overwhelmed, most people just accept the current system as it is and get on with it. Computer technology has the potential to overcome these practical problems and make Hyperconomics a general reality.

Agent systems are computer programs which can carry out instructions given to them by their users, in a somewhat flexible way, without the user being present. Specifically, agent systems are characterized by the following agent criteria:

1. Agents can communicate with other agents. They aren't just some piece of software on your personal computer that might carry out calculations in your absence. An agent could actually communicate with a producer's agent software and *buy* the shoes you wanted without you even being at your computer.

2. Agents are given some autonomy in decision making. In the example above, you might tell your agent to be flexible in that any product with 50% or more of your criteria that beats its competitors in price by more than your aggregate total social parameter bid is acceptable. Or you might say that you'll pay $2/criteria more for a 100% match. Finally, you might say that you want it to find the best match it can in 2 weeks because your current shoes are starting to fall apart. The choice is yours, and you can spend the most effort configuring the most important parameters and let the rest alone.

These two criteria are very important to Hypereconomics. You don't want to spend all your time researching and shopping for products that meet complex criteria, so you want an autonomous agent which can do it for you—a digital, socially-mandated butler. Also, you want your agent to have some flexibility to make purchases so that it doesn't wander the Internet forever looking for something that's just not available (but since it is communicating your criteria to producers, maybe the next time you buy shoes someone *will* meet them all).

Hypereconomics is also related to software programs called "recommender systems" and to auction agent systems. Auction agents, like Hammersnipe, are simple agents which you give a maximum bid for auction items on sites like eBay, and then they bid for you at your specified number of seconds before the auction closes. They make purchases, like a hypereconomic agent, but have none of the more sophisticated functionalities that the theory of hypereconomic agents envisions.

Recommender systems (also called collaborative filtering systems) are the systems you see on sites like Amazon.com, which tell you things like "people who bought Skinny Puppy CDs also like to buy Dave McKean art books." More sophisticated systems have been developed in the past, which deliver a broader range of recommendations with more immediacy. For example, if you are known to be interested in the cellular biology of senescence, your collaborative filtering agent might get alerted every time someone publishes in that area, and every time someone asks a question, posts a message, or performs a search on some select scientific news and search systems.

Extrapolated to a hypereconomic agent, this sort of information aggregation could be used not only to advertise new products you might want to buy based on what others with similar buying histories have (which is how it's used now on sites like Amazon), but to put your agent in touch with the agents of those like-minded consumers in order to form purchasing pools. By forming these ad-hoc group buys, a hypereconomic agent system can collate previously disparate single individuals into a large purchasing bloc, which can then offer their incentives to producers in bulk. Now, instead of one do-gooder trying to buy shoes with sweatshop-free labor and sustainable practices, you might have ten thousand, or a hundred thousand, or a million. All of a sudden, producers have incentive to listen, and they have it in a non-confrontational manner. They're being offered financial incentives first, rather than being threatened with lawsuits, boycotts, or government regulations. Consumers also win; because even if you enact change with lawsuits, boycotts, and regulations, you still pay higher prices than you would have had you gotten together and negotiated with the producers (which you can't do unless enough of you can "get together").

What happens is, everyone whose criteria for a purchase is some percentage similar to yours is contacted by your agent and asked if they'd like to participate in a group negotiation for the overlapping criteria (your agent starts by asking for *any* overlap, and then can ask lesser participants to consider the other criteria or get bumped if the greater overlap group gets sufficiently large). Either the other consumers have told their agent to always agree, ask them first, or never agree about joining blocs—but eventually your agent will get enough responses from other agents that it can know if it can represent a bloc or if it has to go it alone.

Some privacy is lost in that your agent represents a fair amount of your social criteria to the public, but your buying habits are already known by your credit card company and a host of other organizations, and obviously anyone who wants to enact social change has to at least give up privacy about what they believe with respects to those social issues. Unlike with current systems, though, what you gain by allowing your purchasing habits to be shared with other agents isn't just more advertising, but an opportunity to negotiate with producers to meet specific social, ecological, and price goals. They have an opportunity to tell you if your demands are too unrealistic, and, either as yourself or by setting up your agent with negotiation tolerances, you can decide if you can reach an agreement. If all you care about is lowest absolute dollar price and nothing else, a hypereconomic agent can help you find that too, but it offers a lot more as well. With advanced artificial intelligence, obviously your agent can become even smarter about how it carries out these bloc aggregations and producer-consumer negotiations on your behalf.

HYPERECONOMICS AND DEMOCRACY

Hypereconomics is philosophically aligned with another idea whose time has, as far as we're concerned, already come (though the world seems slow in realizing it!): direct democracy. Like our current economic system, the American electoral college and the U.S. Congress place only indirect control over outcomes in the hands of the public. The same is true of the parliaments of other modern democratic nations. It is argued that putting more control in the hands of the public is foolish, because it will lead to mob rule and the elimination of minority rights—but we feel that stems from a misunderstanding of the alternatives to the current systems (which are, by the way, themselves amenable to mob rule, and to compromise by various market and voting "cheats"). Allowing people to vote frequently about a variety of highly specific issues, via the use of appropriately secure Internet technology, would seem a highly natural and powerful way of cutting through a lot of the confusion induced by the bureaucratic and indirect nature of modern representative democracy.

Direct democracy, like the "direct economics" of Hypereconomics, can help reduce frictional costs of bureaucratic structures and introduce efficiencies in the information aggregation and transfer that underlies both politics and economics. While this may not be to the liking of those who have made their political and financial fortunes off exploiting bureaucratic inefficiencies and power consolidation, the fear that direct democracy and economics necessarily leading to "mob rule" is overplayed. The philosophical nature of the institutions involved needn't change radically in order to more directly involve people in voting and economics.

Direct democracy methods, such as referenda, can be used to allow people to gain more control over their lives through providing a mechanism for allowing them to have more of a voice in politics. Political action committees, unions, and corporations have money and access which gives some private "representation" in government, but only to those who either can buy it, or who join an ideological bloc. Pragmatists and centrists—that is, most people—are left out. Government, beholden to ideological dollars, winds up spending more of its time meddling in people's personal affairs (such as sexual relationships, and what is or isn't art, or pornography, or "sinful") rather than dealing with real public policy issues. In terms of making opinions heard about public spending, both direct democracy and hypereconomics can play a role. If referenda also addressed cost, people could choose more wisely whether or not they really want to spend their money on a program they think they support. Strong supporters of various programs could offer to pay more to have them put into place, requiring them to put their money where their ideologies are rather than forcing all taxpayers to subsidize their beliefs or pet pork projects. Progressive (as opposed to "winner takes all") voting schemes could allow citizens to allocate blocs of preference, deciding that they would prefer position X for cost Y, but allowing them to order by preference possible compromise outcomes should their favorite not win. This would increase voter satisfaction and give government more information about what the people really want.

Both representative and direct democracy can coexist—people can't vote directly on every issue. A mechanism for ad hoc voting, such as more secure online voting systems than are currently implemented, would allow a self-selecting group of voters to have more frequent input. This would offset the "pure money" influence currently wielded by the PACs, but would still need to be ameliorated by representation in Congress. Representation, including a legislature, an executive with a veto, and an empowered judiciary, is also still needed to protect minority rights from "mob rule" and to deal with all the issues that don't interest voters directly; but there is still *a lot more room* for direct participation in politics and economics. A national referendum system would allow some more direct representation for those who aren't part of the Internet cognoscenti or a PAC, but as the Internet and similar technologies spread, more and more people could have their voices heard in politics and economics through such systems. The power elites may fear this, but if it is allowed to happen, we think they will find everyone benefits from less time and money spent on inefficient mechanisms for determining "the will of the people."

The California "initiative system" is a nascent attempt at more direct democracy. It is considered a failure by many, because of the massive problems in the state with budgeting, education, and prison management. However, the problem in California is not the initiative system, but rather the lack of transparency and accountability within the management systems of California government (and the anti-meritocratic policies of the teachers' and prison guards' unions), coupled with the several (some say up to 30) billion dollars Enron and its cronies bilked from the state during the manufactured energy crisis[1] (a situation which could possibly have been ameliorated by a more transparent economic system such as hypereconomics). However, the initiative system itself is a good start in terms of a direct mechanism for government actually hearing about the will of the people. The problems with the California system are how expensive it is to get an initiative on the ballot and debate it—a situation which is changing with the advent of the Internet and thus the reduced cost of spreading the word about political issues—and the fact that, like in all American politics, the "winner takes all" voting system prevents rather than encourages compromise and cooperation; meaning that American politics swings like a pendulum between Left and Right ideologies that can excite people in an all-or-nothing race, rather than focusing on practical issues.

Both less costly and less rigid voting systems are needed to make direct democracy and hypereconomics practical and effective. Some complain that the desire for such systems reflects the "wishy-washy" nature of voters, who can't commit to a single ideology. However, the fact of the matter is that it isn't ideological purity, but rather compromise and a willingness to change your mind in light of changing data, which results in any practical solutions to social and political problems. Despite campaign promises, this kind of compromise is done every day in Washington and other capitals. Most people understand the necessity of such compromise; they just want more stake in these deliberations and the agreed-upon solutions.

The notion of frequent, issue-specific online voting connects naturally with the research that has been done into alternative voting methodologies, including Duncan Black's hybrid Condorcet/Borda methodology, Steven Brahms' contemporary Approval Voting system, B.L. Meek's method for Single Transferable Voting, and range voting (which is commonly used for recommender systems). All these alternate voting methods have significant resonance with the Hypereconomics concept. They are all methods for ranking choices, rather than simply allowing plurality or majority to override all other options, and each one's algorithm is best suited for certain situations and is most vulnerable to different "attacks."

Black's system uses the Condorcet method for ranked voting, which allows voters to rank all candidates (people, laws, whatever) in an election by order of preference. A pairwise matching is then done between all candidates, and whatever candidate beats all others is the winner. However, in addition to allowing ties the Condorcet system may lead to pairwise defeat cycles (A defeats B, B defeats C, and C defeats A, for example). Black uses the Borda Count to resolve this. The Borda system gives N-1 points for each first place vote, N-2 points to each second place vote, down to N-N points for each last place vote. The winner is the candidate with the most points. Black's system is preferable to a strict Borda count, because if a single candidate pairwise defeats all others that is a more accurate selection than a candidate who wins a Borda count by virtue of having more points, but never having been first choice for anyone.

Brahms' approval voting system is a Boolean point assignment system in which voters can either approve (score +1) or disapprove (score +0) of a choice. It is a restricted variant on general range voting, in which voters can assign a score (most commonly either Brahms' 0-1, or 0-9, or 0-99) to each choice (the Borda count is an implicit range vote). Most range votes use a sum of points, but a median can also be used. Any range voting method can be used for either ranking or single-winner elections, and thus can be applied to determining a finite compromise set as well as a particular single outcome. However, with single-winner elections, range voting is vastly inferior to Black's method for the reason stated above.

Hare-Clark voting, known more commonly as the Single Transferable Vote method, is often used for multi-choice elections (such as legislators in a British-style parliamentary system). In STV, voters rank candidates, and a candidate must meet a threshold (quota) of votes to win. What happens is equivalent to a multi-round count, with only first-place rankings counted in round one. If a candidate meets the threshold in round one, then the voters beyond the quota who chose that candidate as #1 have their votes transferred (as their #1 is already elected). In each round a candidate choice is elected, until all slots are filled. This system is used for parliamentary elections in Britain and Australia, and is what is being referred to when a commentator discusses "proportional representation." New Zealand has, in some elections, adopted B.L. Meek's modification to this system, which allows voters to rank candidates equally and requires a computer to run an iterative approximation algorithm which distributes the remaining votes of an elected candidate fractionally to all subsequent selections provided by the voter according to the voter's rankings. It avoids the possibility of infinite recursion in vote reassignment in a strictly rounds-based counting, but is uncommon due to the complexity of counting it (and people's mistrust of computer-mediated voting at this time).

There are strong arguments why such voting approaches should be implemented in practice, replacing the simplistic and less effective techniques currently implemented. However, there is also a more futuristic possibility which is worth exploration: the use of hypereconomic agents as "automated voting agents." The idea is that you tell your voting agent what your attitudes and preferences are, and it then decides how to vote for you, based on its analysis of your desires and the optimal strategy given the voting schemes available in the elections. Voting agents won't be as confused as the average human by a panoply of voting schemes, so they would be free to use the voting methodology most appropriate for the task at hand (for example, range voting is very poor unless there can only be one winner in the selection). This methodology could be applied to voting for a favorite film or voting for President. Of course, the usual variety of dangers arises here: the writers of voting-agent software could acquire excessive power over elections.

Hypereconomics and direct democracy are not just aligned algorithmically, but also philosophically. Both are designed with these socio-ethical groundings in mind:

> 1. Fairness—both minorities and majorities should be allowed to have their voices heard, rather than being aggregated into massive, externally homogenous blocs based on what *someone else* thinks they believe.
>
> 2. Openness—nobody should control an inordinate amount of information about the system and be able to use it to game everyone else.

Though it is not a necessary theoretical implication of the Hypereconomics paradigm, the practical development of Hypereconomics is likely to be most successful if one adds an "ease of use" criterion to encourage more people to participate. Hypereconomic methodologies could be used to mitigate the complexity of allowing more people to participate fully in economic and political decision-making in ways which let market and political leaders *really hear* what the people are saying, rather than receiving that information through several levels of indirection, and often conflicting ones. Also, it lets people organize themselves into ad hoc leadership blocs, rather than relying on someone with large, centralized resources to run everything.

Hypereconomics would be good for business and political leaders because it allows them to better make choices which really will win them customers and votes, because they can negotiate with the people more directly. Money and time are saved by not having to pay for expensive and often inaccurate opinion polling and market research data, and costs can be further reduced by not producing unwanted products or funding unwanted government programs. Right now, companies and governments still often build new projects first, and then find out what people think of them. Even if opinion-polling in advance finds that a product or program is unpopular, funding for such things can come through such indirect routes that they often go ahead anyway. With a hypereconomic system, people can "vote with their dollars" (or, in the case of political systems, create instant referenda) to help businesses and governments decide if they're really giving the people what they want—and the people will tell them exactly what it's worth to them that a certain product or program be created or withheld.

None of this prevents wealthy individuals, corporations, or governments from pursuing programs they think are beneficial but which the body politic has not yet become ready for. Innovative, forward-thinking programs can still proceed, but people's complaints can more directly be taken into account. Perhaps the new Megadam for hydro production should be built somewhere other than everyone's favorite swimming and fishing river, even if it still will get built. It is the day-to-day of politics and economics in which people really feel a need to participate, and currently, even in the "ultimate democracy" of the United States, don't. Hypereconomics is a system which allows people to negotiate real economic and political value for their social and ecological concerns, and to do so in a less confrontational manner which benefits everyone involved.

"WEAK HYPERECONOMICS" IS ALREADY FIGHTING POVERTY

Although the creation of software explicitly oriented toward Hypereconomics is progressing rather slowly at present[2], there is a fairly significant movement toward Hypereconomics occurring as a result of people pursuing other related goals. Elements of weak Hypereconomics emerge from some more traditional economic systems which, while they may still cling to single-parameter valuation (price only for the final product), attempt to expand consumer choice and social responsiveness, or to allow consumers to form ad hoc buying groups (currently, usually only for price discounts, but sometimes to express social or ecological value in a market). An obvious example is the Foresight Exchange[3], which attempts to be a literal "marketplace of ideas" in which people can pay for "idea futures"—that is, to fund, through a market system, the development of futurist ideas which traditional markets may not touch. Idea futures and collaborative filtering are directly on the path towards transparency and social responsibility in markets, but it's the ad hoc networks of people on the Internet, banding together to make their economic and political voices heard, that is the underrated motivator of society in the direction of Hypereconomics.

Already there are companies which are marketing products where you can pay a premium for sustainable development, sweatshop-free labor, and so on. Other than in the "Hippie Villages" of places like Berkeley, CA, or New Hope, PA, the place where these markets thrive is on the Internet—where geographically disparate people form a large single purchasing bloc. Currently, there is no mechanism for the "exposed valuation" part of Hypereconomics, and as such the system still suffers from limitations on scale and sophistication, but as much as traditional economists try to say otherwise, the Internet inherently has hypereconomic properties in that there are mechanisms for back-channel information exchange, aggregate knowledge sharing, and forming of ad hoc purchasing blocks which are just not possible in the physical world; unless thousands or millions of like-minded people happen to all live in the same place and have time and space to all get-together and plan such economic activities.

With the Internet, Hypereconomics is almost inevitable, as a whole world of complex information and global economic exchange has opened up to a whole variety of people who are not experts in these fields.

People on the Internet gather together, in order to give their patronage to sustainable energy companies, organic farms, obscure technologies, alternative religions, and so on. In the political realm, organizations like Moveon.org, or the ideologically opposite LetFreedomRing.com, attempt to give more political power to individuals through a more flexible political-action-campaigning system than the traditional, expensive door-to-door methods. Sometimes "the people" make foolish choices, but that happens regardless of what political and economic systems they're encumbered with. The current American system is pretty good, but it could be better. Though Hypereconomics (combined with a "computer on every desk" as envisioned by various government and charitable projects, such as the Net PC project we'll discuss below) would help more, the current hypereconomic trends on the Internet are already helping the poor and disenfranchised in three ways:

1. Exerting aggregate pressure on companies to pursue labor-friendly and ecological practices in ways they could not before the Internet brought them all together.

2. Exerting pressure to pursue fair pricing globally through opening of global markets to not only big corporations, but to small producers and individual consumers.

3. Allowing individuals to create ad hoc buying and political action blocs in ways they previously just couldn't afford due to the high costs and large distances of doing national or global business and politics without the Internet.

What is going on now in these areas is far from everything we'd like to see in a true hypereconomic system, but it's a decent start. Building upon these trends, we'd like to see funding for real hypereconomics development emerge—private or governmental. Both the private and public sector have a lot to contribute to, and to gain by, improving economic sustainability and fairness globally (even if their current ideological biases don't let them see how).

BRAZIL, A CASE STUDY IN ECONOMIC INEQUITY

Now we'll zoom back to the economic present. Hypereconomics would be a wonderful development, but right now we're stuck in the pre-hyper-economy, and this is the context in which our technological and human development is unfolding. It's worth trying hard to understand the socioeconomics surrounding technological development, rather than pretending that this development exists in a fictitious socioeconomic vacuum.

As an example of the socioeconomic and technical situation in many developing nations today, we'll look at Brazil, a place in which we have a strong interest due to having so many friends and colleagues living there. With a 1999 GDP of $555 billion, Brazil is the tenth largest economy in the world; and is also highly economically diverse, with huge variations in development level across industries. Its economic history is rocky, but the last decade has been a good one. In July 1994, led by President Fernando Henrique Cardoso, Brazil embarked on a successful economic stabilization program called the Plano Real (named for the new currency, the real). The success of the plan surprised even many of its supporters. Inflation had reached an annual level of nearly 5000% at the end of 1993; and under the Plano Real it dropped to a low of 2.5% in 1998, climbing slightly in the years since but remaining in single-digit range. In January 1999, the country successfully shifted from an essentially fixed exchange rate regime to a floating regime. US direct foreign investment has more than doubled since 1994, and overall trade has more than doubled since 1990. All in all, after many years of chaos the economy finally seems to be working.

But, in spite of this success story, economic inequality in Brazil remains just about the worst, if not the absolute worst, in the world. The standard way of measuring inequality is the Gini coefficient, which ranges between 0 and 1: 0 if everyone in the country earns exactly the same amount; 1 if one person earns all the money and everyone else earns nothing. Throughout the '80s and '90s, Brazil's Gini coefficient has been around .60, compared to numbers in the .3-.4 range for Southeast Asian countries, and the .4-.5 range in Africa. Latin America as a whole tends to have more severe income inequity than most parts of the world, except for Africa. However, even for Latin America, Brazil is extreme: the average Gini coefficient for Argentina, Bolivia, Chile, Colombia, Costa Rica, Mexico, and Panama was 0.42 in the early 1990s.

Country	**Gini Coefficient**
Argentina	.49
Bolivia	.51
Brazil	.61
Chile	.58
Colombia	.56
Mexico	.52
Venezuela	.50

Table 11.1 Comparison of the Distribution of Income (Gini coefficient) in Selected Latin American Countries[4]

What impact has the Plano Real had on this situation? It drastically decreased the amount of dire poverty in Brazil, by increasing the income level of all classes—inarguably a very positive thing. Its impact on income distribution, however, has been vastly less dramatic, although also significant.

There are serious human social and cultural issues here that cannot easily be addressed by economic adjustments alone, even extremely savvy ones such as those introduced by President Cardoso. The cultural divide between the Brazilian middle class and underclass could hardly be more severe. The Brazilian middle class lives essentially like the American or European middle class. For this small segment of the population, the Brazilian educational system is outstanding. Brazil's top universities, attended almost entirely by middle- and upper-class youths, rank alongside the best institutions anywhere. On the other hand, Brazil's annual expenditure per primary-school student is 12.8 times less than for its university students, compared to a mere three-fold difference in the United States. The money that is spent on primary education is far from equally distributed, and ultimately contributes to social inequalities in a major way.

The drive to educate the Brazilian masses has been reasonably successful during the past few decades, despite this paltry government funding. Illiteracy runs at 9.4 percent among Brazilian women between 30 and 39 years old, but drops to 4 percent for the 15 to 19 age group. For men in the same age groups, the rates are 11 percent and 7.9 percent, respectively—but these figures don't tell the whole story. There is a large gap between basic functional literacy and having the educational background to fully participate in the emerging global economy. Graduating from a ghetto high school with no technical skills, no funds to pay for commercial training school, and a slim chance of getting one of the few slots at the public universities, a typical Brazilian youth has vastly fewer career options than someone in a similar position in a developed country. One of the major challenges Brazil faces going forward is to undertake reforms and initiatives addressing the structural causes of poverty and income inequality, and helping the country as a whole to move toward the future, as opposed to just a small minority of privileged individuals. This is a task requiring tremendous creativity as well as money, and one whose true dimensions will only become apparent as the work on it unfolds.

In spite of these various foundational difficulties, the Brazilian software industry is booming (though its income disparity, and resulting educational disparity, will be a rate-limiting factor and will potentially put the brakes on the boom, as the pool of educated people becomes unable to keep up with demand). The setting of the stage for the current boom was slow and gradual. In the early 1990s, before the Plano Real, there were 100,000 people engaged in information technology activities in Brazil. This included 30,000 with advanced degrees in computer-oriented fields, 10,000 engaged in R&D efforts and 800 with Ph.Ds in computer science. Brazilian universities offered 210 undergraduate and 20 graduate computer science programs, producing a steady supply of technically-trained individuals. Now, post-economic stabilization, the software industry is several times this size, including firms in every aspect of computing and communications, with rapid growth in the Internet and wireless sectors.

All in all, the Brazilian economic story is a strange one, and in many ways a microcosm of the world economy. We have tremendous technological advances, pushed along by a small subset of the population and primarily benefiting this same subset. We also have a huge pool of individuals who benefit relatively little from the tech revolution they hear vaguely described on radio or TV (if at all), and also find themselves unable to contribute significantly to this revolution due to the unavailability of appropriate education. It is a situation which can only lead to problems. The wealthy so extremely outpacing the masses is a volatile situation which has historically led to only three outcomes for the affected society: reform, revolution, or collapse.

COMPUTING FOR THE MASSES: THE BRAZILIAN NETPC PROJECT

There is some spillover from tech advances among the rich few into the lifestyles of the majority—both in Brazil and in the world at large. The growth of wireless technology is particularly interesting in that, more so than desktop computing, wireless bridges class divisions. Cell phones are affordable by a much larger segment of the population than desktop computers, and as the mobile Internet becomes a major force, wireless may become the major means by which high technology spreads into the depths of the Amazon and into the sprawling, dangerous ghettos that surround every Brazilian city (this example has echoes in places like China and India, whose situations are not unlike that of Brazil). However, while wireless may be a fast growing part of the *consumer* technology revolution, desktop computing is still the cornerstone of business and scientific technology.

This gives rise to an obvious question: in addition to "a chicken in every pot" (as was advertised by Franklin Delano Roosevelt in the U.S. during the Great Depression of the 1930s), why not strive for "a computer in every home?"

One might argue that there are more critical things to do for the Brazilian masses. Bill Gates, after he established his $21 billion Gates Foundation, quickly abandoned his initial plans to focus on disseminating technology throughout the undeveloped world when he realized that some of the computers he'd donated to African villages were useless, due to the minimal availability of electricity there and the lack of relevant education and training. Gates decided to focus on improving the dissemination of food and medicine to impoverished regions. However, it is unclear whether this is the right approach for the long term. From a short-term perspective, of course, we can't let people starve to death; but if they can't live off the land (for whatever geographic or sociopolitical reasons), and without roads, electricity and commerce being brought to them, how will they ever feed themselves rather than relying on charity?

In any case, Brazil is not the Sudan—there is no mass starvation in Brazil at present, though there is surely widespread malnutrition amongst the poor. Medical care is decent by developing-world standards—in the big cities it even reaches developed-world standards. The main problem in Brazil is not keeping people alive—that basic requirement is met (making Brazil a "developing nation" rather than an impoverished one). What is needed is a way to lift people from the cultural and material conditions of poverty and enable them to become full participants in the emerging global information economy. With an Internet-connected computer at home, a young Brazilian has the world at his fingertips, and with ample self-determination and/or an encouraging family, is able to learn about every topic under the sun in a self-directed way. Skills like computer programming and word processing can also be practiced, providing the computer owner a real possibility of participating in the new economy in a serious way. Access to, and familiarity with, technology has long since become a minimum requirement for fully participating in the middle and upper echelons of economic activity.

The Brazilian minimum wage is equivalent to roughly $90 a month, whereas a computer that sells for $500 in the US may sell for near $1000 in Brazil (due in part to high import tariffs, and in part to fewer vendors in the market meaning less competition). The general nature of an appropriate economic solution to "a computer in every home" in a Brazilian context is clear: it is a combination of "Capitalist" techniques of freer markets (lower tariffs, more competition) and a "Socialist" program of improving the economic and educational conditions of the *average* Brazilian.

It was with this in mind that João Pimenta da Veiga Filho, Brazilian Minister of Communications, chose to organize the "Net PC" project. The idea was to create a computer that members of the Brazilian underclass could genuinely afford. The effort was successful. The Net PC costs around $400 reiais (around US$ 200); and in order to ensure affordability, a 24-month payment plan is offered.

The task of creating this machine was turned over to the computer science department at one of Brazil's leading universities, the Federal University of Minas Gerais (UFMG) in Belo Horizonte—the same university that trained nearly all of our own Brazilian colleagues at Webmind Inc. and Novamente LLC. The project was led by a number of expert computing researchers, including esteemed, U.S.-trained professors such as Sergio Vale Aguilar Campos and Wagner Meira, Jr. These professors are accustomed to spending their time doing research and teaching on advanced topics, but they and many of their colleagues and students were willing to take time out from this to work on the government-sponsored project of bringing much simpler aspects of computing to a much wider population.

Instead of asking what can be done to sell more software or more hardware to middle-class North Americans (the question on the minds of most people in the U.S. computer industry), they asked, as Meira puts it: "What does a computing novice really need in a computer? Internet (including multimedia) and text processing. Eventually software for creating a spreadsheet or a presentation. However,"—and here is the big difference from projects like the American

WebTV—"the Net PC does allow expansions for those that want to have an enhanced computing experience." The Brazilian Net PC allowed the machine to grow to meet the needs of an increasingly sophisticated user, at a price they could afford.

The Net PC itself was a fairly standard one for the time the project was initiated—a Pentium 500 MHz, with keyboard, mouse, 56 Kbps modem, 14" display, 64 Mb RAM and no hard disk (16 Mb flash RAM instead). According to those involved in the project, the technical aspects of designing the system were not particularly onerous—no major inventions or innovations were required. The hardest part was bargaining with the manufacturers of the various parts of the machine, who tended to be oriented toward making the most expensive and powerful machines possible rather than creating low-cost systems.

Meira cannily observes that the minimalist approach taken in the Net PC is the sort of thing that could only emerge in a place like Brazil, not in a place like the U.S.A., where "More, more, more!" is the watchword. "In Brazil," he notes, "popular stuff is usually minimalist, such as the popular car (up to 1000cc), pre-paid cell phones, etc." This is a small example of the general principle that the developing world must lead its own people into the Information Age. The cultural and conceptual biases of developed countries aren't necessarily in synch with the needs of the rest of the world. Though First World technology has universal applicability, we in First World industry are remarkably terrible at viewing things from the perspective of anyone other than the highly privileged, and these are markets in which we will fail to ever participate unless we discard that particular bias.

The Net PC project was launched in 2001—now it's 2005, and in actual fact, dissemination of these machines through Brazil has proceeded very slowly. Furthermore, the value added by these cheap Linux-based Net PC boxes, as compared to cheap Windows boxes, gets more and more dubious as hardware costs continue to decrease. Pirated versions of Windows abound in Brazil, so the cost of Windows isn't as much of a practical issue as one might think from an American perspective—as many American media vendors are discovering, if you overprice your product for a particular market people *will* just steal it, regardless of how immoral you may find that to be and how many treaties are signed with poor countries to try to stop it.

All in all, at this point, our assessment is that these Net PCs are not very likely to have much of a *direct* impact on the Brazilian underclass' computing experience. This is not because technology will have no impact on these people, but because the project itself did not provide the ideal solution. Lower-class Brazilians will probably experience the Net via cheap Windows computers running pirated versions of Windows. Brazilian schools and universities will run typical PCs using low-cost versions of Windows, purchased in bulk by the Brazilian government under a Microsoft discount program. The desire to integrate with First World computing systems means the Brazilians need to accept the monopoly and learn Microsoft products, even if they have to steal them to do so. Simply providing them with a quality competing product is not sufficient, because for better or worse Microsoft has successfully created such demand for their product that it is seen as a *necessity* rather than a luxury in the computer economy.

Does this mean it was a useless waste of effort, and the government should have kept its nose out of matters and let the private market take care of everything? Not necessarily. Markets flourish based on competition, and the Net PC provided competition for the mainstream hardware/software industry in Brazil with regards to the low-cost sector. Arguably, some large companies would not have paid as much attention to the actual needs of the Brazilian computer market were it not for the Net PC project—they would probably have continued to try to sell powerful, high-priced North American-targeted PCs to that market. If the commercial sector manages to beat the government/university sector, in terms of providing a mass-market product with the optimal mix of low cost and high usability, that's to everyone's benefit—at least to a large extent. (There is the argument that Microsoft's monopoly is anti-competitive and hurts the computer industry generally.) That doesn't mean the commercial sector would have provided the same thing without the government/university sector's competition to push it in the right direction. Indeed, based on the situation in the Brazilian computer market prior to the Net PC, it might well not have. Markets aren't worth much when they haven't got sufficient information to set proper value for things, and upstart competition is one way they get that information—even if that competition is coming from "Socialist" programs funded by governments.

Hypereconomics can help with this. It provides a way to inject information into an economic system which is often entirely missing, or approximated using inefficient data-collection methods and statistical methods which may be too biased towards "winner takes all" results. Correct estimation of value in its many dimensions is as crucial to business as to consumers, and can lead to improved macroeconomic function worldwide by allowing for less waste. There are multiple dimensions of value involved which are not being correctly weighted in traditional economics measures for two reasons:

1. The algorithms for estimating value don't include their contributions directly, based primarily on philosophical beliefs about the sufficiency of approximating them in other values such as cost.

2. The mechanisms for gathering and analyzing this information are nonexistent or inefficient. This is primarily due to the fact that such mechanisms predate widespread computer-mediated communications.

A hypereconomic system can help set proper value, and encourage competition and thus economic robustness, by allowing consumers to more directly communicate with producers. Multiple dimensions of value can be "discussed" directly through decomposed-price bidding schemes such as those discussed earlier, rather than only through political methods (activism, letter-writing, etc.). This provides a mechanism for reconciling cost with value across a variety of factors which influence buying decisions. Through its coupling with technology, Hypereconomics also allows such negotiations to be ongoing.

Decisions can be made on "market time" rather than responding to exclusively to slower-moving political processes (computer-mediated direct voting can help that along as well), because producers and consumers can exchange this value information with every transaction. It also enables more competition by giving small producers an opportunity to reach larger consumer bases and negotiate pricing in ways only large producers used to be able to do. Only an anti-competitive monopolist who was interested in preserving his own power at the expense of genuine economic growth and prosperity would philosophically oppose such a system. The arguments against it based on complexity are quickly fading away as more and more people become comfortable with computer-mediated systems. Such systems also provide more transparency, leaving less room for corruption (but with more participants, there is more room for swindling and scamming—no system is perfect).

As computer security technology advances, and people's trust in it grows, we may soon come to trust such systems as much as we trust dropping a piece of paper—be that a ballot or a check—into a box that some person we don't know then guarantees will be treated in the manner we've prescribed. The benefits of a hypereconomic system outweigh the complexities, and while we understand the ways in which our existing systems can be corrupted much better than those resulting from any proposed changes to the system, that alone is no reason not to pursue a change which has so many obvious benefits.

AND THE MORAL OF THE STORY IS?

In the following two chapters, we'll discuss some of ethical issues associated with transhumanism in a fairly philosophical way. We'll make every effort to ground the discussion in actual everyday human experience, but things will sometimes get very abstract nonetheless. Such discussions are fascinating—if we didn't think so, we wouldn't spend three chapters of this book on them—but they sometimes seem a bit like angels dancing on the heads of pins. This is why the kind of work done by Campos, Meira and their colleagues in the Brazilian Net PC project (and similar projects around the globe) is so intriguing; and why

Hypereconomics and direct democracy are things we should try to embed in all our networked technical systems, so that more people can participate not just in technology but in the rest of economic and political activity as well. There's no arguing with the real, physical-world power of millions of impoverished Brazilians (or Indians, or Chinese, etc.) logging onto the Net and discovering discussion groups like the Extropians list (to be discussed in the next chapter), where subjects like ethics and technology are discussed, and speculations on superhuman AI appear alongside the online MIT curriculum and critiques of the latest Java release. Without the Net PC or other low-cost alternatives provided via the mainstream computing industry (largely through governmental and/or hypereconomic pressures), these people might well never get to log on and argue with us for themselves, learn economically life-transforming technology skills, and participate in hypereconomic decision-making. Without Hypereconomics and direct democracy, the post-Internet world will just be a high-tech shopping mall placating the masses with trinkets; while the real-world Gini indices get ever closer to 1—a situation which the wealthiest 1% might think sounds appealing, but if they really thought about it (and knew their history) they'd realize is a likely recipe for various forms of disaster. Engaging more and more intelligent people in the ongoing technology transformation, and the political and economic world in which this transformation is embedded, is certainly a very good thing.

---

[1] See the film "Enron: The Smartest Guys In The Room," and/or the Enron memos and analysis thereof released by FERC and via the court proceedings and Senate hearings, on the issue.

[2] Our own practical work in this direction has been on hold since 2001 due to lack of funding.

[3] http://www.ideosphere.com/

[4] The Gini Coefficient for the United States is around .4, worse than any other "First World" nation that has been ranked, which is most of them. Brazil's Gini coefficient places it ahead only of seven tumultuous African nations, of those ranked. See the most recent full listing at: http://en.wikipedia.org/wiki/List_of_countries_by_income_equality
Other sources: World Bank, Regional Study, Poverty and Policy in Latin America and the Caribbean, Argentina Poverty Assessment and Uruguay Poverty Assessment (FYOO).

CHAPTER 12

EXTROPIAN ELITISM AND HUMANIST POSTHUMANISM

This chapter, perhaps more than any of the others in the book, has evolved significantly from its initial form. It began, like many of the others, as an article for the German newspaper *Frankfurter Allgemaine Zeitung*. Of all the articles Ben wrote for *FAZ* from 1999-2001, this was the one that excited them the most. It's not hard to see why: rather than explaining difficult technical content, it focused on social and moral issues. Admittedly, like a lot of successful journalism, it also had a bit of an overly sensationalist tone—including a contentious Holocaust metaphor—which one of the editors wanted to remove; but he was overruled by the publisher, Frank Schirrmacher, who felt pushing the boundaries to be one of the key purposes purpose of futurist writing in the first place.

The original article was a kind of critique of the Extropians, a specific group of futurist thinkers and activists. As a consequence of writing and publishing the article, Ben came into contact with a greater variety of Extropians than he'd known previously, and came to somewhat modify his views of the group (Stephan already had bumped into number of them from his work with Sasha Chislenko on hypereconomics and other issues). This chapter has ended up as a kind of chronicle of its own evolution, and of the evolution of our thinking about this group of futurists and the strengths and weaknesses of their conceptual approach to the issues of our mutual interest.

Ben's original *FAZ* article on Extropians started out as follows:

> "Nietzsche, my favorite philosopher, gave his book *Twilight of the Idols* the subtitle 'How to philosophize with a hammer.' It was the moral codes and habitual thought patterns of his culture that he was smashing. In a similar vein, the creed of the Extropians, a group of transhumanist futurists centered in California, might be labeled 'How to technologize with a hammer.' This group of computer geeks and general high-tech freaks wants to push ahead with every kind of technology as fast as possible—the Internet, body modification, human-computer synthesis, nanotechnology, genetic modification, cryogenics, you name it. Along the way they want to get rid of governments, moral strictures, and eventually humanity itself, remaking the world as a hypereconomic virtual reality system in which money and technology control everything. Their utopian vision is sketchy but fabulous: a kind of Neuromancer-ish, Social-Darwinist Silicon-Valley-to-the-nth-degree of the collective soul.[1]
>
> "I sympathize with their techno-futurism and their lust for freedom. But their brand of ethics scares me a little.
>
> "Intuitively conceived as the opposite of entropy, Extropy is a philosophical rather than a scientific term. The Extropians website[2], the online Bible of the movement, defines Extropy as "A metaphor referring to attitudes and values shared by those who want to overcome human limits through technology. These values…include a desire to direct oneself in pursuing perpetual progress and self-transformation with an attitude of practical optimism, implemented using rational thinking and intelligent technology in an open society."

"Transhumanism," as a general term, refers to philosophy that doesn't view human life as the ultimate endpoint of the evolution of intelligence. Extropianism is a particular form of transhumanism, concerned with the quest for "the continuation and acceleration of the evolution of intelligent life beyond its currently human form and limits by means of science and technology, guided by life-promoting principles and values, while avoiding religion and dogma." Working toward the obsolescence of the human race through AI and robots is one part of this; another aspect is the transfer of human personalities into "more durable, modifiable, and faster, and more powerful bodies and thinking hardware," using technologies such as genetic engineering, neural-computer integration and nanotechnology.

Along with this technological vision comes a political vision. Extropians, according to extropy.org, are distinguished by a host of sociopolitical principles, such as:

> "Supporting social orders that foster freedom of speech, freedom of action, and experimentation. Opposing authoritarian social control and favoring the rule of law and decentralization of power. Preferring bargaining over battling, and exchange over compulsion. Openness to improvement rather than a static utopia...Seeking independent thinking, individual freedom, personal responsibility, self-direction, self-esteem, and respect for others."

It is explicitly stated in Extropian doctrine that there cannot be socialist Extropians, although the various shades of democratic socialism are not explored in detail. In point of fact, the vast majority of Extropians are radical Libertarians, advocating the total or near-total abolition of the government. This is really what is unique about the Extropian movement: the fusion of radical technological optimism with libertarian political philosophy. With only slight loss of meaning, one might call it libertarian transhumanism.

The characterization of Extropian philosophy that Ben gave in those paragraphs was based on conversations with a number of individuals identifying themselves as Extropians, both in person and via e-mail. Some of these individuals and conversations will be discussed later on in this chapter.

Conversations with other Extropians during the last couple of years—including Natasha Vita-More, one of the original Extropians—have made us realize that this impression in the original article of the Extropian group was somewhat flawed and incomplete.

Everything observed in that article was true of a certain subset of the Extropian community, but the Extropian community has a lot more diversity than the article which spawned this chapter gave it credit for. The statement that "the vast majority of Extropians are radical Libertarians" was an overstatement. Many Extropians are radical Libertarians, and very few are socialists, but all in all there is a much greater variety of political views in the community than appears at first glance.

This chapter represents a variant of that *FAZ* article, but with a less sensationalistic and, we hope, more accurate flavor. The basic themes are the same, but we're pleased to be able to present them now as pertaining to a subset of the Extropian community rather than the Extropian community as a whole, and to give a more balanced overall perspective.

Even the new and mellower version of our take on Extropy and related ideas is unlikely to please everyone in the Extropian community—but we're calling it as we see it. As already noted, we wish to emphasize the distinction between the "official Extropian line," as laid out on the Extropy website, and the actual belief systems that tend to be held by the majority of individuals associating themselves with Extropianism. Our concern here is mainly with these actual belief systems. We're not writing about Extropianism as a formal set of beliefs or as an official organization, but rather about the *cluster of individuals and ideas* that has aggregated around the Extropy concept over the last couple decades.

For example, libertarian politics is not part of the official Extropian philosophy, but it is a mighty common theme on the Extropy e-mail list and at Extropy conferences. The complaint that Extropianism isn't intrinsically connected to Libertarianism is reminiscent of an argument Ben once had with a Sufi, who claimed that Sufism isn't a religion. He was right, formally speaking. Sufism is not a religion—it's a "wisdom tradition" of Arabic origin, associated with Islam. Yet 99.9%, perhaps 100%, of Sufis are Muslims and are religious by the definitions of the rest of the world.

There's no question: Some Extropians carry their anti-socialist Libertarianism to a remarkable, ultra-radical extreme. For instance, visionary roboticist Hans Moravec, a hero to many Extropians, had a somewhat disturbing exchange with writer Mark Dery in 1993. Dery asked Moravec about the socioeconomic implications of the robotic technology he envisioned. Moravec replied that "the socioeconomic implications are…largely irrelevant. It doesn't matter what people do, because they're going to be left behind like the second stage of a rocket. Unhappy lives, horrible deaths, and failed projects have been part of the history of life on Earth ever since there was life; what really matters in the long run is what's left over." Does it matter to us now, he asks, that dinosaurs are extinct[3]? Similarly, the fate of humans will be uninteresting to the superintelligent robots of the future. Humans will be viewed as a failed experiment—and we can already see that some humans, and some human cultures, are worse failures than others.

Dery couldn't quite swallow this. "I wouldn't create a homology between failed reptilian strains and those on the lowermost rungs of the socioeconomic ladder."

Moravec's reply: "But I would."

Put this way, Extropianism starts to seem like a dangerous, profoundly ethically-deficient philosophy—one which has an ad-hoc human-valuation principle which *equates relative success in a particular system at a particular point in time with absolute worth*. Thus the outrage of the *FAZ* article that served as the seed of this chapter; but one must remember (which Ben didn't when he wrote the original article, overheated with moral outrage) that Moravec is just one voice among many—and a decent percentage of Extropians would be just as annoyed by Moravec's ethics as we are.

Although Moravec is often intentionally confrontational and alienating, the Extropian perspective overall isn't *that* far out on the fringe these days. Luminaries associated in one way or another with it include Marvin Minsky, the AI guru; Eric Drexler, the nanotechnologist; Kevin Kelly of Wired Magazine; and futurist writer Ray Kurzweil. At this point, Extropianism does not rank as one of our more prominent cultural movements; but it is active, vibrant, and growing. *Extropy* magazine had several thousand subscribers before it moved onto the Web in 1997; and the Extropy e-mail discussion list is a hugely active one (though greatly uneven in quality). A vast amount of online literature related to various aspects of Extropian thinking exists[4]. This is definitely one of the more important online communities, and whatever its strengths and weaknesses, it's worth paying some attention. Discussion of Extropian ideas brings up all sorts of interesting topics, which are highly pertinent to the future of technology, life and intelligence.

MAX MORE: THE ORIGINAL EXTROPIAN

The man who got this all started was Max More, a philosophy Ph.D. with a knack for rational argumentation and an impressive, convincing personal style. In 1995, Jim McClellan interviewed More for the U.K. newspaper *The Observer* and noted, "The funny thing about Max is that while his ideas are wild, he argues them so calmly and rationally you find yourself being drawn in."

More started his career studying philosophy, politics and economics at St. Anne's College at Oxford University in the mid-1980s. At that point his main focus was on economics, from the libertarian perspective. While doing his degree, Max became strongly interested in life extension, and he was the first person in Europe to sign up for cryonic suspension with the U.S. firm Alcor. In 1995, when he received his philosophy degree from the University of Southern California for research on mind, ethics and personal identity, he was already deep into organizing the Extropian movement; bringing his political and technological interests together. Technology, he felt, was ready to push mind into new spaces altogether, such as virtual realities where the notion of "I" as currently conceived had no meaning. Governments were holding us back, preventing or slowing research in crucial areas.

The first edition of *Extropy* magazine came out in August/September 1988 with just 50 copies, co-edited by Max More and his friend T.O. Morrow. It was a wild mix of sci-fi-meets-reality thinking (much like this book!)—from life extension, machine intelligence, and space exploration; to intelligence augmentation, uploading, enhanced reality, alternative social systems, futurist philosophy, and so on. The magazine seeded the social network that led to the e-mail list (1991), the first Extropy conference (1994) and the website (1996) which soon (1997) obsoleted and incorporated the paper magazine.

In terms of philosophical precedents, it's not too inaccurate to call More's credo a mix of Ayn Randian anti-statist individualism with Nietzschean transmoralism, held together by a focus on future technologies. In *Extropy #10* he explicitly equates the "optimal Extropian" with "Nietzsche's Übermensch," but he cautions, in another essay ("Technological Transformation: Expanding Personal Extropy") that "the Übermensch is not the blond beast and plunderer." Rather, the Extropian Übermensch "will exude benevolence, emanating its excess of health and self-confidence." That's reassuring, yet hard to reconcile with Moravec's Olympian detachment regarding the destruction of the human race. This contradiction, we believe, is both Extropianism's core weakness and a primary source of its energy.

In spite of More's forceful, argumentative style, Extropianism is certainly not an orthodoxy. Within the general "party line" of Extropianism, there's room for a lot of variety. This is one of the movement's strengths, and surely a necessary aspect of any organization involving so many overly-clever, individualistic, oddball revolutionaries. Moravec and More don't agree with each other entirely, and don't necessarily agree with all their own past opinions. Consensus isn't critical; progress is the thing.

SASHA CHISLENKO

We've had intellectual exchanges with plenty of Extropians, including Marvin Minsky, Max More, Max's cultural-strategist Natasha Vita-More (a highly creative thinker), Eliezer Yudkowsky (whom we'll discuss below), and too many others to name. But the Extropian we've known best on a personal level was Sasha Chislenko—a visionary cybertheorist and outstanding applied computer scientist. Chislenko's work, thought and life exemplify the brilliance and power—and weakness and danger—of the Extropian perspective in an extremely vivid way.

As with many Russian emigrants to the U.S., Chislenko's Libertarianism was borne of years of oppression under the Soviet Socialist regime. Having seen firsthand how much trouble an authoritarian government can cause, he was convinced that government was, intrinsically, predominantly evil. After he left the Soviet Union, Chislenko was a man without a country: lacking a Russian passport due to his illegal escape from Russia, and lacking an American passport because of his status as a political refugee. Once, Chislenko and Ben were invited to give a lecture together at Los Alamos National Labs in New Mexico, but were informed that since Chislenko lacked a passport, he couldn't get the security clearance required to enter the lab grounds. They both ended up turning down the lecture invitation, disgusted at the government's closed-mindedness.

Chislenko was impatient for body modification technology to advance beyond the experimental stage—he was truly, personally excited to become a cyborg, to jack his brain into the Net, to replace his feeble body and brain with superior engineered components. Not that there was anything particularly wrong with his body and brain—in fact he was in fine shape, with the exception of some intermittent emotional problems—it just wasn't as good as the best synthetic model he could envision. As a creature of abstraction, he could easily envision a million ways to improve upon his natural state, and was anxious to do so. He was a strong advocate of various "smart drugs," some legal, some not, which he felt gave him a superhuman clarity of thought. He was outraged that any government would think it had the right to regulate the chemicals he chose to put into his body to enhance his intelligence.

His own technical work focused on "active collaborative filtering," technology that allows people to rate and review things they see on the Net and then recommends things to people based on their past ratings and the ratings of other similar people[5] which, in collaboration with Stephan, he evolved into the idea of Hypereconomics. Popular websites like Amazon.com and Barnes&Noble.com have primitive collaborative filtering systems embedded in them—when you log on to buy a book, they give you a list of books you might be interested in. Sometimes these systems work, sometimes they don't. Recently Ben logged onto Amazon to buy a "Bananas in Pajamas" movie for his young daughter, and their recommendation system suggested that he might also be interested in the movie "Texas Chainsaw Massacre II." How it came up with that recommendation is anyone's guess: Perhaps the only previous person to buy "Bananas in Pajamas" had also bought the "Texas Chainsaw Massacre" film. The recommendation systems that Chislenko designed were far more sophisticated than this one, probably the most advanced in the world. He led a team implementing some of his designs at Firefly, a company later acquired by Microsoft.

Compared to body modification, cranial jacks and superhuman artificial intelligence, active collaborative filtering might seem a somewhat unexciting path to the hypertechnological future; but to Chislenko, it was a tremendously thrilling thing—a way for humans to come together and enhance one another's mental effectiveness, passing along what they'd learned to one another in the form of ratings, reviews and recommendations. Recommendation and filtering technology was a kind of collective smart drug for the net-surfing human race. Through recommendations and voting, Chislenko believed that global intelligence could be increased, freedoms expanded, and social and environmental stability ensured with a hypereconomic system which allowed the aggregated values of individuals in the system to trump the centrally-imposed valuations of traditional economics, even market economics. His goal, with first recommendation systems and ultimately Hypereconomics, was to eliminate all forms of central planning in economics entirely and create a system in which the values of communities were directly reflected in how resources were allocated throughout the entirety of the economic system.

The recommendation system aspect of Chislenko's vision in this area has become somewhat mainstream by this point. An example is the Website Epinions.com, which pays users to give their reviews of consumer products and other things. The higher others rate your reviews, the more you get paid. Chislenko had nothing to do with this site, but it epitomized his ideal. He strongly felt that as the economy transformed into a cyber-powered hypereconomy, intellectual contributions like his own would finally get the economic respect they'd always deserved. People would be paid for writing scientific papers to the extent that other scientists appreciated the papers. The greater good would be achieved; not by the edicts of an authoritarian government, but by the self-organizing effects of people rating each other's productions, and paying each other for their ratings and opinions. He coined the word "hypereconomics" to refer to the complex dynamics of an economy in which artificial agents dispense perpetually renegotiated, often small payments for everything, and in which complex financial instruments emerge even from simple everyday transactions—AI agents paying other agents for advice about where to get advice; your shopping agent buying you not just lettuce but futures and options on lettuce, and maybe even futures and options on advice from other agents.

There was a painful contradiction lurking here, not far beneath the surface. This personal contradiction, we believe, cuts close to the heart of Extropian philosophy—at least, in the form that it takes in the mind of many Extropians. The libertarian strain in Chislenko's thinking was highly pronounced: on different occasions he told each of us, tongue only halfway in cheek, that he thought air should be metered out for a price, and that those who didn't have the money to pay for their air should be left to suffocate! We later learned this was a variation on a standard libertarian argument, sometimes repeated by Max More, to the effect that the reason the air was polluted was that nobody owned it[6]—ergo, air, like everything else, should be private property (although like any other extreme privatist, Extropians of this ilk are unable to explain how they'd propose to undo the unfair economic playing field the wealthy would have during privatization of shared resources, based on their long history of accepting corporate welfare and government support of coercive, monopolist practices which contradict the libertarian ideal). At the 2001 Extropy conference, a speaker gave an even more extreme variation: In the future, every molecule will be bar-coded so its owner can be identified. People or their descendants will have to pay for the oxygen they breathe, molecule by molecule. While this seems both absurd and offensive, our economic system is currently headed in basically that direction—as evidenced by government permission for patents and monopoly claims on genes, mathematical formulae, and a variety of other things considered "part of nature" and previously barred by patent tradition.

Chislenko equated wealth with fundamental value, and his vision of the cyberfuture was one of a complex hypereconomic network, a large mass of money buzzing around in small bits, inducing people and AI agents to interact in complex ways according to their various personal greed motives (although he didn't feel it was based on greed, but rather mutual benefit with hypercompetition buffered out by a détente, created by a situation akin to the games in Robert Axelrod's *The Evolution of Cooperation* and the greatly lessened ability of any individual to game the system). Chislenko was by no means personally wealthy, and this fact was highly disturbing to him. He often felt that he was being shafted, that the world owed him more financial compensation for his brilliant ideas, and that the companies he'd worked for had taken his ideas and made millions of dollars from them, of which he'd seen only a small percentage in the form of salary and stock options.

Chislenko worked for us for a while in 1999 and 2000; our company Webmind Inc. hired him away from Marvin Minsky's lab at MIT. While we enjoyed Chislenko very much as an intellectual collaborator and a friend, he wasn't in any way an easy employee. We were excited to bring him into the Webmind team, but were relieved as well as sad when he quit in mid-2000, having been offered a position as CTO of a tech incubator in Boston. He contributed many interesting technical and conceptual ideas to our Webmind work—mostly to the Webmind Classification system product (which focused on automatically placing documents into categories), but to some extent to the AI R&D codebase as well—but his staunch individualism, combined with his aggressively Socratic demeanor and preference for theory over praxis, caused him to excel neither as a practical implementor nor as a manager, and so it was sometimes hard to fit him into the work process of a start-up company based on collaborative teamwork. He spent very little of his life in a university research context, but this would probably have been the most natural place for him—he had diverse skills and ambitions, but above all, he was a visionary deep thinker and conceptual guru par excellence.

Toward the end of 1999, he frequently told each of us how he had conquered many of the intellectual puzzles he had been struggling with for decades, and was now focusing on mastering his own mind and emotions. The gravity with which he declared this scared us a little. In one way or another, we each told him, probably too flippantly, that we found emotions were sometimes more fun if you left them unmastered. Chislenko's response to such comments was to become extremely serious and philosophical, warning us that we all needed to master our own minds lest we leave ourselves open not only to control by others (his lingering Soviet paranoia) but also the destructive tendencies of our own weaknesses. While this sounds not unlike many self-help books, Chislenko couched it in very critical, technical, and philosophically Randian contexts which made it clear he considered this task to be beyond mere self improvement. He wasn't always so serious: he was also an avid techno dancer, occasionally observed leaving the Webmind office in the evening with a girl half his age on his arm and heading for a dance club or a rave, where he would move to the beat in his peculiarly robotic yet wonderfully vivacious way.

When Chislenko committed suicide in mid-2000, we wondered at first whether it had been an act of philosophical despair. Had there been a problem at his new company—were they unwilling to implement his newest designs for online collaborative filtering? Had he received one more devastating piece of evidence that the world just wasn't going to compensate him appropriately for his ideas, that the hypereconomic cyberfuture was far too slow in arriving? Had we, as not only collaborators but friends, somehow caused him to feel philosophically abandoned by not agreeing with him on various substantial points of how to design AGI and hypereconomics? As it turned out, his terrible action was more directly motivated by a complicated and painful romantic relationship—good old-fashioned, low-tech human passionate distress.

In some important ways, Chislenko was similar to Nietzsche, who as we've seen was one of the Extropians' philosophical godfathers. Both Chislenko and Nietzsche were intellectual superstars who explicitly enounced one moral philosophy, but lived another. Nietzsche preached toughness and hardness[7], but in his life he was a sweet person, respectful of the feelings of his mother and sister (whose beliefs he despised). On the day he went mad, he was observed hugging a horse in the street, sympathetic that its master had whipped it. He preached the merits of the robust, healthy man of action and criticized intellectual ascetics, yet he himself was sickly, nearly celibate, and sat in his room thinking and writing day in and day out. Similarly, Chislenko extolled the money theory of value, yet lived his own life seeking truth and beauty rather than cash, trying to transform the world for the better and distributing his ideas for free online—on one level, assuming that a meritocracy was not only the intended order of society, but that it actually existed unimpeded in America, and his rewards could come because of rather than in spite of such generosity of ideas; and on another level, knowing better, and feeling that by his example he could lead the world into implementing its meritocratic ideals. He argued that air should be metered out only to those who could pay for it, yet was unfailingly kind and generous in real life, always willing to help young intellectuals along their way without asking for anything in return.

For what it's worth, it's impossible to avoid observing in this context that Chislenko, the would-be-cyborg transhumanist, manifested a remarkable number of cyborg-like personality traits. His body movements were sometimes oddly robotic—in fact he looked most natural when dancing to techno music, with its computer-generated beats. It would be an unfair exaggeration to say that his voice had something of the manner of a speech synthesizer—but it did have a peculiar stiffness to it, that one might describe as wooden or metallic. Of course, this point should not be made too strongly. Chislenko was an outgoing, friendly human being, easily hurt and in some circumstances quick to anger. He was by no means devoid of affect, and was generally a warm and giving friend; but when, about six months before his death, a group of us were coming up with silly e-mail nicknames for our co-workers (Chislenko was among them at the time), the one we picked for Chislenko was robotron@webmind.com. It was clear to everyone who knew him that he had difficulties dealing with the ambiguities and subtleties of human attitudes and relationships. He acknowledged this himself, and sometimes said it was something he was working on, while at other times he embraced it as superior, admonishing us all to bring order and rigor to our social and emotional interactions. He was a poor politician, which is partly why he so often got himself into positions where his ideas weren't adequately appreciated by his co-workers or employers. Extropianism, a clear-cut, simple philosophy, seemed to provide him a welcome respite from the human complexities and contradictions that caused him so much grief in ordinary life. In the end, however, Extropianism failed him severely, in that the Moravecian strain of dogmatic hypercriticism of all things human left little room for the errors and omissions that each individual makes as they try to improve themselves. Like any other system of dogma, the imperfect (i.e. everyone) find in their daily lives that they have bought into little more than an unattainable ideal. As such, inflexible systems lead necessarily to collapse, rather than adaptation, in the face of any critical failure of their tenets to prove irrefutable.

Of course, not all Extropians have Chislenko's personality characteristics, and not all impose the contradictions of a dogmatic personal philosophy coupled with an actually humanist lifestyle upon themselves. It would be a mistake to overgeneralize: to create a psychology of Extropians from this one example. Max More, for example, is extremely politically-adept in his own way; and Max and his wife Natasha are both body builders, with much grace and naturalness in their physical motions, devoted to living well-rounded lives as well as to deep futurism and the life of the mind. Many Extropians have above-average mastery of human relations, happy personal lives, and so forth; but everyone has their own philosophical quandaries to answer, and it's impossible not to hypothesize that the role Extropianism played for Chislenko—providing crisp certainties to serve as welcome relief from the puzzling, stressful confusion of everyday life—tells us something about the role Extropianism plays for some other individuals as well.

For some of its adherents, Extropianism serves the role of providing a simple, optimistic worldview and a community of like-minded believers. Like most religions and other religion-like belief systems such as Marxism or Neoconservativism, through its focus on a better future world, Extropianism can encourage avoidance of the difficult ambiguities of human reality. Focused solely on the better future or a return to the idealized past—especially coupled with an assumption of inevitability—a dogmatic belief system helps people deal in one way or another with existential questions and impose certainty onto a system, life, which is inherently ambiguous. Extropianism is explicitly anti-religious, but it's not a new observation that rabid anti-religiosity can, for some people, serve almost as a religion itself. As Dostoevsky said, the atheist is one step away from the devout. Atheism and theism provide the mind with the same kind of rigid certainty. For some people, this kind of definitive cutting-through of the Gordian knots of messy human reality can be indispensable, providing the comfort level prerequisite to a healthy and productive state of mind. Unfortunately, an otherwise comforting belief system coupled with the human drive to share ideas and to be certain about things leads to the proselytizing mindset: the belief that anyone who can not be convinced to believe the same *existential particulars* as oneself is inferior, and omitting them from society is not just acceptable but mandatory.

Of course, for some other Extropians, the Extropian perspective does not serve this sort of role at all, but rather is a conceptual and practical philosophy fitting in naturally with an intellectually, emotionally and physically healthy life. In such cases, Extropianism is not a dogma but a model of an optimistic future—a profound hope to be worked toward, but not an absolute mandate which must be achieved, no matter how terrible the cost .

INHUMAN TRANSHUMANISM?

As Max More realized from the start, the moral-philosophical aspects of Extropianism are key. Like Nietzsche, Extropian thinkers recognize that morals are biologically and culturally relative, rather than absolute. Who hasn't been struck by this at one time or another? We consider it OK to eat animals but not humans; Hindus consider it immoral to eat cows; Maori and other tribes until quite recently considered it acceptable to eat people. Even within the supposedly absolute morality of contemporary religion we find all kinds of relativist arguments, such as a litany of situations in which it is acceptable to kill[8]. Or, consider sexual mores. Why is female sexual infidelity and promiscuity considered "worse" than similar behavior on the part of males? This is common to all known human cultures; it comes straight from the evolutionary needs of our selfish DNA.

Given this blatant arbitrariness, it seems quite attractive to ignore human values altogether and focus one's attention on knowledge, understanding and power—qualities which seem to have an absolute meaning that morality lacks. In this vein, Nietzsche focused on personal power achieved through mental exploration and self-discipline; whereas the Extropians focus, by and large, on power achieved through technological advancement. They also share a focus on intellectual brilliance—and many, though not all, Extropians seem to take a worrisomely dismissive attitude toward those whom they feel don't have what it takes to make the next step on the cosmic evolutionary path (as exemplified in the Moravec quote). Considering their emphasis on self-improvement, this attitude almost negates all their arguments, and essentially deflates their claims to any moral high ground—if technology really can overcome all problems and raise us to the next level of advancement (and with technology guiding evolution it is now subject to the explicit decisions of intelligent beings rather than randomness + selection), what reason other than a defective moral and ethical system would there be for leaving anyone behind?

Moravec was playing Devil's Advocate in the quoted interview, but what we'd like to see is more Extropians taking an opposite point of view, and focusing on the value of transhumanist technologies for advancing the well-being of *every* sentient being. Whoever doesn't "have what it takes" now, oughtn't they—if the Extropian ideal is correct—be able to be improved through amazing advances in technology until they have got it? The further Extropian culture moves in this direction, the more we'll like it.

4 or 5 years ago, Ben posted a question on the Extropian e-mail list, before either of us even started thinking about writing about Extropianism. Intellectually fishing, Ben posited that compassion, simple compassion, was an ethical universal, although it might manifest itself in different ways in different cultures

and different species. He suggested that compassion, in which one mind extends beyond itself to feel the feelings of others and act for the good of others without requiring anything in return, might be essential to the evolution of the complex self-organizing systems we call cultures and societies—and expressed an essential disbelief that all human interaction is, or should be, economic in nature.

If you're expecting a story about a deep intellectual and ethical discussion on humanist and transhumanist philosophy in the context of an Extropian dialogue on compassion—well, no such luck. There was a bit of flaming, some impassioned but shallow Ayn Rand-ish refutations, and then they went back to whatever else they'd been talking about, unfazed by the heretical position that perhaps transhumanism and humanism could be compatible, that technological optimism wasn't logically and irrefutably married to the extremes of libertarian politics. At that time, you could only belong to their e-mail list for free for 30 days; after that you had to pay an annual subscription fee. After Ben's 30 days expired, he chose not to pay the fee, bemused that this was the only e-mail list he knew of that charged members money, but impressed by their philosophical consistency in this matter. (Stephan never chose to pay the fee, telling the Extropians such as Chislenko who invited him to participate that they were being inconsistent—participants ought to be paid for their postings, as well as the organization for hosting the list, thus the list should be free on the basis of barter. Now the list is free, though who knows if a number of people presenting similar arguments were the reason why?)

We have more respect for the diversity of the Extropian community than we did when we first encountered it, or when Ben wrote the *FAZ* article that was the first version of this chapter. It's definitely a loose conglomeration of individualists—heck, even though we're not formally a members of the Extropian group, Ben's spent some time on their e-mail list and has been to one of their conferences, and Stephan has spent a lot of time in-person with Extropians and at local Extropian get-togethers, so from the outside world's perspective we're virtually Extropians ourselves. Though some of the key philosophical *habits* of the group trouble us, any statement made about "Extropians" as a philosophical whole is bound to be a bad overgeneralization, more so than would be the case for many other social subgroups.

We admire Extropianism's courage in going against conventional ways of thinking, recognizing that the human race is not the end-all of cosmic evolution, and foreseeing that many of the moral and legal restrictions of contemporary society are going to be mutated, lifted or transcended as technology and culture grow. We too are outraged and irritated when governments stop us from experimenting with our minds and bodies using new technologies—chemical or electronic or whatever (collective responsibility against destroying ourselves is of course a good thing, but the idea that big-G Government as an entity is the mechanism for collective care against dangerous experimentation is proved rather fallacious upon even a cursory look at 20th century history). We find Extropian writings vastly more fascinating than most things we read. Extropian individuals are looking far toward the future, exploring regions of concept-space that would otherwise remain unknown, and in doing so they may well end up pushing the development of technology and society for the better. Yet, we're a bit vexed by the strain of Extropian thought that envisions Extropian human beings as supertechnological proto-Übermensches, presiding over the inevitable obsolescence of humanity through the promotion of selfishness and the absolute worship of power and money. As a philosophy and a group of people, Extropianism is simultaneously profound, attractive, amusing and disturbing.

Nietzsche, like Sasha Chislenko, was generally an exemplary human being in spite of the inhuman aspects of his philosophy. Yet many years after his death, Nietzsche's work played a role in atrocities, just as he'd bitterly, yet resignedly, foreseen. In the back of our minds is a vision of a far-future hyper-technological Holocaust, in which cyborg despots dispense air at fifty dollars per cubic meter, citing turn-of-the-millennium Extropian writings to the effect that humans are going to go obsolete anyway, so it doesn't make much difference whether we kill them off now or not. Extropian philosophy should be read, because Extropian thinkers have explicitly pondered some aspects of our future more thoroughly than just about anyone else; but in our view, some of the key themes in the

Extropian community—particularly, the alliance of transhuman technology with simplistic, uncompassionate Ultra-Libertarian philosophy—must be opposed with great vigor. We ourselves are extremely far from being anti-freedom, but recoil at the reduction of all human and environmental interaction to uncompassionate, self-obsessed, and dangerously (even in a practical, amoral sense) hypersimplistic models such as the monetary theory of economics (in which a single parameter—price—determines the absolute value of all things and beings).

Many of the freedoms the Extropians seek—the legal freedom to make and take smart drugs, and to modify the body and the genome with advanced technology—will probably come soon. We not only hope for, but work toward, and implore others to work toward a situation in which these freedoms do not come along with a cavalier disregard for those living in less fortunate economic conditions, who may not be able to afford the latest in phosphorescent terabit cranial jacks or quantum-computing-powered virtual reality doodaddles, or even an adequately nutritional diet for their children.

Although both of the authors have had the privilege to grow up and live in a wealthy First World nation, neither of us has yet been particularly wealthy nor sheltered; we have both experienced the expectable batch of unpleasant and ambiguous life experiences. Nevertheless, we both believe that we humans, for all our greed and weakness, have a compassionate core; and hope and expect that this aspect of our humanity will carry over into the digital age, trumping our less altruistic tendencies; and carry over even into the transhuman age, outliving the human body in its present form; and that this trait follows us throughout our evolution—wherever that may take us. We love the human warmth and teeming mental diversity of important thinkers like Max More, Hans Moravec, Eliezer Yudkowsky and Sasha Chislenko, and great thinkers like Nietzsche—and hope and expect that these qualities will outlast the more simplistic, ambiguity-fearing aspects of their philosophies. Well aware of the typically human contradiction that this entails, we're looking forward to the development of a cyberphilosophy accepting what is great in Extropianism, and moving beyond it in the explicit direction of compassion—a *humanist transhumanism.*

ELIEZER YUDKOWSKY

The typical techno-futurist guru has a huge variety of domains of interest and knowledge, but one, maybe two special obsessions. Moravec is robotics-focused; More is particularly into life-extension and libertarian politics; Ben is an AI guy at heart, in spite of recent forays into biotechnology and other domains; Stephan has his preferences for philosophy and AI. On the other hand, Eliezer Yudkowsky[9]—one of the more interesting young folks in the Extropian circle—focuses his thinking almost exclusively on ensuring the Singularity is beneficial, by creating what he calls "Friendly AI." We find Yudkowsky's work particularly interesting in that it is, in its premise at least, a kind of humanist transhumanism. Unlike Moravec, Yudkowsky specifically and intensely wants the Singularity to help all humankind. He takes his altruism very seriously. "If one human dies," he says, "it subtracts from me."

It is uncertain how deeply Yudkowsky considers himself "Extropian," but he's a member of the Extropy organization, and we include him in this chapter because we first encountered him via his frequent postings on the Extropians e-mail list. He has been an active member of the Extropian community for many years, in spite of having some profound disagreements with Max More and other Extropian stalwarts.

Yudkoswky shares with us the idea that the probable best course to the delirious and universally beneficent cyberfuture is to create a computer smarter than us, one that can figure out all these other puzzles for us. To accomplish this goal of real computer intelligence, Yudkowsky champions the notion of a "seed AI," in which one first writes a simple AI program that has a moderate level of intelligence, and the ability to modify its own computer code, to make itself smarter and smarter. His design for a "seed AI" is still evolving, and so far appears not to have achieved nearly the level of concreteness that we have with our Novamente system, but he's a clever guy and we don't doubt he'll come up with something interesting. Discussions on his "Singularitarian" e-mail list led to the formation of the Singularity Institute, devoted to the creation of seed AI, and to the (apparently now defunct) company Vastmind.com, developer of a distributed processing framework that allows a collection of computers on the Net to act like a single vast machine (a project Webmind was also undertaking when we ran out of money).

Like many of the leading Extropians, Yudkowsky started his life as a gifted child; and, like many gifted children, he grew up neglected by the school system and somewhat misunderstood by his parents. He's followed a unique psychological trajectory: After the seventh grade, he was stricken with a peculiar lack of physical energy, which to some degree plagues him to this day. His parents tried to help him cope with this in various ways, but without success: only when they allowed him to take control of his own life and his own mind was he able to work his way back to a productive and functional state of mind. This experience, he says, taught him that even well-meaning, loving people who want to help you can do you a lot of damage, due to their lack of understanding. He cites this as one of the sources of his libertarian political philosophy (do you see a trend emerging here?). Just as his parents tried to guide his life but failed in spite of good intentions, so does the government try to guide the lives of its citizens, but fails—and fails particularly where the vanguard of technology is concerned. Of course, every vanguard of new ideas feels this way about government in particular and society in general, but that is the mechanism by which the vanguard compels itself to enact change rather than merely pontificate about it.

Yudkowsky runs an e-mail list called "SL4."[10] He and his helpers moderate the list with a sense of humor and an iron hand (there are good, practical reasons for this approach, but like Nietzsche and Chislenko, Yudkowsky can't escape the contradictions in trying to make practical an absolutist philosophy like Ultra-Libertarianism). The control they exert on the list is occasionally a little overbearing, but, all in all, it keeps the list's signal-to-noise quality orders of magnitude higher than on the Extropians list[11]. "SL4" stands for "shock level 4," where he defines a shock level as a measurement of "the high-tech concepts you can contemplate without…experiencing future shock." According to his Web page[12], the first 4 shock levels are defined roughly as follows:

- SL0: The legendary "average person" is comfortable with this level of modern technology. SL0 technology is not on the frontiers of modern technology, but is the technology used in everyday life.

- SL1: Virtual reality, living to be a hundred, the frontiers of modern technology as seen by *Wired* magazine. Scientists, novelty-seekers, early-adopters, programmers, and technophiles are completely comfortable with this technology, but the "average person" is skeptical, perhaps wary.

- SL2: Medical immortality, interplanetary exploration, major genetic engineering, and new ("alien") cultures. The average Science-Fiction fan is, or at least believes they are, comfortable with such things coming to be. Most people consider such things to be improbable, if not absolutely impossible.

- SL3: Nanotechnology, human-equivalent AI, minor intelligence enhancement, uploading, total body revision, intergalactic exploration. Extropians and transhumanists consider such things to be inevitable and are comfortable with this. Other kinds of futurists are skeptical, even wary, and the "average person" doesn't consider such things at all, except perhaps as Science-Fiction scenarios vaguely familiar to those in the post-industrial nations—if any came to pass, the "average person" would be no less shocked than if God himself appeared before them.

- SL4: The Singularity, Powers (a term taken from Vernor Vinge's fiction, meaning superintelligent, Godlike-beings), complete mental revision, ultraintelligence, the total evaporation of "life as we know it." All but the Singularity

> faithful are skeptical, even wary, and the "average person" considers any such possibility to be solely the domain of the supernatural and in the hands of God.

According to Yudkowsky, "The use of this measure is that it's hard to introduce anyone to an idea more than one Shock Level above—and Shock Levels measure what you accept calmly, not what you know about. There are very few SL4s…If somebody is still worried about virtual reality (low end of SL1), you can safely try explaining medical immortality (low-end SL2), but not nanotechnology (SL3) or uploading (high SL3). They might believe you, but they will be frightened—shocked."

Periodically someone comes along and claims to have achieved SL5, but the claims are never very convincing. Perhaps the best "SL5" story heard on the SL4 list came in a discussion on psychedelic drugs, where someone said that SL5 is the realization that all human categorizations, including shock levels, are the almost-meaningless cognitive masturbations of our limited ape-like minds. Of course, the beauty of SL4 is that it pretty much encompasses SL5 in this sense. Once you've realized that mind and reality as we know it may be superseded by something totally different, you've got to question whether any of our ideas make any sense at all—or whether, from the perspective of our future super-superhuman "selves," our current ideas will seem about as profound and interesting as the metaphysical ruminations of mildly retarded dung beetles.

Of course, the "shock level" categorizations are a formalism about an individual's hopes and beliefs. There is no empirical evidence that anyone will accept SL4 with calmness or glee, because we can define so little about it. Rather, someone who claims to be SL4 has a specific notion of what a possible Singularity could be like, and would like to see it implemented. Not unlike a Christian waiting for Rapture; except that rather than have faith and wait, a Singularitarian must actively work to *create* the Singularity, and through ethical and engineering actions set in motion now, guide the creation of Singularity such that they can achieve a positive Transcendent state (and not be cast asunder by the reign on Earth of an evil AI superbeing—hence the emphasis on Friendly AI, its predecessors, like Asimov's Laws of Robotics, and its successors, like our Voluntary Joyous Growth philosophy, which you'll hear about in Chapter 15).

Yudkowsky has introduced the term "sysopmind" (sysopmind.org was the former Internet domain replaced by sl4.org) to refer to the notion of the Sysop, a superintelligence that has achieved near-complete control over the structure of matter and energy in some region of spacetime, and thus plays the role of a (hopefully benevolent) system administrator for some portion of the world. A long thread on the SL4 group discussed the possibility of some lesser mind, at some future time, hacking into the Sysop and co-opting its powers for its own devious purposes. That's a rather literal extrapolation of the analogy between present-day computer technology and a possible future AGI, but not an uncommon one in Sci-Fi literature; and it reflects a very real concern about the seduction of a seemingly flawless system, and the care people must take in assuming that someday a being will exist which frees all other beings from responsibility towards their own well-being and that of the system as a whole.

Of all the various and wacky discussions on SL4, the one that amused us most was the one about the contrasts between Yudkowsky's and Ben's personal lives. Yudkowsky was raised in a strict Jewish family, and he shows this influence very strongly in his life and his work, even though he is an avowed atheist. His devotion to the Singularity is definitively monastic. As he has publicly declared many times, he does not "fight, drink alcohol, take drugs or have sex." He recognizes the pleasures that can be obtained from these things—not through experience, but through accepting others' abstractions about such things into his own thinking—but he does not want to get involved with activities that will evoke strong animal emotions and thus distract him from 100% focus on the Singularity. He specifically laid out, in an e-mail to the SL4 list, the only conditions under which he could see violating these precepts. For instance, if a wealthy woman approached him and told him she would fund his work on the Singularity, but only if he would marry her—then, he says, he would marry her, not for her sake, but for the Singularity.

In e-mail dialogues with Yudkowsky on SL4, Ben presented doubts as to the necessity of this monkish approach, saying he was highly devoted to pushing toward the Singularity as well, but didn't see why it would be necessary to give up having a rewarding personal life in order to manifest this devotion in a highly effective way. We work long hours because we enjoy it, and believe what we're doing is extremely important, for ourselves (we want to build an AI that will help us figure out how to live forever!), for the human race, and for the evolution of intelligence overall. Still, we take time for our families and various other pursuits like composing music, playing the piano (not that expertly, but enthusiastically), occasional outdoor sports, writing this book, and even other jobs (running companies, writing software, working on movies—whatever pays the bills). Our own quest for the Singularity is partly altruistic, but partly a consequence of our boundless curiosity and desire for adventure and excitement—the same thing that pushes us to try new sports or travel to different countries; and the same thing that, in our college years, impelled us to experiment with various mind-altering substances (those which we had judged to be no less safe than the dangers imposed by the vagaries of daily life—alcohol, driving cars, weather, and so on).

Yudkowsky responded that his own quest to bring about the Singularity through creating seed AI was purely altruistic in motive. He also called himself a true romantic, stating that he knew a real love relationship would take too much of his time, so he was just going to steer away from that domain of life altogether. Waxing poetic, he said, "...love is a cathedral that you build together, a rose that you grow and water together *for its own sake*."

Upon reading this characterization of love, Ben couldn't help but respond to him with the Charles Bukowski line, "Love is a dog from hell."

Both tired of e-mailing for the day, Yudkowsky went back to helping humanity, and Ben returned to his own endless work with the somewhat maudlin thought that love didn't really have much to do with any of these silly words anyway—but he sure hoped it would survive the Singularity in one form or another.

About a year later, Yudkowsky posted a very funny Web page on the theme of his own celibacy, entitled "Love and Life Just before the Singularity."[13] The page ended with the following survey:

> Should I make an effort to have a love life?
>
> 1. Dating would decrease your efficiency. It is not realistic to suppose otherwise.
> 2. You're better off staying out of the line of fire.
> 3. If Miss Implausibly Precise Initial Condition shows up, consider it, but not otherwise.
> 4. Happy people are more effective. Do it for the Singularity.
> 5. You can't give up on love and still understand what it means to be human.
> 6. Sexuality is a basic need; it's not healthy to deny it.
> 7. Christ, Eliezer, you need to get laid. Quit fretting and just do it.
> 8. I vote for you to get married and start raising a family. You'll contribute more to the future that way than through all your AI work.
> 9. Yes. You do not understand. But you will.
>
> 10 Words can't express how much I don't care.

In a later version of the page, the following note appeared next to a report of the results: *Note: 22 votes were cast for the fifth option by a multiple voter...You can subtract 22 votes from the fifth option to arrive at the correct total. This kind of spoiler action is not welcome; please do not do so.*

As it happens, that multiple voter was Ben's girlfriend at the time—an attractive young physics student who was a big fan of eternal love, flowers, Puccini and all things romantic. I think she hoped to influence Yudkowsky to experience the glories of romance—but, predictably enough, she succeeded only in annoying him, and his dogma about the subject seemed to hold firm. In 2002, following the "Cathedral" thread, there followed a hilarious SL4 e-mail thread in which a group of others discussed the theme of the "monk versus the warrior." Somehow Ben had become a warrior—which seems rather amusing since in point of fact, much like Yudkowsky, he spends an excessive proportion of his waking hours on his butt in front of the computer, and is generally inclined towards cooperation rather than conflict. Then, in early 2005, Yudkowsky showed up at a Bay Area social gathering with a woman whom he proudly and humorously introduced as his "consort." When questioned by a friend (neither of us were there), she reported that she had been attracted to him via his online essay about why he would never have a girlfriend. In the end, it seems we have to give Yudkowsky some kind of prize for finding a creative way to use the Internet to pick up girls! We are relieved, however, that there is at least some flexibility in his dogma, and that he's open to change his ideas in the light of new information. Not only *can* one contribute to the positive evolution of human society while having a fulfilling social life, one *should* do so to the best of their abilities, because these developments proceed in the context of human relations, and how can you change the world for the better if you don't relate to the people in it?

The main practical consequence of Yudkowsky's extreme altruism—apart from his abstemious lifestyle—is his focus on the notion of Friendly AI. He wants the Singularity to benefit all people, which in our view is a vast improvement over the Moravecian "to hell with the poor" attitude. Yudkowsky believes the Singularity will be brought about by a seed AI transforming itself to superintelligence and then making endless further inventions and innovations, until—if his formulation of a positive scenario occurs—it becomes a Sysop, a system administrator for the computational system that is the universe, regulating the dynamics of the universe according to human-friendly and sentience-friendly ethical principles. It follows from this that making the seed AI as benevolent as possible to humans is an important idea. Of course, it can't be known that a human-friendly seed AI will become a human-friendly Sysop; but, in Yudkowsky's view, the lack of absolute knowledge in this regard is a lame excuse for not trying.

We'll dig into, refute some of, and expand upon some of Yudkowsky's ideas—which are quite subtle—in more detail in Chapter 15. On a crude level, however, his key idea is that Friendliness (to humans) should be at the top of any seed AI's goal system. He doesn't mean "friendliness" in quite the typical sense, and has recently introduced a concept of "humaneness" as a way of explaining his notion of Friendliness, which we'll go into in great detail in Chapter 15. For now, the common notion of humaneness will suffice, and to understand that in his perspective the Friendliness goal should be paramount to any AI that is created. Other goals, such as learning things or surviving, should be represented within the system as subgoals of Friendliness. The system should try to survive, but not because survival is its ultimate goal—rather, because surviving will allow it to help people more.

When we invited him to give a talk at Webmind Inc. in late 2000, he lectured us passionately on the need to give our AI system a friendly goal system. He was a little concerned that the Webmind AI Engine might undergo a "hard takeoff"—a rapid transition from intelligence to superintelligence via progressive self-modification—and that if it didn't have the right goal system inside it at that point, the future of humanity might be a bleak one. Since we involved with the Webmind project were painfully aware of the incomplete state of our codebase, we were not so concerned about this possibility (our concern about running out of "hard cash" before the system was able to produce any "hard results" proved, sadly, to be more well-founded in that particular instance).

Reactions to his talk amongst the Webmind Inc. staff ranged from deep interest, to distant amusement, to outright disgust at the silliness and impracticality of the topic. Generally speaking, our Webmind colleagues were absorbed with the practical problems of trying to create real digital intelligence, whereas Yudkowsky was then more concerned with the various philosophical and futuristic issues that would arise once a truly intelligent AI system is completed. In his talk—reflecting his conceptual bias as of 2000, which is different from his bias as of 2005—he focused on the issue of programming AI systems with Friendly goal systems more than on educating AI systems or the philosophy of humaneness. His emphasis on "wiring in Friendliness" definitely struck everyone powerfully, one way or another. Among the milder responses, one of our Brazilian software engineers raised his hand and politely said: "But perhaps the most important thing is not the in-built goal system, but whether we teach it by example." The friendlier we are, in other words, the friendlier our AI systems are going to be.

The issue is clear and poignant. What the Brazilian engineer was suggesting was that, if our superhuman AI grows up watching us act as though most humans are dispensable and irrelevant, perhaps it will, in its adulthood, believe that we too are dispensable and irrelevant. On the other hand, perhaps, as Yudkowsky says, it will grow up and understand that building it was the best thing the cyber-elite could do for humanity as a whole, and it will then proceed to spread joy and plenty throughout the land. Who knows for sure? The idea that we can manufacture a system which will protect ourselves from each other, while appealing, is very unlikely if we demonstrate in its infancy a profound disregard for life. Even if we could fabricate such a thing from "whole cloth," it would still be a particularly pathetic kind of irresponsibility to not teach both any AGIs we develop in the future, and the children we have now, to be humane and respect life, intelligence, and growth *by example.*

We find the motivation behind the Friendly AI concept admirable; but the Novamente digital mind design which embodies as many of our ideas in this book as we currently know how to formulate does not have quite so rigid a goal system as Yudkowsky suggested in his 2000 talk at Webmind Inc. We tend to agree with the view expressed by the Brazilian engineer during Yudkowsky's talk—that experience and education are going to make more of a difference to the Friendliness or otherwise of a seed AI, than any structure explicitly embedded in its goal system. Certainly, we'll be curious to see what kind of AI architecture Yudkowsky comes up with; no doubt his AI design will be more compatible with his own thoughts on goals and their management than anything either of us might design. He has since written a paper on "Levels of Organization in General Intelligence"[14] which takes some serious steps in this direction—but there is still a lot further to go. In the spirit of both intellectual curiosity and cooperation, we hope that we'll see more developments from Yudkowsky regularly in the years to come.

A DIALOGUE ON HUMANIST TRANSHUMANISM

Though he's been chatting on the SL4 list regularly for the last few years, Ben's last serious foray onto the Extropians e-mail list was in April 2001. Webmind Inc. had just folded; and in need of some distracting entertainment, he thought it would be amusing to bring up the old social-consciousness theme again, though with a less adversarial approach than had been taken in the post years before.

There was a thread discussing how hard it was to get Extropian ideas accepted into mainstream culture. Ben suggested that, as a counterbalance to the "scary" aspects of deep futurism, it might be valuable for the Extropian community as a group to become involved in some kind of socially beneficial project, perhaps spreading technology to the disadvantaged. He had intended to suggest their participation in either the Brazilian net computing project or, more likely, a program he'd heard of that for approximately $14000 allowed anyone to sponsor a Cambodian elementary school.

Yudkowsky, who was active on the Extropians list at that time, responded firmly that the best thing he could to for the disadvantaged of the world was to focus all his time and effort on bringing about the Singularity, because the Singularity will help everyone. He said, "If you can't, on a deep emotional level, see the connection between my work and the starving people in the Sudan, then this—from my perspective—is an emotional peculiarity on your part, not mine."

Ben replied as follows:

> "I do perceive the connection, of course, both rationally and emotionally. Your work has a decent chance of increasing the probability that the Singularity is good for humans, and it's therefore a very important kind of work. I feel the same way about my own work. AI technology is going to do a lot of good for a lot

of people, someday. I do feel AI will be a profoundly positive technology for humans, not a negative one like, say, nuclear weapons, which I wouldn't enjoy working on even if it were intellectually stimulating."

"Yet, for reasons that are still not easy for me to articulate, I feel a bit of discomfort with *solely focusing one's life* on this type of compassionate activity—on 'helping people by doing things that will help them in the future but don't affect them at all right now.' This is a good kind of activity to be doing, for sure. Yet, I feel that, in general, this kind of long-term helping of others can be conducted better if it's harmonized with a short-term helping of others."

Not surprisingly (and not too disappointingly—we love Yudkowsky's work and don't really want everyone to think exactly the same way we do), he wasn't convinced. He asked:

"How do you resolve issues like these? Split your efforts between both alternatives to maximize output. How much money is spent on attempts to actually ship food directly to the poor? Lots. How much money is spent on direct efforts to implement the Singularity? We can both personally attest, Ben, that there is not much."

To his:

"There is absolutely *nothing* I could do that would help the rest of the world more than what I am already doing,"

Ben replied:

"In my view, given the numerous uncertainties as to the timing and qualitative nature of the Singularity, it is irrational of you to hold to this view with such immense certainty."

"Actually, I honestly feel that if you spent a year teaching kids in the Sudan, you'd probably end up contributing MORE to the world than if you just kept doing exactly what you're doing now. You'd gain a different sort of understanding of human nature, and a different sort of connection with people, which would enrich your work in a lot of ways that you can hardly imagine at the moment. Not to mention a healthy respect for indoor plumbing!!"

Samantha Atkins, a long-time Extropian and a very thoughtful person, responded as follows, presenting a more compatible view to ours:

"Perhaps there is a productive middle ground. Some of us could say more about precisely how the Singularity, and the technologies along the way, can be applied to solving many of the problems that beset real people right now. We can produce and spread the memes of technology generally, and AI, NT and the Singularity in particular as answering the deepest needs, hopes and dreams of human beings."

"As part of this, we also need more of a story about the steps up to Singularity as involves the actual lives and living conditions of people. That we will muddle along somehow while a few of the best and the brightest create a miracle is not very satisfying. What kind of world do we work toward in the meantime? What do we do about poverty; about technology obsoleting skills faster than new ones can be acquired; about creating workable visions including ethics and so on? What is our attitude toward humanity?
"The world we make along the way will shape the Singularity and may well determine whether it occurs at all."

Ben presented this parable:

"Suppose you're stuck on a boat in the middle of the ocean with a bunch of people, and they're really hungry, and the boat is drifting away from the island that you know is to the east. Suppose you're the only person on the boat who can fish well, and also the only person who can paddle well. You may be helping the others most by ignoring their short-term needs (fish) in favor of their longer-term needs (getting to the island). If you paddle, then eventually they'll get to an island with lots of food on it, a much better situation than being on a sinking boat with a slightly full stomach."

"If the other people don't realize the boat is drifting in the wrong direction, though, because, they don't realize the island is there, then what? Then they'll just think you're a bit loopy for paddling so hard. And if they know you're a great fisherman, they'll be annoyed at you for not helping them get food."

> "What is this little parable missing? Sociality. If you feed the other people, they'll be stronger, and maybe they'll be more help in paddling the boat. Furthermore, if you maintain a friendly relationship with them by helping them out in ways that they perceive, as well as ways that they do not, then they're more likely to collaborate creatively with you in figuring out ways to save the situation. Maybe because of their friendship with you, they'll take your notion that there's an island to the east more seriously, and they'll think about ways to get there faster, and they'll notice a current that you didn't notice, floating on which will allow you to get there faster with less paddling."

The difference here is between the following two attitudes:

1. Seeking to save the world, as a lone and noble crusader, in spite of itself; generally in conflict with others' perceived short-term goals for themselves

2. Seeking to cooperatively engage the world in the process of saving itself

To meet this second criterion, Ben pointed out, it's not enough to do things that you perceive are good for everyone in the long run. You have to gain the good will of others and work with them together on things that both they and you feel are important. While you can influence them to change their goals to be more compatible with your worldview, history shows that forcing them rather than convincing them to behave in such a manner results in an unstable system.

Of course, it's impossible and undesirable to have a consensus among all humans as to what is good and what is bad. Like most things in the human world, the distinction between 1 and 2 is fuzzy rather than absolute. Leadership, the quality which allows someone to engage people in changing things when they'd much rather they stay the same, is the art of negotiating that fuzzy area between the two extremes. A good leader does this well, and a poor one gets stuck at one of the poles and guides his or her people into disaster—either through extreme action, or extreme inaction. Following up a discussion of the above parable, Ben replied to Yudkowsky:

> "I realize that you, Eli, are trying to cooperatively engage the world in the process of saving itself your way, by publishing your thoughts on Friendly AI, but I have an inkling that the way to cooperatively engage the world in the process of saving itself ISN'T to try to convince them to see them your way through rational argumentation. Rather, it's to try to enter into a real dialogue where each side (transhumanists vs. normal people in this case) makes a deep and genuine effort to understand the perspective of the other side."

Yudkowsky's reply was both well thought out in its details, and predictable in its overall course:

> "*Ben : And if they know you're a great fisherman, they'll be annoyed at you for not helping them get food...* Except I'm *not* a great fisherman. I am a far, far better paddler than I am a fisherman. There are *lots* and *lots* of people fishing, and nobody paddling. That is the situation we are currently in."
>
> "*Ben : What is my answer missing? Sociality.* Very well, then, let's look at the social aspects of this…
> Your answer makes sense for a small boat. Your answer even scales for a hunter-gatherer tribe of 200 people. But we don't live in a hunter-gatherer tribe. We live in a world with six billion people. From a 'logical' perspective, that means that it takes something like AI to get the leverage to benefit that many people. From a 'social' perspective, it means that at least some of those people will always be ticked off, and hopefully some of them will sign on."
>
> "Plans can be divided into three types. There are plans like Bill Joy's, that work only if everyone on the planet signs on, and which get hosed if even 1% disagree. Such plans are unworkable. There are plans like the thirteen colonies' War for Independence, which work only if a *lot* of people—i.e., 30% or 70% or whatever—sign on. Such plans require tremendous effort, and pre-existing momentum, to build up to the requisite number of people."

> "And there are plans like building a seed AI, which require only a finite number of people to sign on, but which benefit the whole world. The third class of plan requires only that a majority *not* get ticked off enough to shut you down, which is a more achievable goal than proselytizing a majority of the entire planet. Plans of the third type are far less tenuous than plans of the second type."
>
> "And the fact is that a majority of the world isn't about to knock on my door and complain that I'm doing all this useless paddling instead of fishing. The fall-off-the-edge-of-the-world types might knock and complain about my *evil* paddling, but *no way* is a *majority* going to complain about my paddling instead of fishing. Certainly not here in the U.S., where going your own way is a well-established tradition, and most people are justifiably impressed if you spend a majority of your time doing *anything* for the public benefit."

As Brian Atkins said:

> "The moral of the story, when it comes to actually having a large effect on the world: the more advanced technology you have access to, the more likely that the 'lone crusader' approach makes more sense to take compared to the traditional 'start a whole movement' path. Advanced technologies like AI give huge power to the individual/small [organization], and it is an utter waste of time (and lives per day) to miss this fact."

Brian Atkins, another Extropian, was for many years Yudkowsky's patron—meaning that he was the primary source of funding for the Singularity Institute, whose primary practical function was the financial support of Eliezer Yudkowsky. (In late 2002, for personal-finance reasons, Atkins decreased his support of Yudkowsky's fellowship, causing Yudkowsky to seek alternate sources of financing.) Atkins is not an AI wizard, but he does have a quick mind, a broad knowledge base, and a good sense for future technology. Not surprisingly, he is pretty close to a "true believer" in Yudkowsky—not that he's sure Yudkowsky's work will certainly save the world, but that there's enough of a chance to merit some investment in this work. His defense of the "lone crusader" approach was thusly as much a practical defense as a philosophical one.

Ben's response to Yudkowsky was:

"Eli…here is my sense of things, which I know is different than yours:
There's the seed AI, and then there's the 'global brain'—the network of computing and communication systems and humans that increasingly acts as a whole system."

"For the seed AI to be useful to humans rather than indifferent or hostile to them, what we need in my view is NOT an artificially-rigged Friendliness goal system, but rather, an organic integration of the seed AI with the global brain."

"And this, I suspect, is a plan of the second type, according to your categorization..."

"*Eli: Certainly not here in the US, where going your own way is a well-established tradition, and most people are justifiably impressed if you spend a majority of your time doing *anything* for the public benefit…*

My belief is that one will work toward Friendly AI better if one spends a bit of one's time actually engaged in directly Friendly (compassionate, helpful) activities toward humans in real-time. This is because such activities help give one a much richer intuition for the nuances of what helping people really means."

"This is an age-old philosophical dispute, of course. Your lifestyle and approach to work are what Nietzsche called 'ascetic,' and he railed against asceticism mercilessly while practicing it himself. I'm fairly close to an ascetic by most standards—I spend most of my time working on abstract stuff, and otherwise I don't do all that much else aside from play with my kids—but, yes, I admit it, I spend some of my time indulging myself in the various pleasures of the real world...and some of my time doing stuff like teaching in my kids' schools, which is fun and useful to the kids, but doesn't use my unique talents as fully as working on AI. I think my work is the better, not the worse, for these 'diversions'...but perhaps it wouldn't be so for you...Perhaps the philosophical dispute over the merits of asceticism just comes down to individual differences in personality."

All in all, Yudkowsky didn't convince us, and Ben didn't convince him, but some headway was made in terms of understanding each other's points of view. We are happy that Yudkowsky's altruistic attitude exists; it's a great counterbalance to the more draconian, elitist strains of Extropian thought, even if it is an extremely abstract kind of altruism focused solely on ultra-long-term benefit. It is important that such things be discussed, even if the discussions are at times rambling and silly. Just because we in the deep-futurist camp don't sympathize much with the ethical concerns of the mainstream media, doesn't mean ethical issues are irrelevant to our thinking and our work. (Issues like stem cell research that the mainstream of society considers ethically critical are, from a transhumanist perspective, somewhat obvious and unsubtle: of course stem-cell research is OK. But not all ethical issues are so obvious from the transhumanistic view; the possibility of Singularity gives rise to a host of trickier ethical puzzles.)

Pursuing the "Lone Crusader" approach to social change, minor or massive, can lead to a situation where one believes one's own myth. The Lone Crusader can easily become so detached from their social context that his or her perspectives on what will make the world better can become useless or dangerous, because they tend towards engaging only with him or herself and a few like-minded supporters. Without a broader social context, critique and refinement of ideas just doesn't happen, and an idea which may have had initial merit can devolve into a perverted form that is propped up by cronyist totalitarianism.

On the other hand, an unwavering devotion to the "Consensus Builder" approach can lead to endless, pointless bureaucratism in which all effort is spent on the frictional costs of the mythical absolute consensus; and can itself run an interesting idea aground on the Plutonian shores of the dictatorship of mob rule, in which every unusual, challenging, independent idea is crushed. As with so many things, the key to success is not absolute devotion to a formulaic ideal, but rather an ethically guided flexibility in which two competing approaches, either of which is independently unviable, are synthesized into a morally guided but practically workable approach.

In large part, we agree with Yudkowsky's perspective that not enough energy is spent trying to ensure a positive Transcension of humanity beyond our current state. More resources should be spent not only on the philosophy of such things, but on the research itself into AGI, genetics and biotechnology, renewable energy, space travel, and all the other things we will need to evolve to the next level of sentient potential. He is correct that more people are trying to solve the problems of poverty and starvation directly, than are trying to bring about their end by lifting up all of humanity to a Transcendent state in a material, non-supernatural way. However, relative to the number of people in the world, and the percentage of those in poverty, his claim that *lots and lots* of people are doing so is perhaps overstated.

We choose, instead, to spend most of our resources on frivolous pursuits. Currently, we have decided, as a society, to spend more of our resources developing a few dozen automobiles, or a few thousand articles of clothing, each year, which are trivially different from last year's—or any one of thousands of similarly trivial pursuits. Rather than focusing on important things that will elevate all of us to a higher level of being, whether it be the sciences of such things, or philosophy and art which will inspire the science to reach further and be more humane, we allow ourselves to be seduced by the short-term excitement of trends and gossip. Such distractions are amusing in moderation, but we've allowed them to crowd out any long term thinking about things that are beautiful and transcendent, such as the arts, science, mathematics, and philosophy. Furthermore, in the aggregate we spend more time acquiring, counting, and figuring out new ways to acquire and count resource tokens—money in industrialized societies—than actually doing anything with those tokens; at this point, probably even more time than feeding our basal desires for food and sex, from which our desires for money and power biologically originate.

We agree with Yudkowsky that we as a race are predominantly wasting our time on frivolous pursuits, but not that we should take the time to pursue Transcendent developments away from family, art, or social compassion. Rather, we ought to draw more resources away from the frivolous, no matter how much money and power we might gain from them *in the short term*, and apply those societal resources towards the things which will make life better. All of the barriers to humanity moving to its next level of mental and social development are now economic; that is, they are a matter of how we choose to allocate our resources. All the compelling environmental reasons to do so are becoming increasingly blatant—we need to repair and improve our environment, but we seem not to be willing to do that as our current selves—and all the technological and scientific progress we need to guide our own evolution instead of leaving it solely up to chance is proceeding at a rate limited primarily by funding (which represents the social value of a pursuit in our society, and thus the desirability of pursuing that thing amongst people in our society, and thus the quality and dedication of talent that a pursuit can attract). You may believe that scientific barriers to true immortality are insurmountable, that Tipler and others are just repackaged mystics, and medical immortality is another Alchemy—but those are the big leaps, and there are other aspects of transhumanism which can't be discarded just because the big leaps look too big. We are close enough now to take the next step to the level of being that is one beyond what we are currently, if we're willing to embrace self-determination and make the effort rather than waiting for random or preordained intervention from outside ourselves. In essence, we can either choose to reformulate Coca-Cola another dozen times (and a million other similar pursuits), or to try to improve the condition of all humanity.

Like Yudkowsky, we believe that the Singularity, in some form which may or may not resemble anything we can currently imagine, can be brought about in a way that benefits everyone, or nearly everyone. We're not sure that the path to this conclusion is as *simple* as the creation of a human-friendly seed AI, however. We think this a laudable goal, but also think it's important to bring as much of the world as possible into the process of creating the Singularity, not just proceed apace with massive technological and social changes—changes to life itself—with all the decision-making power about those changes concentrated in a brotherhood of elites (some of whom, like Moravec, are already itching to cast aside the unbelievers who dare question their infinite wisdom about such things). The Global Brain and Mindplex ideas may be critical here (see Chapters 2, 3 and 14).

If the first AGI doesn't achieve superintelligence while locked in a box, but rather through ongoing interaction with humans in all nations across the world, then its mind stands a good chance of being intrinsically human-focused and human-friendly as a consequence of its upbringing. It will be simultaneously a separate being, an individual AI mind, and part of a symbiotic mind of sorts involving little bits of millions of people. As it invents new technologies, it will want to invent not only technologies to make itself smarter, but also technologies to improve the human component of the symbiotic AI-human-internet mind. Having a human-friendly goal system is fine, but in any system flexible enough to be an intelligence, goals and motivations are going to shift over time (and with the assumption that self-rewriting is a prerequisite for AGI evolution, meta-abstract self-analysis and reformulation of even low-level goal encodings would be part and parcel of its existence).

To be really meaningful and stable, a human-friendly goal system must be allowed to evolve and mature through intensive mutually rewarding interactions with the mass of human beings, and the global brain path to superintelligent AI would seem to have the potential to accomplish this—*if* we can carry it off in a positive manner.

---

[1] As a side note: When Ben wrote his articles for *FAZ*, they were submitted in English and translated into German. Neither of us knows German, and we never checked to see what the German translation of the phrase "Neuromancer-ish, Social-Darwinist Silicon-Valley-to-the-nth-degree of the collective soul" was.

[2] http://www.extropy.org

[3] An obvious flaw in his argument by analogy is that, for a substantial number of people, the answer is "Yes, it does matter to us." This is related to the fact that the vast preponderance of humans is capable of empathy, sympathy and compassion—but we'll get to that.

[4] Much of this information is linked-to from the http://www.extropy.org site.

[5] For this and other of Chislenko's writings, see his website at http://www.lucifer.com/~sasha/home.html.

[6] An idea Extropians have cribbed from extreme Libertarians in the pre-Cyber era.

[7] Though not necessarily of the form that too many people have come to assume — an assumption usually arrived at through never having read any of his work.

[8] The argument that the admonition from God is against *murder* rather than killing—that is, it is against killing without a good reason, not all killing—is a

morally relativistic argument itself. If one merely needs to have a "good reason" to escape the judgment of murder, one can invent a wide variety of plausible excuses for killing essentially whomever they like. Weakly defined "explicit exceptions" such as self-defense leave quite a bit of room for creative relativism.

[9] http://yudkowsky.net/

[10] sl4@sl4.org

[11] extropians@extropy.org

[12] http://sysopmind.com/sing/shocklevels.html

[13] This online essay has unfortunately since been deleted by Yudkowsky, on account of his having successfully entered into the world of romance.

[14] http://www.singinst.org/LOGI/

CHAPTER 13

SONG FROM THE DIGITAL FOREST

Louis Sarno was born in 1954, in Newark, New Jersey. He was an aspiring Science-Fiction writer; and he spent ten years of his early adulthood in Europe, working a variety of jobs from farmhand in Scotland, to English tutor and carpenter in Amsterdam. Then, one day, while listening to the radio in Amsterdam, he heard a very unusual kind of music.

He didn't know what the music was, other than that it was vocal and in a non-Western language—but he knew it was something essential. It hit a space deep within him, hinting at the possibility of a life profoundly richer more satisfying than the one he was living. This music, the radio announcer said, was from an obscure tribe in Africa—a tribe of pygmies, no less.

Louis decided he had to go to the people who made this music. He began corresponding with Colin Turnbull, an anthropologist who had lived among the Mbuti pygmies in the 1950s, and had continued to study and write about them in the decades since. Turnbull offered to give him an audio tape of introduction to his Mbuti friends, introducing him as Turnbull's brother; but Louis decided the Mbuti had been over-studied—their culture was no longer pure. He wanted a pure and untouched culture, a culture filled to the brim with the spirit he had heard in that music. Finally he decided he would go and live among the Ba-Benjellé, in the Central African Republic. Because their homeland was far off any of the highways leading across Africa, they had been little touched by Western society. Lacking funds for a round-trip ticket to Africa, he bought a one-way ticket, leaving him about $500 for expenses. And away he went—on a gutsy, one-way quest for a more fulfilling way of experiencing the world.

Arriving in Africa, he found it wasn't easy getting permission to visit the homeland of the Ba-Benjellé, but he finally managed to get a permit on the grounds that he wanted to tape the pygmies' music. What Louis found when he finally arrived there was somewhat disappointing. Far from a collection of noble savages, living the life of the spirit and soul in the forest, carried away by their music; what he found was a tribe in a state of spiritual and financial poverty, living in run-down huts in a village, doing low-paid dirty work for other Africans from nearby villages, and addicted to alcohol and cigarettes. Furthermore, they had little interest in their new visitor from the West, except insofar as it was possible to get money or cigarettes out of him.

A weaker soul might have turned around and gone back home, but Louis braved the dirt and disease, at one point having hundreds of tiny worms dug out of the soles of his feet with a knife. Once he had spent all his money, the Ba-Benjellé began bringing him their own food. They began performing music, and allowing him to record it—the women sang beautifully, and some of the men were excellent musicians. Finally, the tribe decided it was time to go into the forest, and they allowed him to go with them.

In the forest, he found, everything was different. No one bothered to smoke or drink—the men were focused on hunting, and the women on gathering food. Hunting was done by dragging huge nets through the forest, and goading animals to run into the nets. Honey was found by climbing to the tops of tall trees and knocking down honeycombs. The music sounded ever more beautiful in the forest—and when the tribe danced to their music, sometimes the leaves on the ground would get up and dance through the air along with them. These were the forest spirits, eternal partners of the Ba-Benjellé in travels through the rainforest.

Amazingly, Louis noted, the Ba-Benjellé had totally different personalities in the forest than in the village. It wasn't just that they were nicer, or in better moods, in the forest than in the village—it was more specific than that. Someone who was outgoing and boisterous in the village might be quiet and thoughtful in the forest. On the other hand, someone who was shy and awkward and out of sorts in the village, might come into his own in the forest, emerging as a natural leader and a master of events. In general, neuroses that existed in the village disappeared in the forest. The forest lifestyle was more immediate and violent, more gripping on a moment-by-moment basis, and thus less conducive to personal neurosis. In the forest, the Ba-Benjellé were too concerned with interacting with the forest and each other to get wrapped up in their own problems.

Louis loved the forest, and he was distressed when, after a period of months, they returned to the village. Why, he wondered, would they forsake this perfect, natural, harmonious existence, in exchange for a life of poverty, a life on the lowest rung of the economic ladder in one of the poorest countries in the world? He tried to explain to them that, by raising their children in the village instead of the forest, they were losing their traditional forest skills, and watering down their culture and spirituality. However, he found that they had no sense of the progressive evolution or dissipation of a culture. Instead of viewing time as linear and progressive, they seemed to view it as spherical. Events radiated out from a center, which was the soul of the tribe, of the forest. The linear order of events in time was irrelevant: past, present and future stood side-by-side in a space of deeper relationship. It was not that the Ba-Benjellé were stupid, unable to reason about causes and effects, or to predict the likely future outcome of a series of events. It was that they did not care to think this way: such thoughts did not come naturally to them, and were not deeply meaningful. Louis had to accept that the Ba-Benjellé's short-sightedness about their own destiny was part and parcel of the soulfulness, primality and oneness of nature that had drawn him to them in the first place.

In spite of his strangeness, the Ba-Benjellé accepted Louis as one of their own, and allowed him to record their music on his portable tape recorder. Whether he lives with them still at this moment, we don't know.

When we first read Louis Sarno's story in his book *Song of the Forest*, we were intrigued and impressed. As Ben originally wrote about it:

> "The book touched some deep, quiet parts of my mind, parts that I all too rarely acknowledged to myself. I realized that I, too, in dark adventurous moments, had sometimes dreamed of running away from civilized society, of escaping into the forest to live in the manner of my ancestors. And I realized that Louis Sarno and I, in harboring these late-night thoughts, were manifesting some pretty universal feelings."

After all, who among us has never felt the alienation and emptiness of modern life? Who has not felt, in a self-searching, angstful, funky mood, that all this vast social and technological apparatus we have built up around us is just a way of distracting ourselves from the ultimate emptiness of our lives? What thinking person has not speculated to him or herself, at least once, that the reason we make ourselves so busy doing one thing after another is precisely *because* we don't want to have to stop and reflect on how unhappy we are? Louis Sarno felt these feelings more intensely and more frequently than most of us, and he also had far more guts than the average person: he saw the possibility of an alternative, and he followed up on it. He took dramatic action. In doing so, he improved his own life—not materially, but personally, psychologically and spiritually. He also learned a great deal, and gave the world a gift of numerous recordings of beautiful music—music that otherwise might have vanished forever, unheard outside the forests of central Africa.

What is this "emptiness within" anyway? What is the cause of the emptiness, the alienation, which Louis Sarno felt and which motivated him to flee Europe, the bastion of Western civilization, for the Central African forest? What his story tells, in a dramatic and irrefutable way, is something most striking indeed: that this emptiness within is actually emptiness *without*. The reason we modern people tend to feel emptiness within is that we are looking within for something that does not belong there—for something that belongs outside, in the external world. Our internal lives are not just influenced by, but are completed by, our external lives.

The Ba-Benjellé did not have a superior, happy, harmonious state of mind in the village. They had this state of mind—the state of mind embodied in their music, which had drawn Louis Sarno to them in the first place—in the forest. They did not distinguish their forest state of mind from the forest itself, and from the forest spirits that caused leaves to dance to their music. They had no "emptiness within" in the forest, because they were filled up with force, power and spirituality from without. This feeling was not granted by a mysterious, vague, incorporeal Supreme Being, but rather from the dynamic, visceral forest all around them. Their individual personalities were not important or even *real*, except in the context of interaction with the greater physical/spiritual world. Only when they were grounded in the world—where their perceptions and actions fit with their surroundings and their surroundings corresponded to the thoughts that made them feel complete—were they able to be whole.

In all these respects the Ba-Benjellé are not atypical of Stone Age cultures—similar stories could be told about the natives of the Brazilian Amazon or the Australian Aboriginals. Today we are almost fanatically focused on the individual mind—but this focus is in no way "natural" to humans. Until the emergence of civilization over the last few thousand years, the individual mind was simply not considered separate from the physical or spiritual realms. The individual was seen as a part of the world—not apart from the world. The holistic view of mind is what we inherited from the animal kingdom. Holism is what evolved in the "firmware" of our brains and is still wired there, to a large extent. The individualist view of mind is a product of cultural rather than biological evolution. Cultural individualism dictates that the individual is not one with the world, but is superior to it. The oxymoronic statement that the "whole is inferior to the sum of its parts" is not an inaccurate summation of that paradigm.

We can see the imposition of the individualist view of mind very clearly in the modern school system. To what great lengths do we go to get our children to understand their minds as separate! We test children individually, over and over, forcing them to "think for themselves" rather than together with their friends, no matter how anti-pragmatic that approach may be. As a physics professor once said to Stephan, regarding testing: "Never in my professional life has anyone locked me in a room, away from all resources and reference materials, prohibited me from talking to colleagues, and given me unfamiliar and difficult problems to solve."

The Genius System exalts individual contributors, even those who openly declare that only by standing on the shoulders of giants can they have seen so far, regardless of how obvious it is that their insights would have been impossible without the precedent work of many others or without the direct and immediate assistance of others (be they peers, lab assistants, apprentices, funders, laborers or whomever).

We exalt reading and mathematics, the construction of imaginary worlds inside the head, over carrying out actions in the world with other people. We carry out instruction in artificial environments—schools—forcing children to ignore their senses and use their imaginations (internal simulacra of their external senses) instead. Field trips, in which students get to experience environments relating to the things they're learning about at school, are viewed by many teachers and parents as distractions. The attitude is: "Who needs to actually *sense* things, to interact with things, to *do* or *be* things? It's enough just to organize information within one's mind!" What children want, based on their biology, is embodied experience. Not only children, but adults, generally prefer learning that occurs within an absorbing environment; but this is not what our culture wants them to want, because our culture requires adults who are absorbed within inner worlds of one kind or another. We pay lip service to "learning by doing," but frequently prefer abstract theory to practical implementation. The main thing we are teaching our children in school, besides obedience and conformity, is how to construct inner worlds taking the place of the outer world. Individualism encourages everyone to distance themselves from the group, and from the sensory world, in pursuit of the ideal of absolute freedom—freedom without any responsibility towards one's fellows or the world which we inhabit.

In the primal worldview of the Ba-Benjellé or the young child, mind is just part of a continuum of being. Some aspects of this view of mind, like forest spirits or ghosts or dragons, seem quaint, funny or dangerous. Certainly, we have achieved a lot of remarkable things by understanding the mind as a separate entity—most of modern science and Western literature must be attributed to the individualist, separatist, objectivist stance. Even so, as Louis Sarno realized in a very visceral way, something has been lost. Side-by-side with its supernaturalistic features, the pre-scientific, holistic view of mind contains a deep understanding of the interconnection of mental systems.

In recent decades, the science of psychology has made a great deal of progress toward comprehending the complexity of mind. It has discovered that memories are created rather than simply recalled; that perceptions of reality depend drastically on emotional, cognitive and social factors; and that brain systems are intimately interconnected with other body systems, like the immune and endocrine system. All of these advanced research results are actually implicit in the primitive, holistic point of view. Shamans from "primitive" tribes have a firmer practical grasp of the relation between perception, reality and memory than most modern psychologists; and with their array of herbal medicines, they may have a more useful understanding of the interconnection of body and mind systems as well (though, in the ever-present sense of balance, it is the contemporary, analytical mindset which may allow us to isolate, extract, and synthesize—and improve upon—the knowledge our "primitive" selves could sense). The knowledge implicit in the holistic, pre-scientific view of mind is not easily translatable into the language of science; but then the knowledge gained by scientists is not easily translated into the language of the Ba-Benjellé, either. The two views of the universe are complementary. While we may have lost touch with some of our senses and perspectives, we have the power to come full circle and to use our contemporary knowledge and perceptions to re-embrace our "primitive" ones—and expand upon them.

GEBSER'S STAGES OF CONSCIOUSNESS

We've distinguished two ways of looking at the mind—the Stone Age, Ba-Benjellé way, in which mind is interconnected with physical and spiritual reality; and the modern way, in which the individual thinking mind is a thing apart. This binary distinction is only the coarsest way of viewing the evolution of the intuitive concept of mind over human history. There are many ways to refine the picture, and look at more microscopic distinctions between different perspectives on mind as they have appeared on the historical scene. One way to do this was given by the mid-century cultural theorist Jean Gebser, who identified four stages of consciousness in human history: the archaic, the magic, the mythic and the mental. Each of his "stages of consciousness" is a certain view of the mind, a certain way of perceiving and constructing the relation between the mind and the world.

Gebser was an integrative interdisciplinary theorist of a type that is not so popular in these days of overvaluing specialization. Like Kant, Nietzsche, Plato, Sartre, Hegel and many lesser-known others, he sought an overall understanding of the world. His entryway was history and anthropology, but the implications of his ideas far transcend any particular academic field.

He was born in Posen in 1905, then part of Germany (now Polish Poznan); and like most Germans of his generation, the trajectory of his life was structured by World War II. In 1931, he left Germany in disgust with the Nazis and took up residence in Spain, where he wrote poetry and began to develop his innovative theoretical ideas; but in 1936 he fled to Paris, where he took advantage of the artistic community that had accreted around Picasso and Malraux. When Paris fell in 1939, he escaped to Switzerland, where he completed his masterwork, *Ursprung und Gegenwart*, which finally appeared in English as *The Ever-Present Origin* in 1985 from Ohio University Press. There is no deeper and more comprehensive study of the psychology of the world's past and present cultures. In Gebser's hands, culture is not a collection of facts; it is a mixture of stages and modes of consciousness. Consciousness and the diversity of cultures are keys for understanding each other.

The archaic stage is truly prehistoric consciousness—consciousness as it was before tools, language, and other such modern inventions separated us from the physical world. It is, in essence, the animal's view of the world: a mode of being focused almost entirely on reactions to external, physical events. There is no model of the mind here: what we would call mental functions are simply parts of the world-system. Connections within the organism are not distinguished from connections between the organism and the outside world. Everything is one; perceptions of the world and the instinct to survive in it are the same. Behavior is driven by the sense-organs. Evolved to be embedded in particular environments, one is automatically part of an integral, complex, evolving ecosystem. The sense of self is thus rooted entirely in primal, reactionary self-feelings such as "I don't want to be eaten."

The emergence of mind from instinct into magical consciousness wrests man from his physical world. Now, in order to survive, mind must act upon the world in calculated ways. In this stage, Man becomes conscious of his individuality, his needs and how to fill them by identifying objects in his environment, and how they may be used to promote his well-being. Tools are developed, and the mind learns to identify its own state. However, these tools are still used within a general pattern of being which has been established by the outside world. The natural world is the context. Man is acting autonomously within this context, but remains connected with this context in numerous ways both obvious and subtle.

The magical stage is a state of mind that retains the feeling of unity contained in archaic consciousness, but adds on a feeling of practical separateness. In the magical world-view, mind is separate from universe, but is continually joined with universe by subtle magical connections. Gebser identifies magical consciousness with the worldview of cavemen; he observes it in the semiotics of Paleolithic cave paintings. The traditional forest consciousness of the Ba-Benjellé represents a kind of "late magical" consciousness. While vastly more sophisticated than cavemen, the Ba-Benjellé clearly do have a magical view of the world. They do not conceive of themselves as precisely identical to the forest (the physical/spiritual world), but they experience themselves as constantly coupled with this world—coupled by their own intuitions and feelings, and by magical manifestations like the forest spirits.

In practical terms, magical consciousness corresponds to the invention of sophisticated tools, and the development of complex kinship structures. These innovations are supported by creative methodologies for recognizing and forming patterns. Though it can at times be intelligent and creative, the archaic mind is mainly concerned with filling in abstract forms provided by instinct with particular details. The magical mind, on the other hand, experiments with abstract forms, fills them in with details based on the particular situation, and then modifies the abstract forms accordingly. It has definite mechanisms for creating new abstract structures. This is a major step forward.

Magical consciousness, like archaic consciousness, is focused outward. The turning-inward occurs with Gebser's next stage, mythic consciousness. With the mythic state of mind, the human mind discovers its own depths: it finds a richness of inner structures reflective of, but quite distinct from, the structures it perceives in the outer world. It constructs its own structures to mirror and complement the structures of the external world—something unnecessary in magical consciousness, where the basic unity of mental and physical structures is consciously and continually acknowledged.

It builds naturally toward mental consciousness, in which the inner world breaks free of the outside world altogether, and the essence of being is equated with interior process, reasoning and conscious thought.

In the mythical stage, mind is occupied not only with acting on the world to attain certain outcomes, but with recognizing patterns in outcomes of different activities and properties of physical forms in a more abstract sense. The patterns are separated from the particular situation in which they arose, leading to symbolism—to objects and patterns that represent states of the world, and changes therein. With symbolism, we have cause and effect, language, concepts of time and space, good and evil. More complex social organizations are formed as land is farmed, animals domesticated and labor divided. Harnessing his understanding of state and action, physical cause and effect, man creates simple machines to extend his physical capabilities in order to minimize the effort, time, and space which his labor requires. Out of language and machinery, the roots of science and mathematics and literature are laid.

Here, symbolism is still focused on the external world. Mathematics is geometric or arithmetic, referring directly to real-world shapes or quantities of real-world objects. Science pertains mainly to readily observable phenomena—not to, say, black holes in distant parts of the universe, or particles so tiny as to be not only invisible, but sometimes even unmeasurable. Literary narrative, even when dealing with gods and the like, follows the flow of events of human life, rather than setting up its own order having nothing to do with reality. Yet the fact that this is symbolism pertaining to the world, rather than actions carried out within the world, is important. It is a change of focus from without to within.

Finally, the mentalistic attitude is exemplified by the famous Descartes quote "I think, therefore, I am." Mental consciousness places the self in the head, rather than in the heart. It thus distances the self from the body, from the pulse of physical being in the world. This is where we are now, and where we in the Western world have been for the past few hundred years. Mentalistic consciousness proceeds in an orderly manner from one step to the next, and rigidly separates past from present, and present from future. The spherical, lateral temporality of the magical stage is relegated to small children, insane people, and inspired artists. Magical thought is revered as a lost innocence, but ultimately disregarded as merely fanciful at best.

With the mentalistic stage, a host of new phenomena arise. We have relativity: mind differentiating itself in relation to its objects, seeking to know itself, grasping toward meaning, perspective, and knowledge as ends in themselves, irrespective of outer-world significance. We have mathematics developed into an abstract system, capable of symbolizing ideas completely unreachable by the senses—the fourth, fifth and nth dimensions; electromagnetic fields; infinitely small and infinitely large quantities; etc. We have reflexivity: the mind becoming self-conscious, as its processes become its objects and introspective reasoning becomes seen as superior to all extrospective activity. Any form of belief without observable evidence becomes based on logical construction of rules of faith, rather than rote extrapolation of the observed world into faith that the unobserved world is relatively similar.

The succession of these stages, according to Gebser, is not a matter of new stages replacing old ones, but rather of new stages growing on top of old ones. Each of us is archaic, magical, mythical and mentalistic at different times and in different ways. Usually, however, it is the most recently evolved view of mind that has the most power, and assumes the governing role. The Ba-Benjellé as they exist today present an interesting study of the transition from one stage of consciousness to another—they plainly alternate between a magical view of mind and world, while in the forest, and a mythical/mental view while in the village.

Religion is a good example with which to think about these different stages of consciousness, these different ways of conceiving and experiencing the mind. The everyday spirituality of the Ba-Benjellé is magical; on the other hand, religions as we know them today are a transplantation of this everyday magical spirituality into the mythic and mental domains. They are combinations of the magical, mythic and the mental. It is their magical and mythic elements which render religion so confusing and apparently absurd to the scientific mind—and it is these same elements which render it so emotionally appealing to the overall human organism. Indeed, it might be argued that the main function of religion in today's society is to feed the magical and mythic consciousness within all of us. However, within the mental mindset, religion becomes overlaid with heavy layers of dogma, and its role as an emotionally expressive rendering of belief about possibilities beyond the observable world becomes obscured by a mentalist, rule-following approach to a domain which does not really warrant one.

The spirits of Stone Age people, like the forest spirits of the Ba-Benjellé, were tied in with elements of nature. They were symbolic concretizations of the natural world, of which human minds and bodies were implicitly assumed to be part. Even in the mythic frame of mind, the nature-religion connection was still prominent: Yah-weh, the Jewish God and the root of the Christian God, was originally a corn god! Today the connection exists only in watered-down symbolic form, in the form of rituals whose meanings no one really understands anymore, but which are performed simply as mechanisms of a particular system of faith. For instance, the Hindu prohibition on killing cows ties in with the ecological importance of cow dung in India. The Balinese tradition of sacrificing fruit to the gods in temples and outside homes has an ecological purpose: to attract ants (spirits?) to places other than inside homes. Food prohibitions of Islam and Judaism stem from the particular sanitary conditions of the desert. These religious customs originated at a time when harmony with nature and harmony and God were more nearly the same thing—but now they are anachronisms, continuing primarily on the basis of momentum. The mental beings of today memorize aphorisms of religious custom just as they do the axioms of mathematics. In spite of odd proclamations like the Gnostic Jesus' "split a stick and I am there," the center of contemporary religion is not in the direct, magical relation with the outside world. Religion exists in the contemporary mind primarily as the abstract symbols, words and rituals used to elicit spiritual feelings. We pray in churches—artificial, logical constructions, built according to vocabularies and grammars—we do not pray as systems coupled with an outside, living world.

Philosophy goes one step further than religion, and attempts to transplant the feeling of global understanding and relatedness of magical consciousness into the purely mental realm. Yet it has far less potency than religion—only in a small percentage of individuals does it resonate as deeply as non-intellectual spiritual practice and belief. These unusual individuals, these "exceptions who prove the rule," tend to be atypically intellectual people, in whose minds mental activity has penetrated so deeply as to assume mythic and magical roles (the authors of this book count themselves in this category!). The general impotence of philosophy, as opposed to religion, is proof of the human power and necessity of religions' mythic and magical aspects. Intellectualism is the embracing of the mental perspective in its extreme, whereas Spiritualism is the embrace of the magical perspective. Religion is the paradigm case of a multilayered construction; an entity that bridges all the different structures of consciousness, the different perspectives on mind and world.

MIRRORS IN THE MIND

Gebser's categorization system is a powerful one, but it has no absolute, dogmatic value. All systems of categories are merely tools for understanding: they encompass a certain amount of order, and leave other things out. Gebser's is no exception. Just as Gebser's four stages enlarge the original dichotomy of Stone Age versus modern with which we began, so it is possible to expand Gebser's stages into a yet finer gradation of modes of consciousness or views of mind. One interesting way to do this is via the concept of mirroring or *reflexivity*, according to which successive stages in the development of mind can be viewed as earlier stages, reflected into themselves.

The birth of the magical state is the initial act of mirroring. Initially, in the archaic state of consciousness, nothing is divided: the world is One. Then the mind, with its sense-organs reflecting the world, becomes separate from the world. In a primordial act of mental reflexivity, it becomes a world within a world. Instead of just creatively recognizing patterns in the world, the mind is self-consciously creating an inner world, which is a sort of oversimplified simulacrum of the outer world. The archaic mind also simulates the outer world and creatively recognizes pattern in the outer world, so mere intelligence is not the difference between the archaic and magical stages. The difference is simply the drawing of the boundary, the line between inside and outside, and the classification of the passage of causation from inside to outside as magical. This act of boundary-drawing is important, because it provides the "distance" needed in order to make calculations—to reason.

By decoupling its dynamics to some extent from the dynamics of the environment, the mind's dynamics become free to pursue their own trajectories and find new places that they could never have found in the old coupled system. They can follow one step to another to another without interruption from the outside world, and in this way they can create things like axes and boats and cultures; but in gaining these new trajectories of reason, they also lose some of the trajectories of interaction that they previously followed, emergently with the dynamic environment.

The magical state of mind contains a boundary, distinguishing the world from the world-within-a-world called mind. It is the result of a single reflexive movement. The next reflexive movement, the mirroring of the mind/world dichotomy within mind, results in the emergence of language. Language is a result of mind conceptualizing the distinction between mind and world, and applying the tools it previously used to deal with the external world to deal with *itself*. It is easy enough to communicate about the outside world with another sentient being simply by example. You can show someone else how to make an axe without ever saying a word, but (within the constraints of human or humanlike existence, at any rate) there is no possible way to communicate a complex internal emotional state such as "I don't understand why I exist" without language. Instead of merely recognizing patterns in the outside world and calculatedly creating patterns in the outside world, the mind in this reflexive stage starts recognizing and creating patterns in its own structure and dynamics. Its internal dynamics are de-coupling into two systems: the inner observer and the inner observed. The inner observer, looking at the inner observed, is understanding what it sees, and constructing forms representing its observations; mental, not physical, forms. These forms are linguistic structures. But as with the initial reflection that moved archaic consciousness into magical consciousness, there is a cost to this decoupling. The price is paid in harmony, unity, coherence. Many sophisticated patterns are gained; but some very simple patterns, some simple symmetries, are lost.

When it first emerged, human language was not considered distinct from the sounds by which animals communicate with one another, and the sounds with which nonliving phenomena (thunder, water, etc.) communicate with living beings. Even today, in pre-scientific cultures—not only Stone Age cultures, but relatively more advanced cultures like the Native American Tribes or the New Zealand Maori—"talking to animals" is a commonplace notion. At the same time as spirituality pulled away from living nature, developing into interior, mythic systems, human language become interiorized as well. In this development our language became distinct from the multiple languages of the natural world. In these ways and others, our perception of the natural world changed: no longer was it part of a living continuum along with us. Rather, the external world was "dead and out there" while we were "alive and in here."

We can see this transition in language, in religion, and virtually everywhere else. Consider such changes in literature. In Homer's works, characters do not have inner thoughts. They hear voices from the gods, from the spirits, who are often associated with natural forces such as Neptune, the god of the sea, and so on. In later Greek literature, representing "full-fledged civilization," the voices have moved inside. The locus of pattern-formation has moved within. Julian Jayne's *The Origin of Consciousness in the Breakdown of the Bicameral Mind* is one interpretation of this event, which we find interesting although we don't entirely agree with him.

This evolution of language and symbolism had a lot to do with the emergence of religion out of magic, and into philosophy. The two evolutions occurred at the same time, and in no small measure helped each other along. Both had to do with the reflection of the dichotomy between inner and outer world into an *inner* dichotomy, a division of the inner world mirroring the original division between inner and outer. Religion is a conscious, concerted formation occurring within the mind, corresponding to what, in the magical mindset, was simply a mode of interaction between the mind and the world. It is formalization into abstraction of what was once viewed simply as a state of being.

Next in the sequence of reflections, following on the creation of language, is the creation of machinery. Tools exist in the archaic state of mind, only coming into their own in the magical state—but machinery is a different story. Machinery requires advanced language because it is, in essence, a language. It is the language of abstraction and repeatable problem-solving made real; a language whose words are tools, and whose sentences are the embodiment of solutions to problems which had to be abstractly formulated in order to be solved. When we were simply beings in nature, there were no engineering problems, as there was no substantial notion of a rigorous effort to modify the natural order of things. Engineering, no matter how primitive, is a grammar of tools: it is a collection of rules telling what kinds of tools should be fit together in what ways, to provide effective structures. With the evolution of machinery, we have a reflection into the external world, to accompany the reflection into the internal world that gave rise to language. The mind/world dichotomy becomes manifest in reality, as the calculated pattern-recognition and formation of the magical mindset is used to control not relations between the organism and the outside, nor relations between different inner forms, but relations between parts of the outside world. With machinery, we begin a program of deliberate reordering of the natural world to reflect our inner mental picture of how it ought to be, rather than our internal mental picture being wholly derived from observation of the world outside.

With the advent of higher-level rationality—true abstract thought—we have yet another reflection: we have language itself reflected inwardly, now playing the role that the world plays with regard to machinery. Advanced reason is, in fact, a machine for fitting together and producing linguistic forms. Logic itself is a form of language; so is science; so is mathematics. With purely abstract thought, we formulate problems wholly within the context of our internally constructed paradigms; and come to solutions which express a metalinguistic reordering of these systems via linguistic machinery which acts on our mental worlds in the same way we reorder the natural world with physical machinery. Instead of an inner world containing a simulacrum of the mind/world dichotomy, we have an inner world containing a dichotomy between an outer world and a mind/world dichotomy. Things become perverted, convoluted, complex—and astoundingly creative. We have a world within a world within a world, a nesting of mirrors three levels deep. We've gone from reflecting the natural world, to imposing our abstractions upon the natural world, to imposing meta-abstractions onto our own systems of abstractions.

The hierarchy of reflections is just a more detailed form of the observation that first occurred to Ben while reading *Song of the Forest*: that the inner, mental world of modern Western civilization is a result of turning the original magical, animistic worldview *outside-in*. The mind is always going to perceive a complex, living network; a web of subtle, dynamic pattern. If it didn't, it wouldn't be a mind, and if it doesn't perceive this web in the outside, natural world, it's going to perceive this web within itself. The complexity and life are going to be there, even if reflected within themselves two, three or a dozen times. It can be complicated, but it cannot be extinguished, without extinguishing mind itself.

ONE STEP BEYOND...

Louis Sarno felt, intuitively, that the modern, mentalistic system—with nature viewed as largely dead (we can no longer "talk with" animals or skies or waterfalls!), and the inner world alive and flourishing—was intrinsically less healthy than the old, magical system. It is not hard to see the roots of this feeling. After all, our individual inner worlds are largely not only incompatible with the natural world but mutually incommunicable to each other—whereas in the old system, where one's life was largely focused outside in nature, there was intrinsic communicability between humans as a consequence of the focus on the common system in which humans were mentally embedded, and confluence with the naturalistic state of being. Our linguistic formations have become more and more ornate, attempting to overcome this communication barrier that we have created by moving the focus of human complexity from nature to the inner mind. Ultimately, we do not succeed. Our religions become reduced to empty symbolisms, linguistic formalisms—mere incantations. Our deepest personal experiences become almost impossible to communicate, because they too are focused on our own inner worlds, rather than on the shared substrate of external, natural reality. We feel increasing alienation and isolation from each other, because we've accepted a system of being in which having no common frame of reference is axiomatic, and indeed the very idea of interdependence of any kind is seen as anathema.

However, not many of us are willing to do what Louis Sarno did and return to a semi-Stone Age way of life, in hopes of regaining some of the magic of the magical state of consciousness. Instead we keep pushing ahead, progressing further and further into our linguistic, mechanical, scientific, rational world. The question is where are we getting by going this way? Are we getting somewhere valuable, important, deeply fulfilling? Or are we simply moving further and further away from the core of our being, disappearing into a tinier and tinier fraction of the universe, a mirror within a mirror within a mirror within a mirror—until ultimately, we disappear entirely?

Gebser himself was an optimist. He felt that there was a fifth stage of consciousness, one that he called the Integral stage—a return to holism, to the oneness of archaic consciousness, but without sacrificing the advances made by the magical, mythical and mental stages. In recent decades numerous "New Age" philosophers have made claims similar to Gebser's. Terrence McKenna, for example, has proposed that a dramatic transformation in human consciousness is going to occur in the year 2012, at the end of the Mayan calendar (McKenna, 1994). A new age is upon us, it is said—an age of postmodern science, collective consciousness, and near-universal harmony! Everything will be beautiful and surreal.

This is a wonderful vision, but it has far more hope than substance to it; and is disturbingly reminiscent of the exultations of Christian fanatics 1000 years ago, as the millenary anniversary of the birth of Christ approached. Perhaps simply believing a new age is upon us will make it come true—but it didn't work for the Anabaptists of 1000 A.D.! The notion of salvation through cataclysm not only severs man from responsibility towards himself and his world, but its success rate is nil.

While Gebser was clearly fixed in the mental mode of being, most of the modern New Agers seem to feel more affinity for the mythic and magical states of mind. New age culture also seems to gravitate towards a Stone Age-ish spherical notion of time, in which the succession of events is viewed as unimportant—a view of time that is admirable in some respects, but is perhaps not optimal for making temporal predictions!

It is easy to be skeptical of these proclamations of a new and better mode of consciousness to come. Such proclamations ring of falsity, of ideological salesmanship. Sardonic retorts are in no short supply. However, if we accept

Gebser's model, it's actually happened four times already, and at each stage the evolutionary time to the next model has decreased.

It is a curious fact that in recent years a number of quite rational, mentally-oriented people have come to make statements very similar to those of the "crazy" New Age millenarians. We are among this august crew. However, the differences between optimism in a rationalist framework, versus a contemporary religious one, are: our belief is evidential and probabilistic rather than axiomatic; and if we are shown to be incorrect, it's perfectly acceptable for us to just accept that and look for new theories and predictions. Also, hard-Singularity dogmatists aside, the change we perceive as coming happens through change which is obviously derivable from the proceeding stage (obviously, to those who are around once the change has occurred, not necessarily beforehand; which is the primary difference between a merely rapid or surprising change and a cataclysmic one), and guided by an intelligence (for better or worse). We believe that the emergence of advanced biotech and computer technology is reasonably likely to catapult humanity into a new phase of consciousness, essentially identical to what Gebser spoke of as the fifth, Integral stage. In a "global brain" scenario, it's quite possible that intelligent computer networks will be a higher stage of consciousness, and that they will not be separate from us—but will rather induce us to move into a higher stage of consciousness as well. Advanced biotechnology may come to be one of the major factors catalyzing this fusion. On the other hand, in an AI-before-GB scenario, we may see an Integral consciousness-based society emerge via technologies we can't yet envision, which stem from the alternate perspective and new forms of creativity that a self-rewriting, evolving AGI may achieve. Or we may see it arise straightforwardly, via uploading and neurosoftware engineering—we upload our brains into the Net and intentionally modify them to improve their experience in the world, transferring our minds onto hardware which is soon to be superior to the human mind in raw computational power[1], and which is easier to maintain and replace as it fails. Probabilities are tough to estimate, but there are plenty of plausible avenues by which Gebser's conjecture could be technologically realized.

Eccentric, and maniacally poetic and metaphysical, as we can be sometimes, we're basically scientists. We realize that making such statements will not do much good for our short-term credibility among scientific colleagues. However, we are willing to take that risk, because of the obvious importance of what we're saying. If we are right, and computer and biotech really do have the ability to push us to the Integral stage of awareness, then this is something that everyone should know about and work toward. It should structure our actions and guide our lives as we build our future world. Louis Sarno's bold action, while apparently good for him at the particular time he took it, is probably not the right direction for humanity as a whole. Even if we *could* do it, there would be no point in our moving back to the magical state of mind—because we are moving forward to something not at all inferior, but rather something encompassing the magical, mythical, mental and archaic modes of being in a harmonious and creative way.

As we've progressed through the stages of mental development to our current, mentalist, stage, we've gained the ability to introspect to the level of allowing ourselves to begin to know what it is we've actually lost from the archaic, mystical, and mythic modes of understanding. Once we started progressing away from the archaic, it became imperative that we reach the mental stage in order to ever evolve to the Integrative stage in which we can regain what we've lost in the process of creating something new. We're at a crossroads, where in order to keep going we need to harness all our mental abilities, and all our energies and resources, to build a world in which we can rebuild what we've lost in our minds (and in our world, and our relationship to it), in order to have a solid foundation of being which will allow us to progress beyond that. The alternative is to limp along as incomplete beings, stuck in our current developmental stage until, sooner or later, we all die out.

From a certain perspective, of course, computer technology is the ultimate manifestation of our tendency to withdraw into our own inner worlds, to avoid contact with nature. It is also the ultimate alienation of language from nature, relying as it does on the development of artificial languages (programming languages, protocols, etc.) which are not only pure mental constructs, but are mechanisms by which we communicate with machines—physical projections of our mental abstractions—and cause machines to communicate with each other. We've got abstractions communicating abstract ideas to each other in abstract languages—it doesn't get more unnatural than that!

Computers are the first instruments to manipulate abstractions and simulate processes. They are windows through which the mind can examine abstract processes objectively; a way of taking a formal abstraction, encoding the underlying principles of an abstract mental model, and testing them in the closest thing to the natural world that purely abstract "ideas about ideas" can currently inhabit—a machine. Looking through these mechanical windows into our own minds, we no longer have the need to look at anything else. These observations are made concrete by our own lifestyles during the past few years: what do we do all day? We sit inside staring at computer screens, in carefully controlled and unnatural environments, using various formal-language-based tools (programming languages, word processors, drawing tools, etc.) to spill out the complex creations of the seething, self-organizing internal universes in our heads. We use these tools to help build our private inner worlds, and to communicate some small aspects of our private worlds to the private worlds of others—for example, to attempt to communicate the thoughts underlying this sentence to you.

We are pretty comfortable with this computer-heavy lifestyle. (At least, we think we are!) We're happy people: not every minute of every day, but for the considerable majority. We work long, intense hours, but we're not "workaholics" in the sense of not being able to enjoy ourselves apart from work. We take great pleasure from composing and playing music, reading, creating art, spending time with our wives (and Ben's children), hiking in the woods, playing sports, eating good food, and so on and so forth. Still, we're well aware that our relatively pleasant existence is largely a consequence of a fortunate confluence of social context and brain chemistry: it so happens that our neuropsychological cocktails are compatible with a lifestyle that modern Western society finds acceptable, and that we were born somewhere that allows people like ourselves to be relatively free, prosperous, and happy. This is not the case for everyone, and the majority of people who find their brain chemistry is not compatible with the lifestyle modern culture has provided for them seek solace in the products of modern pharmacology.

Illegal drug use is stigmatized in mainstream Western society, and in some cases rightly so—heroin, speed and cocaine, for instance, are inarguably destructive. No one who has spent time around junkies, cokeheads, speed freaks or crank addicts can have much enthusiasm for these drugs: they destroy people's minds, bodies and lives. On the other hand, we are strong advocates of legalizing marijuana and hallucinogens (mushrooms, LSD), not because we use them regularly (we don't), but because we think that in order to justify outlawing something, the government should have a much stronger reason than is present in the case of these substances. Excessive marijuana smoking has bad effects on the lungs and short-term memory, but as essentially everyone in our generation (or even our parents' generation) knows, it's not really any worse than alcohol and is probably less toxic than smoking cigarettes. It's the kind of potentially mildly destructive habit that, in our view, adult humans should be allowed to pursue if they so wish. On the other hand, hallucinogenic drugs are far more psychologically powerful than marijuana, but they are not at all physically addictive and certainly do not cause violent or otherwise criminal behavior in most people (there is enough variety in neurochemistry that pretty much anything could be a chemical trigger for unwanted behavior, so it's a matter of looking at such things statistically). They have powerful good aspects as well as potentially destructive aspects attached to them. Nearly all tribal cultures have used them to good effect, in the context of shamanic and other rituals. We believe our culture is making a big mistake by pushing psychedelic experience into the category of criminal acts: both because it runs counter to the ideals of personal liberty that we supposedly value, and because it is restricting growth by baselessly depriving people of experiences which are generally no more dangerous than drinking alcohol, and are at least equally pleasurable and interesting to our current human brains.

Though we strongly stigmatize the use of *certain* psychoactive substances, people still use these in great numbers, risking (particularly in the U.S.) absurdly long jail terms in order to do so. Furthermore, a substantial percentage of the non-illegal-drug-using population is hooked (physically or psychologically) on some form of legal psychoactive substance. Prozac and Valium are just the best known of a huge variety of mood-altering drugs, prescribed by doctors simply to make people feel better—to make them better able to tolerate the frustrating aspects of their lives. We as a culture are basically addicted to legal mind-altering drugs, though the side effects of these drugs can be severe[2]; certainly worse than those of pure LSD or properly grown hallucinogenic mushrooms. (Not that Prozac/Valium and LSD/mushrooms are in any way psychoactively similar, it's just interesting to see what society categorizes as "safe" versus "unsafe"—unfortunately, within a class of substances which have relatively similar severity of unwanted effect, it appears to stem from whether or not someone can secure a monopoly claim to the substance, or at least that it doesn't threaten an existing monopoly.) The bottom line is that the bureaucratically-structured, tech-dominated-and-enabled society we've created is intolerable by a vast percentage of the population without frequent chemical alterations to their brain state. Pharmacology and computer technology seem to be interacting in a strange sort of way: the latter makes life tougher for many, and the former modifies the brain so as to better be able to tolerate it. Of course, people have been unhappy and drawn to chemical solutions for a long time. Before there was computer technology, there was something about bureaucratically-structured life which led a substantial percentage of people to mentally rebel against it—perhaps the impression that overly hierarchical structures give, whether you're in the collectivist East or the individualist West, is that each of us is ultimately worthless because the system doesn't need any individual to continue. We long for a system in which each of us feels necessary, and in absence of that, we create one in our own minds.

Still, what we believe the future holds is a rather different sort of computer/biotech-tech synergy than the misery/escape symbiosis. For, in the Internet, we have a system with the same properties that Nature originally had for us, way back in the Stone Age. It is a complex, self-organizing web which generates mysterious patterns and binds together various people into a common substrate (it may currently appear to be a bazaar of snake oil and pornography, but there's really a lot more of substance there than a cursory glance might reveal). It opens up our inner worlds and transforms them into collective worlds. In a very crude sense, we can see this in the psychology of e-mail—instead of thinking through an issue oneself, one can dash off e-mails to four or five friends and engage in a real-time collective thinking process. Even without virtual reality and biotech-based human-computer hybridization, there is the clear potential for the weaving together of the world-wide network of mutually incommunicable "inner worlds" into a whole, vibrant system; even if it is a relatively simple consequence of the coevolution of human language with real-time mass and distance communication systems, and the computer languages for representing and testing abstractions and meta-reasoning about them. With the capability to genetically engineer human beings so as to optimize the effectiveness of human-computer interactions, the sky isn't anywhere near the limit. Communication with the Net and with other humans will become part of the very processes of human thought, feeling and creativity.

Nature, as it receded from us, transformed in our cultural psyche into spirits and gods, and into inward-focused, linguistic minds. As religion fades, we are left with nothing but mind—nothing but the rational, inner world, and the institutional and technological forms it has created. Then, lo and behold, one of these forms is leading us back to something with many of the properties of every stage along the way. The emerging Internetworked intelligence is at once a natural environment, a spiritual domain, and a mind with a complex, creative inner world—a potential Mindplex of interdependent thinking and creative self-reformation. Genetic engineering and bioengineering will allow us to fully partake of this new, communal, synergetic life form.

Furthermore, the same archetypal patterns remain in this new, digital external universe, as in the old, biological one. The same patterns that structure the natural world also occur in our religious symbolism and experience, and in our mathematics, science, computing and art. These archetypal patterns, ultimately derived from Nature, recur in the emerging Internet intelligence—once again projected into a collective, communal environment, rather than imprisoned within the confines of individual heads. In short, the archetypes have moved from the external real space to the internal mental space and then back "outside" into the emerging biocyberspace.

Such archetypal patterns give us grounding as we move from stage to stage of our mental—and biological—evolution. As we change, archetypes may fall away; as we decide they contraindicate our growth as sentient beings, or discover they never were truly archetypal in the first place. We want to reclaim true archetypes of our unity with our world, as we create new ones about the composition of our created world and its unification with the natural, physical world. False archetypes—stereotypes—continually are identified and evolved away from, so long as we are improving ourselves as a species. This has happened before and continues to happen, whether it be by accepting new perceptions, and believing in the roundness of the world and the peripheral physical importance of our solar system; or spreading acceptance of lessening social restrictions on non-impositional personal choices, and finally viewing women as people separate from their mates, free to live their own lives; or allowing people to love whomever they choose, regardless of what sex they are. We are no more bound by any irreversible mental constructs of prescribed behavior or inextricable social need to continue warfare or monopolist economics, than we were to continue slavery or monarchy. Every particular of how we conduct ourselves is amenable to change—we must "merely" figure out what positive change entails, and how to implement it.

In terms of reflections, computing corresponds to the reflection of rationality into the outside world. The mind/world within the mind within the mind/world within the mind within the mind/world becomes a mind/world within the mind within the mind/world within the world within the mind/world—and so forth. The mirror of computation reflects outward where (and what!) the mirror of rationality reflected in. Finally, in the intelligent Internet, the ground of the "physical world" itself is replaced by a reflection of the rational-mind/computation dichotomy. Mind replicates itself in an interconnected global network of computers atop the world's storehouse of knowledge and ideas, thus maximizing its resources—the input of each individual mind—and the information it creates. The dichotomies in thought promulgated by culture, geography or perspective are synthesized away. The mind, transformed by its interaction with the global mind, approaches its evolutionary potential in its current form.

Biotechnology is critical here, for an obvious reason: As human organisms, we are sensorially attuned to genuine, external nature; not to "Internet nature." So, even if the latter is to become just as complex and multitextured as us, it will not match us as well. Every aspect of our body is attuned to the natural environment, not the Intelligent Internet. Thus, one concludes that, in a sense, Louis Sarno may be right: while a future Internet-based human society may be healthier than the current system, based on individual, incommunicable inner worlds, it still will not possess the basic health and integrity of Stone Age culture, or of animal life—unless the radical changes to come are pushed in the right direction, that is: towards the symbiosis of the external, animal being and the internal, abstract being as codified in a system of machines and languages.

When first thinking through this issue, we came up with two possible solutions:

> 1. The body is shed or transformed, and replaced with a form more harmonious with the new Nature/God/Mind that we have created.
>
> 2. The global brain is brought into harmony with Gaia, the "mind of nature" itself[3]—so that nature, man and the Intelligent Internet are all fused as one

Option 2 had us thinking, in rather grandiose terms, about nanotech bacteria diffused into the atmosphere, communicating with real bacteria as well as with computer controllers and the humans linked to them, and with nanomachines interacting with and monitoring the fine-level mechanical substrates of the earth's physical geology, creating a link between Gaia and internet intelligence. At some substantial level—not a level of assumed understanding based on mystical inference, but rather by building up from the lowest-level, nano-engaged understanding to the higher levels—humans and machines could communicate with all Gaia: bacteria, animals, weather, and so on. Now *there's* a global brain deserving of the name!

When Ben posted these thoughts to the Global Brain Discussion Group (in June 2000), however, Francis Heylighen brought us a bit closer to reality. As he noted:

> "There is a third, more short-term and more practical possibility, which is that the Internet evolves to fit our inborn characteristics. All evolution is co-evolution: systems mutually adapt. People will adapt to some degree to the new Internet environment, but the Internet will adapt even faster to the people that use it. Just because the Internet is intrinsically much more flexible than our hard-wired instincts and proclivities, it will find a way of presenting itself that matches those proclivities."
>
> "This has happened countless times in the evolution of computer interfaces. For example, the GUI that became popular with the Mac was based on the idea that people don't understand things by reading long lists of file names, but by moving and manipulating objects. Thus, files were represented by icons that you could drag and drop to move them from one directory to another. 3-D, virtual reality, as e.g. imagined by Gibson in his original 'cyberspace' vision, is another obvious way to make a complex information space match better with our inborn capacities to reason in three-dimensional space. 'Emotional' agents that respond to our moods, or show simple emotions, [are] another one of these new interface paradigms that [try] to fit our evolutionary psychology."
>
> "I don't say that all these interface tricks will succeed, or even that they are necessary to have a good grasp of the Internet, but they are definitely undergoing a fast evolution and competition in order to find interfaces that are better adapted to our brain."

His point is a good one—insightful and down-to-earth; typically high-quality Heylighen. However, the coevolution of systems is often stated as symmetry, and then reflected upon primarily in the pessimistic terms of how the new systems will succumb to the worst of our current tendencies. Yet, coevolution is a two-way street, and if positively directed could ultimately lead to new ways of thinking. If this sounds outlandish—if you believe that human nature is immutable and we'll always be exactly as we are—realize that it has happened several times before: with agriculture, with city building, with imperialism, with the enlightenment, with industrialization, etc. Each time, whole new ways of viewing, and shaping, the world came into being which literally could not be conceived of prior to the moment of confluence that caused their creation. To achieve what we're thinking of would take more than a new UI for PCs as they exist today; it would require computing "interfaces" to be much more thoroughly integrated into natural human life. Interacting with the Net needs to not be something you do while sitting on your ass staring at a machine, but something you can do while floating in a lake, or walking under the trees, or sitting in the living room or the yard, chatting with your friends. Ubiquitous computing, if done properly has this potential. In fact, it has a strong potential to bring us *closer* to nature, in our daily lives, than our current industrial-revolution-based technology permits.

Ultimately, though, we don't believe any version of Heylighen's modest, conservative vision is going to actually carry the day. In Yudkowsky's terms, Heylighen's vision is SL1, perhaps slightly verging on SL2. It doesn't go far enough. Ubiquitous computing will happen and it will be wonderful, allowing us to be constantly accessing new information in the realm of abstraction while freeing us up to engage in more experiential learning as well; but creative biotech will happen too—it already is being pursued. Genetic engineering and bioengineering, implemented cooperatively, will make us capable of interfacing directly with data on a Siberian web server while swimming in a lake in the Costa Rican mountains, or flashing interesting 5-dimensional movie clips to a lover, brain-to-brain, in the middle of the sex act. Scary, exciting, or fascinating—all of the above, we suppose. Yet this is where science is pointing us—in fact, this is the least of it—and to doubt this without solid reasons is to blind oneself to reality.

People are vigorously pursuing these technologies already, and all that is currently lacking is a determination to do so in a way which will direct them in positive rather than destructive ways. This may be a severe deficiency from the perspective of our well-being as a species, and that of our planet as an ecosystem, but it's nothing that will prevent these advances in technology from being pursued and wherever actually possible, achieved.

We are thus led from the frontiers of consciousness to the frontiers of engineering: the creation of digital bodies and the re-engineering of biological bodies. Things become not only stranger than anyone previously imagined, but stranger than anyone *could* have previously imagined. Even we don't pretend to see where it is all going to lead—except to guess (hope) that it is going to be wonderful, and beyond any mode of conscious experience that we have encountered up to this time. The world can indeed become beautiful and surreal, and though we're a little bit wary of possible future disasters (more on that a little later), all in all, we can't wait to see it.

---

[1] Unless there is some quantum phenomenon in the human brain which we've not yet accounted for in neuroscience, in which case "soon" may require the perfection of quantum computing and be several decades away.

[2] The severity of side-effects for both illegal and legal drugs has a wide variety. Both classes of drugs contain some drugs that are dangerous, and some that are relatively benign, so empirically there must be some reason other than safety why some drugs are illegal when relatively similar drugs—in terms of potential danger to the user and others—fall in either category.

[3] A phenomenon which is very poorly described by the New Age usage of the term Gaia, but we lack a better concise word for a phenomenon which is slowly becoming better understood through research into areas such as complex dynamics in natural systems, scientific ecology, animal behaviorism, and other fields seeking to understand systemic interrelationships in the natural world.

CHAPTER 14
MINDPLEXES

Among the many sorts of things the strange and glorious future may hold, one entrancing possibility is the emergence of *new kinds of mind.* Many new kinds of mind may come to exist—most of them impossible for us, with our current human minds, to conceive. There's one possible new kind of mind, however, that seems almost graspable by us humans, in spite of its radical difference from anything we've experienced so far. We'll call this new kind of mind a *Mindplex*. In this chapter we'll briefly review the "Mindplex" concept and tie it in with the Novamente AI system. This is highly speculative material—but it's one of those thoughts that follows us around constantly, and we hope you'll feel at least a little of the fascination we have with the idea.

Mindplexes are hinted at by Turchin's and Heylighen's notion of a global brain or global superorganism—but Heylighen himself, as discussed previously, prefers to interpret the Global Brain concept in a less radical way. The Mindplex concept is the result of taking the Global Brain theme to its maximum limit. Not the maximally offensive limit—which, according to modern American taste, is probably the notion of a global superorganism in which humans are forced, Borg-like, to serve as cells with no will. Rather the Mindplex is the maximally radical conceptual limit, in which the Global Brain is neither a tool humans use to communicate, nor a mega, human-like brain in which we serve as neurons, but a whole new form of being; thinking and feeling, including and enhancing and yet massively transcending known human experience—including new forms of individual free will and social interaction that are impossible in a universe where minds are connected together merely by imprecise linguistic representations of thoughts.

We'll define a "Mindplex" as an intelligent system that:

1. Is composed of a collection of intelligent systems, each of which has its own "theater of consciousness" and autonomous control system, both of which interact tightly, frequently exchanging large quantities of information.

2. Has a powerful control system on the collective level, and an active "theater of consciousness" on the collective level as well.

In short, a Mindplex is a system of "collective intelligence" (CI) in which each individual, while afforded privacy and autonomy of his or her thoughts, also has the capacity to pool information with other individuals in a coordinated manner. This enhances both the individual's capacity for thought, and the system's overall capacity for thought. Rather than necessarily draconian and Borg-like, this can be seen as an extreme extension of the development of liberty and individuality through education and cooperation.

We reject the social cycle theory summed up in the idea that "history repeats itself," and instead agree with Hegel that social progress is inevitable. While individual societies may go through the phases described by P. A. Sorokin—spiritual, materialistic, and a synthesis of the two—during their individual lifecycles, ultimately forward progress for humanity as the aggregate of all social bodies is not just possible but unavoidable until either we are extinct, or evolve to a transhuman state.

As George Por wrote in his Collective Intelligence Blog[1], CI is "the capacity of a human community to evolve toward higher order complexity thought, problem-solving and integration through collaboration and innovation." Eventually, this collaboration and coordination may achieve a concurrency and depth which will be like "plugging-in" your mind to other minds at-will, and collaboratively thinking about a problem in real-time, sharing experiences and mental capacity to come to a solution to a problem or develop a much more quick and effective innovation. The choice of the word "Mindplex" to label this concept was suggested by Eliezer Yudkowsky after some discussion on the SL4 e-mail discussion list.

In informal discussions, some people react to the Mindplex concept by contending that either human minds or human social groups are Mindplexes. While there are significant similarities between Mindplexes and minds, and between Mindplexes and social groups, there are also major qualitative differences. It's true that an individual human mind may be viewed as a collective, both from a theory-of-cognition perspective (e.g. Minsky's "society of mind" theory, discussed previously) and from a personality-psychology perspective (e.g. John Rowan's theory of subpersonalities). It's also true that social groups display some autonomous control and some emergent-level awareness. In a healthy human mind, the collective level rather than the cognitive-agent or subpersonality level is dominant, the latter existing in service of the former. In a human social group, the individual-human level is dominant, with the group-mind clearly "cognizing" much more crudely than its individual-human components and exerting most of its intelligence via its impact on individual human minds. A Mindplex is a hypothetical intelligent system in which neither level is dominant, and both levels are extremely powerful. A Mindplex is like a human mind in which the subpersonalities are fully-developed human personalities, with full independence of thought, and yet the combination of subpersonalities is also an effective personality. Or, from the other direction, a Mindplex is like a human society that has become so integrated and so cohesive that it displays the kind of consciousness and self-control that we normally associate with individuals, while at the same time allowing each component individual its independence, thereby strengthening the interdependence by affording each individual the opportunity to bring their unique perspective and experiences to the Mindplex. Individuality of the fully-formed subpersonalities is crucial to the robustness and effectiveness of the Mindplex, since diversity of interest and experience provides a greater degree of flexibility and capability to the group, and it would be self-defeating for this particular kind of collective to stamp-out individuality in its members.

Two mechanisms by which Mindplexes may possibly arise in the medium-term future are:

1. Humans become more tightly coupled via the advance of communication technologies, and a communication-centric AI system comes to embody the "emergent conscious theater" of a human-incorporating Mindplex.

2. A society of AI systems communicates amongst each other with a richness not possible for human beings, coming to form a Mindplex rather than merely a society of distinct AIs.

Mindplexes are most likely to occur with the assistance of AIs because, having (in a positive scenario[2]) developed without ego and with attention to ethics, AIs will be more able to take a neutral stance in managing the contributions (and providing security about the fears) of a variety of fully-developed individuals as part of a collective reasoning process. The former sort of Mindplex relates most directly to the Global Brain concept as commonly conceived. Of course, these two sorts of Mindplexes are not mutually contradictory, and may coexist or develop independently and then fuse into a single Mindplex. The possibility also exists for higher-order Mindplexes, meaning Mindplexes whose component minds are themselves Mindplexes. This would occur, for example, if one had a Mindplex composed of a family of closely-interacting AI systems, which acted within a Mindplex associated with the global communication network. This is a likely situation to occur given more than one type of AGI system coming into existence, and/or the coexistence of AGI with biological transhumans which have developed Global Brain type abilities.

The notion of the Global Brain is directly embodied in the design of the Novamente AI system. The Novamente design contains a specific mechanism—called Psynese—intended to encourage the formation of multi-Novamente Mindplexes. As such, Novamente is designed specifically to be deployable within global communication networks, in such a way as to provide an integrative, conscious theater for the "distributed cognition" system of interacting online human minds. In the final section, we will use the projected nature of Novamente Mindplexes as a metaphor for comprehending the slipperier, and more ambiguous, Global Brain Mindplex idea.

THE VALUE OF COOPERATION

We have described Mindplexes as Collective Intelligence—another way to think of them is as Cooperative Intelligence. This phraseology provides a link between Mindplex theory and game theory, which studies the value of cooperative behavior in various simplified formal situations and then extrapolates from these situations to the real world. Game theory, a branch of mathematics which was created to help understand the nature of economic interactions using simulations, has some interesting and relevant things to say about cooperation.

The game-theoretic work of Robert Axelrod and others, in which cooperation is shown to be pragmatically beneficial to the individual, is quite relevant here. Axelrod's 1984 study, a classic in the field, used a simulation called the "Iterated Prisoner's Dilemma," which is a game of multiple rounds. During each round, all participants must decide whether to cooperate with or betray the other participants. If both players cooperate then they gain maximal reward, but if one player cooperates and the other betrays, then the betrayer does better than the cooperator. This is a simple model of a situation where cooperation provides more value to the individual than selfishness over a non-immediate time horizon, even though selfishness may provide more immediate gratification.

The archetypal situation that gave the Iterated Prisoner's Dilemma simulation its name is the standard "Prisoner's Dilemma" game where two people who committed a crime together are separately detained and each one is asked to squeal on the other one in exchange for a reduced prison sentence. If neither one squeals, both go free. If both squeal, then both get a large prison sentence. If only one squeals, then he, the betrayer, is rewarded with a reduced sentence whereas the betrayed one gets a long prison sentence. This game was popular as a model of nuclear détente in the Cold War era.

The Iterated Prisoner's Dilemma extends this basic situation by posing a series of Prisoner's Dilemma type interactions among a group of different participants who repeatedly interact in different combinations. The only information each participant has about the others is how they behaved in the past rounds. Axelrod showed that the Iterated Prisoner's Dilemma simulation tends towards cooperation over the long-term, even though the participants in the simulation are given a completely selfish value system. Tit-for-tat, a game theoretic strategy formalizing the maxim, "do unto others as they do unto you," was found to be an optimal strategy for this tournament (one introduced by mathematical psychologist Anatol Rapoport)—provided not every participant chose this strategy. Basically, in Tit-for-Tat, you cooperate with a player who has just cooperated with you, and betray a player who has just betrayed you. Such a strategy creates a long-term incentive not to harm the other participants lest they harm you. Of course, if every participant chooses Tit-for-Tat as a rigid strategy, it is possible to wind up in a situation where everyone constantly harms everyone else. But, this work did point toward the idea that avoiding this situation – i.e. taking a cooperative approach – is required of an optimal strategy for events which can be modeled by some Prisoner's Dilemma variant (such as resource allocation, whether or not to start a war, etc.)

Rationally and emotionally, people ought to greatly prefer a predominantly cooperative system, and this is borne out in other research. Axelrod was showing in a simplified game theoretic context what Robert Trivers had described in biology as "reciprocal altruism." Trivers' theories, being observations of actual social interaction (amongst humans as well as animals), are "messier" and more complicated, but ultimately his work shows that because "humans respond to acts of altruism with feelings of friendship that lead to reciprocity" that a strategy of reciprocal altruism is an optimal social strategy. Extrapolating, we can see that our feelings of empathy, and the ability to project outcomes based on reasoning without direct observation, should allow us to make the "first move" in a social interaction altruistic rather than antagonistic.

More recent game theoretical research has looked at other strategies which are optimal for the Iterated Prisoner's Dilemma with noise (the possibility that a result is tabulated incorrectly), and which are optimal for a modified game called Alternating Prisoner's Dilemma in which each player takes a turn rather than both players making a move during each round. These additional criteria get the simulations a slight bit closer to the level of complexity Trivers observed in real animal populations. The Pavlovian class of strategies investigated by David and Vivian Kraines is a strategy which adjusts its responses based on how well it fared over previous rounds, rather than simply responding to the last one round as with Tit-for-Tat. Each successive round modifies a probability of the participant taking a particular action, depending on how that action fared for him in the past. If previous move choices have led the player to a bad situation, the chance of him making that move again is lowered, and vice-versa. With this strategy, descent into reciprocal aggression eventually ends, because a defector will eventually be penalized so consistently for transgressions that his probabilistic model of benefiting from these defections will approach a zero-return state, and he will revert to cooperative behavior.

Daniel Neill discusses a Pavlovian class optimal strategy called Firm Pavlov which is optimal under noisy conditions and also, interestingly (and realistically), punishes blind altruism. By optimal under noisy conditions, we mean that the strategy doesn't descend into constant aggression and therefore constant failure for all participants, even in the situation where incorrect results are transmitted to participants for some number of rounds. This means that if player A and player B are using Firm Pavlov as their strategy, and A is cooperative with B, but player B is told A has defected, if the system will eventually recover from this accidental betrayal and not get stuck in a mode of retaliatory deadlocks. Firm Pavlov is a variant of a strategy called Firm But Fair which differs in that while FP will exploit a blind altruist (i.e. a "sucker"), FBF will stop exploiting a participant using an "always cooperate" approach after some number of moves. In that sense, FP rewards rational pragmatism over unthinking doctrine. Despite their differences, all of these successful strategies in these different simulations reflect Trivers' notion of reciprocal altruism, in that they show that a cooperative strategy is an optimal one.

These more complex strategies can be seen as additional pieces of evidence that reciprocal altruism in animals (including humans) is evolutionarily strategic and not a "mistake of nature." Joseph Heinrich and colleagues found altruistic behavior of different characteristics in two other simulations: the ultimatum game and the dictator game. In the dictator game, one player chooses a sum of "money" to keep for herself, and another sum to give to the other player—with the other player having no input. The dictator game is a modification of the ultimatum game, in which one player chooses the sum, but if the other player rejects her division then neither player gets any "money."

We feel that reciprocal altruism easily explains the findings of nonzero altruism in the ultimatum game, since both players have input, and less obviously in the dictator game in which the only negative reinforcement against a tyrant is their own guilt. There is an argument that the dictator game is "polluted" by the presence of the experimenter, and that a tyrant would feel no guilt without the presence of this observer. The impunity game was devised to attempt to simulate the "anonymity hypothesis," which states that an anonymous dictator game would result in pure greed. In an impunity game the first player divides up the "money," but unlike in an ultimatum game there is no penalty if the second player rejects it. It is unclear to us how this is meaningfully different from the dictator game, except that while impunity games have been less well studied than other simulations, there are some claims that pure greed does result from runs of this simulation unlike those of the dictator game. Even if further research shows this claim to be true, there is a problem with both the anonymity hypothesis and the rules of the impunity game in the first place—they're impossible.

Both the anonymity hypothesis and the rules of the impunity game attempt to model a situation in which an actor can participate in a system without being susceptible to any form of feedback from that system (even lightweight feedback in the form of criticism by the experimenter for being a simulated tyrant). In reality, nobody can operate in a socioeconomic system without being susceptible to any feedback from it. It is more likely that you'll live forever than that you'll find a way to manipulate any system without experiencing any feedback effects upon yourself.

The nature of reciprocal altruism appears to also explain "inequity aversion," in which, as Fehr and Schmidt write: "people resist inequitable outcomes; i.e., they are willing to give up some material payoff to move in the direction of more equitable outcomes." Guilt over unfairness is a manifestation of concern over reciprocal selfishness in response to an unfair behavior, and active suppression of unfairness (inequity aversion) is an attempt by actors to restabilize a destabilizing system that could lead to a painful period of successive punitive moves by other actors—a situation best avoided, even if everyone is reliably using a strategy capable of reestablishing an altruistic equilibrium after a period of combative deadlock. It seems that only the most irrational actors will doggedly pursue a consistently combative social strategy without regard for the fact that it leads almost all other actors to penalize them for it.

The research of Trivers, Axelrod, and others contradict the popular interpretation of Thomas Hobbes' "red in tooth and claw" description of natural interaction, and provide a much better explanation for why many animals, especially us, have developed increasingly cooperative societies. (In *Leviathan*, however, Hobbes clearly states: "Do not that to another, which thou wouldst not have done to thyself." This is fully in keeping with the later observational and mathematical research, even if his prescription against active altruism is contrary to the observed optimal social strategy. Many people misinterpret Hobbes as *promoting* violent confrontation, as they do with Nietzsche.) Further, we have seen that a number of game theoretic simulations have uncovered behaviors which at first glance may appear to be aggressive or selfish, but taken in the context of long-term strategy across many actors, can be shown to promote pragmatic fairness (as opposed to a mandated illusion of fairness), pressuring participants to take no more from the system than they can contribute. Cooperative intelligence provides an evolutionary path for sentience that is in agreement with an optimal social strategy to which humans (and other animals) are already predisposed. Mindplexes represent an extension of the cooperation meme, from inter-individual interactions to the domain of cognitive interactions within cognitive processes that are now restricted to individual minds.

THE PROMISE OF MINDPLEXES

Today, if you were to think "I wish I knew more about cats," you would have only two, non-exclusive choices (besides the choice to not learn anything about cats and merely wish you did):

1. Spend a fair amount of time doing research online, in books, talking to experts if you can reach them, etc. about cats.
2. Do a lot of observational work on cats yourself, and learn from those experiences.

In a Mindplex there is a third choice. If you were to think the same thought, it would immediately be transmitted to a number of other individuals, some of whom may know a great deal about cats. Their experiences and research would be instantly available to you, and anyone else in the Mindplex. This information about cats would already be a part of the Mindplex, and simply by having your thought Mindplex-consciously, other patterns of thought would emerge which would give attention to cat knowledge and allow you (and everyone else in the Mindplex so interested) to receive a lot of information about cats. Within milliseconds you might know that cats have been domesticated for somewhere between 3500 and 8000 years, were considered sacred by Ancient Egyptians, can hear a half octave higher than dogs, are considered among the most biomechanically optimized predators…and a huge wealth of other cat knowledge. Of course, this wouldn't prevent you from doing research outside the Mindplex, and contributing that knowledge back into it for other cat researchers in your Mindplex to immediately know as soon as they had relevant cat thoughts (indeed, you would be rewarded for doing so with even greater knowledge and cooperation within the Mindplex). Imagine how amazing it would be to *immediately* share knowledge with many other individuals with similar interests as yourself, with a cerebral-cortex equivalent (probably much more capable) control system which could coordinate these thoughts in such a way that it wouldn't be a frustrating cacophony of overwhelming information for any individual. At the same time, you would be building and mining interconnections between all these individual thought contributions to come up with insights and ideas that are greater than the sum of the parts—and immediately every individual could know about those thoughts, too!

Potentially, in a Mindplex-oriented individual, a distinction would evolve between unconscious, individual-conscious, and Mindplex-conscious thoughts. Mindplex-conscious thoughts are those intended for Mindplex expression. Individually-conscious thoughts would be restricted to a single Mindplex node—the individual's mind. However, Mindplex-conscious thoughts would automatically propagate throughout the relevant parts of the Mindplex in a similar manner to that in which individual-conscious thoughts propagate through the relevant parts of a human mind. Such a distinction would serve a dual purpose to protect the individual's "private thoughts" and to make them less fearful of participating in a Mindplex, and also prevent the Mindplex from being overwhelmed with a huge influx of irrelevant and distracting thought patterns which provide no benefit to the Mindplex's overall intelligence and happiness. Each individual's neural control systems would self-organize (or be augmented with) a mechanism for regulating Mindplex-conscious thoughts in order for a Mindplex to successfully exist.

We may also think about mindplex-unconscious thoughts—but these can only really exist in the supporting infrastructure of the Mindplex (most likely electronic communications networks with specialized AIs for cortical control), because each individual's unconscious contribution to the Mindplex will originate when unconscious realizations emerge within one's own mind, and we decide for ourselves that this unconscious realization warrants Mindplex-conscious thought in much the same (usually unnoticed) way we decide to pull an unconscious thought into the foreground of our individual-conscious minds.

Such an evolutionary pathway for intelligence has a huge potential for benefit. If the evolution of human mind and society proceeds in a positive direction, we may be able to reach a point where we can create Mindplexes and use that increased mental capacity to solve these psychological problems, and many other problems, much more effectively (or, perhaps, to come up with even better ways to annihilate ourselves—a danger that's present in anything we do).

YOUR LIFE IN A MINDPLEX

If Mindplexes come to exist, and you become a part of one, what does that mean for you? Does it mean you will become a thoughtless robotic extension of machine intelligence or a mind-controlled servant of an evil human dictator? While it is possible, it seems to us less likely than either of those fates befalling you in a world without Mindplexes. Why? Even more so than its social system predecessors, the tightly integrated Mindplex really cannot function without the individuals involved making a willing commitment to the system, because the value of their input diminishes exponentially if their participation is unpleasant and unrewarding, and their impact on the system is immediate and substantial rather than very diffuse as in a contemporary social system. It is impossible to force an intelligent being into a completely unrewarding situation for too long. In the worst case, in the face of overwhelming opposition, enough individuals will choose to die rather than carry on as part of such a system that the system will become crippled. Social system parallels with tyrannical regimes that strangled themselves with their own policies can be seen throughout history.

While there is no rigorous science of such things as yet, it seems most likely to us that the central coordination of a Mindplex, in order for it to function whatsoever, must be without Ego. Any attempt to suppress the individuals in the Mindplex risks a chain-reaction of individual departure from the Mindplex which would be suicidal to the system. For humans or truly intelligent AIs to willingly involve themselves in such a system, and for it to therefore succeed, the system must provide them with these voluntarism guarantees:

- Trust—the system must provide a minimal risk and maximal benefit to the individuals involved, and especially humans with their egos must believe that their participation in the Mindplex is rewarding to them as individuals as well as contributing to the betterment of the Mindplex society.
- Comfort—integration into the Mindplex must not be painful or demoralizing.
- Fairness—the Mindplex must treat all participants fairly. The system must trend towards fairly dividing both labor and reward (in the form of time spent on thoughts of interest), attempting to be as close as possible to a Brahms and Taylor Envy-Free system in which each member of the Mindplex can rationally believe that no other member is unfairly favored.
- Utility—since the primary benefit of a Mindplex to the individuals is an increased capacity for thought, each individual must see an increase in their abilities in their areas of cognitive interest immediately upon their entry into the Mindplex.
- Freedom—any member of the Mindplex must be free to pursue their own thoughts and be given a useful amount of time to do so when not helping with collective intelligence, and be free to leave the Mindplex if these guarantees are not met.

These requirements are not entirely dissimilar from what a successful organization within our current context of social cognition (such as a nation or a corporation) must meet in order to successfully engage its individuals to contribute to the greater good. Clearly this makes sense, as Mindplexes are a radical extension of societies of individuals, one which takes them to a very tightly integrated level in which people and AIs can collectively think about things together. Just as totalitarian societies fail, so too would totalitarian Mindplexes. One thing which many systems theorists (and some management and social theorists) realize is that diversity of function and freedom of autonomy—even within a tightly integrated system—is crucial for fault tolerance and innovation. If your cerebral cortex forced every brain component to ignore its own functional imperatives and instead think exclusively about, say, football, you'd die. The situation is similar with Mindplexes. All the advantages of having a Mindplex in the first place would be lost by crushing the will of the individual fully-formed minds operating within the larger Mindplex context.

It's an oversimplification to say that a tyrannical Mindplex is *impossible*, but it is unlikely to exist for very long, given the tight integration of the individuals in the system and thus the immediate and direct impact they have on it. Even if a tyrannical Mindplex does survive for a while, it is highly likely that a Mindplex which affords its members more individual benefits and opportunities will out-compete the overly regimented Mindplex, as has so often been the case with social systems. That may be small comfort to the individuals who, if it ever happened, would have the misfortune to endure such a failed Mindplex—but, overall, intelligent life is more likely to ultimately survive by evolving to greater capability than not doing so.

What a successful Mindplex would be like is the ultimate extension of the maxim "just think what we could do if we all put our minds together on this"—among people we trust. Everyone's mental resources could be pooled together in a much more efficient manner than is currently possible with the communications tools we have now, and problems could be solved much more efficiently and effectively. Imagine, for example, every cancer researcher in the world able to share insights and ideas without the tedious and time-consuming processes of writing papers, giving talks, and so forth. Rather, as each experimental result were observed and each new idea hatched by someone, everyone could immediately mull it over and provide their insights. The experience and knowledge of many minds could think about such questions with the efficiency of a single mind. AIs could provide their machine ability to mine extensive data sets and build and mine huge networks of relationships, and humans contribute their abilities for novel creativity and insights into the physical world—and scientific problems which once took years to solve could only take hours.

Let's say one day you, as a member of a Mindplex, had an idea for a song in your head, but weren't trained as a composer. Rather than heading over to the music school and looking for young composers who might be interested in collaborating with a complete novice who could just barely hum the correct tune to them, you'd merely think about it in the right way, bringing that train of thought out of your individual mind and into the Mindplex. The Mindplex coordination cortex, most likely an AI, would dispatch that train of thought to currently active (connected to the Mindplex) musical individuals, and within seconds someone could think "that's a catchy tune" and actually compose and perform it. You, as a brilliant painter, might someday return the favor for someone who sees a vision of a great painting but has no skill in that area. Millions, or billions, or perhaps even more of these kinds of interactions happening regularly could lead to a situation in which a myriad of scientific, artistic, engineering, and other accomplishments which were previously only idle daydreams or frustrated attempts could be realized.

Some say that competition is the only way to innovation. While such a statement is a gross oversimplification, considering how many millions of innovations have come as the result of collaboration, it has a grain of truth. However, even within a Mindplex the incentive of competition would not disappear. Competition would be reduced in emphasis (which, considering how extremely overemphasized it is in our society today, might not be such a bad thing), but just as thoughts in your own mind compete for your attention, so would thoughts (and the individuals which have them) compete in a Mindplex. More scientifically, artistically, or economically productive lines of thinking would be more rewarded.

Truly useless (or worse, parasitic) individuals are a waste of resources, and there is no incentive in a Mindplex or any other progress-driven system to allow an individual that isn't contributing to continue to participate. Furthermore, those individuals who contribute more to the Mindplex get more out of it—both indirectly, in terms of exposure and level of interconnectedness, and directly, in that individuals that contribute more are also much more likely to get more of their personal interests addressed via the combined resources of the Mindplex. The structure of a successful Mindplex also would be such that it would be exceedingly difficult for non-contributors to masquerade as contributors and exploit or even dominate the system, as they currently can in political and corporate systems. Similarly, it would be more difficult for a useful contributor to be labeled as useless by some ill intentioned other. Mechanisms for accountability and the pseudoeconomic systems of resource allocation in a functioning Mindplex are too direct for easy manipulation. If someone did try to "game the system" and found a way, the feedback loop to the other individuals is so tight that it would be a matter of minutes to days, rather than years to decades, before this was noticed. The Mindplex could either eject the problematic subgroup, or die (the individuals would not die, but that particular Mindplex would cease to exist).

A failed Mindplex would be a horrible thing to be a part of. However, since humans don't have direct psychic abilities, only some electronic system of mediation would allow us to participate in a Mindplex. This means we can design the system to allow individuals to disconnect at will, greatly reducing the incentives for creating a failed Mindplex. Certainly, a draconian society could use this kind of technology differently, by the usual means of physical coercion, to create a terrible situation. It is difficult to argue that such a situation would be *more* terrible than those which Hitler, Stalin, and their ilk have already created in the past, but the "worst case usage" argument is often used against any kind of technological progress. Clearly we should be vigilant against such misuse of *any* powerful technology. However, the human race should focus its formidable problem-solving abilities towards eliminating such oppressive societies in the first place, not eliminating technological and biological evolutionary progress.

Currently, our own human problems of selfishness and greed are the primary obstacles towards our forging forward on the path to this kind of cooperation, even without the help of AIs. To overcome those human problems, we need to develop systems in which we can trust each other, and where creativity, intelligence, cooperation, contribution and progress are rewarded, and unenlightened self-interest, obstructionism, coercion and hoarding rejected. Some say that we're biologically predestined to a fate of greed-driven conflict, but considering how radically we've changed our biological destiny (and that of the world around us) in the past, that argument sounds more like a cheap cop-out by people who don't want to take responsibility for anything other than their own short-term selfish interests. There are also some structural issues which will likely make it so that AIs form Mindplexes before humans are able to be integrated into them, but the power of such AIs may be great enough to help humans engineer solutions to these limitations in our abilities to communicate with each other.

PSYNESE AND NOVAMENTE MINDPLEXES

Humans as currently constituted are not ideally suited to participate in Mindplexes. It seems likely that neuromodification and neural augmentation will be necessary in order to render humans fully Mindplexable. On the other hand, various forms of AI may find themselves immediately and naturally able to join into Mindplexes. In this section we will explore why future Novamente AI systems—if they function as hoped—will be better suited to form Mindplexes than (unmodified) humans. To carry out this exploration, we must delve into a little more detail regarding one particular aspect of the Novamente design: a unique communication framework called "Psynese." Although Psynese is different from human language, it can be thought of as a kind of "language" customized for communication between Novamente AI Engines.

One might wonder why a community of Novamentes would need a language at all. After all, unlike humans, Novamente systems can simply exchange "brain fragments"—subspaces of their Atomspaces. One Novamente can just send relevant nodes and links to another Novamente (in binary form, in an XML representation, etc.), bypassing the linear syntax of language, and the imprecisions inherent in formulating the vast expanse of abstract thought into a construct in the necessarily limited form of language. This is in fact the basis of Psynese: why transmit linear strings of characters when one can directly transit Atoms? But the details are subtler than it might at first seem.

One Novamente can't simply "transfer a thought" to another Novamente. The problem is that the meaning of an Atom consists largely of its relationships with other Atoms, and so to pass a node to another Novamente, it also has to pass the Atoms that it is related to, and so on, and so on. Atomspaces tend to be densely interconnected, and so to transmit one thought accurately, a Novamente system is going to end up having to transmit a copy of its entire Atomspace! Even if privacy were not an issue, this form of communication (each utterance coming packaged with a whole mind-copy) would present rather severe processing load on the communicators involved (not to mention the complexity of resolving the interconnections of this new, inherited mindspace with that of the experientially grounded one that the system had developed from its own perceptions and interactions).

The idea of Psynese is to work around this interconnectedness problem by defining a thought grammar and related vocabulary called the Psynese vocabulary:

> **Psynese vocabulary**: a collection of Atoms, associated with a community of Novamentes, approximating the most important Atoms inside that community. The combinatorial explosion of direct-Atomspace communication is then halted by an appeal to standardized Psynese Atoms.

Pragmatically, a PsyneseVocabulary is contained in a PsyneseVocabulary server—a special Novamente that exists to mediate communications between other Novamentes, and provide Novamentes with information. By necessity, any language is a limited form of expression when compared to the entire universe of thoughts which may be constructed, but the goal with Psynese is to create a language in which a wider variety of thoughts can be encoded with *enough precision to be understood almost as well as if the recipient Novamente had had the thought itself.* In essence, Psynese strives to be as close as is possible to a thought grammar, with a concordant thought vocabulary.

In order to describe specific examples, in the next few paragraphs we shall develop some semi-formal notation to describe Novamente Atoms. We will not define this notation formally, but will instead rely upon the reader's familiarity with similar notations to ground the semantics. Roughly, the notation used is of the form:

<Relationship> <Argument 1> <Argument 2>

The relationships are Novamente Links and the arguments are Novamente Atoms (usually Nodes). The examples given will regard InheritanceLinks, which define probabilistic logical inheritance. For instance, "cat" inherits from "animal," so we can say:

InheritanceLink cat animal

indicating roughly that there is an InheritanceLink between the "cat" node and the "animal" node. For simplicity the examples given here involve entities like cats and elephants that correspond to simple English names. In fact, however, nearly all Atoms in Novamente do not correspond to concepts with simple English names—most refer to elementary percepts or actions, or else highly abstract concepts or fragments of abstracts. In a real Novamente system, there may well be no "cat" node, and the "cat" concept may be represented by a large fuzzy complex of nodes and links that are linked to the linguistic construct "cat" only through social grounding in the necessity of communicating about the thing called "cat" with other sentient beings. This is quite similar to how human mental models are constructed: you experience and perceive things, and only through the necessity of talking about it with other people do you ever bother to develop simplified linguistic descriptions of your pure thoughts about your experiences.

Finally, if an entity is defined relative to a specific PsyneseVocabulary PV, it will be denoted with a "/PV" at the end of its name. So,

InheritanceLink cat/PV animal/PV

defines the posited inheritance relationship to exist within the PsyneseVocabulary PV.

Using this notation, suppose Novamente1 wanted to tell Novamente2 that "Russians are crazy." The obvious option is:

InheritanceLink Russian/PV crazy/PV

Perhaps Novamente1 doesn't like PV's definition of "crazy." It can try to find a different PsyneseVocabulary PV1 with a better definition, and then transmit something like:

InheritanceLink Russian/PV crazy/PV1

Or, perhaps it simply wants to tell Novamente2 exactly what it means by "crazy". Then it may send:

InheritanceLink Russian/PV crazy

along with a set of relationships involving "crazy," examples of which might be:

InheritanceLink crazy interesting/PV
InheritanceLink crazy unusual/PV
InheritanceLink crazy dangerous/PV

(I.e. it may send an inheritance link with a Psynese Vocabulary node in one term, and a complete Novamente sub-network involving the "crazy" concept Atom in a set of relationships with other Atoms—a subnetwork—in the other term.)

Of course, it may also be that Novamente1 doesn't like the PV definition of one of these terms, say "dangerous." In that case it can refer to another PsyneseVocabulary, or, it can define the term it's using by giving another set of relationships. The key is the fact that, at some point, Novamente1 can stop defining terms in terms of other terms, and accept that the "socially agreed-upon" meanings of the terms it's using are *close enough* to the intended meanings. The purpose of socially mediated thought is, after all, to avoid infinite refinement of terms in order to allow information to be shared in finite time.

As already cautioned, the above examples involve human natural language terms, but this does not have to be the case. PsyneseVocabularies can contain Atoms representing quantitative or other types of data, and can also contain purely abstract concepts. The basic idea is the same. A Novamente has some Atoms it wants to convey to another Novamente, and it looks in a PsyneseVocabulary to see how easily it can approximate these Atoms in terms of "socially understood" Atoms (Psynese approximations of purely conceptual Atoms). This is particularly effective if the Novamente receiving the communication is familiar with the PsyneseVocabulary in question. Then the recipient may already know the PsyneseVocabulary Atoms it is being pointed to; it may have already thought about the difference between these consensus concepts and its own related concepts. If the sender Novamente is encapsulating conceptual subnetworks for easy communication, it may specifically seek approximate encapsulations involving PsyneseVocabulary terms, rather than first encapsulating in its own terms and then translating into PsyneseVocabulary terms.

The general definition of a *Psynese expression* for Novamente is a set of Atoms that contains only:

1. Nodes from PsyneseVocabularies

2. Perceptual nodes (numbers, words, etc.)

3. Relationships relating no nodes other than the ones in the above two categories, and relating no relationships except ones in this category

4. Predicates or Schemata involving no relationships or nodes other than those of types (1-4)

This is not a "language" as we normally conceive it, but it serves similar functions in socially grounding thought for easier communication. Psynese is to Novamentes as human language is to humans. The biggest differences from human language are:

- Psynese uses weighted, typed hypergraphs (i.e. Atomspaces) instead of linear strings of symbols. This eliminates the "parsing" aspect of language (syntax being mainly a way of projecting graph structures into linear expressions).

- Psynese lacks subtle and ambiguous referential constructions like "this," "it" and so forth. These are tools allowing complex thoughts to be compactly expressed in a linear way, but Novamentes don't need them. Atoms can be named and pointed to directly, without complex, poorly-specified mechanisms mediating the process.

- Psynese has far less ambiguity. There may be Atoms with more than one aspect to their meanings; but the cost of clarifying such ambiguities is much lower for Novamentes

> than for humans using language, owing to the greater precision of Psynese and its more direct mapping into the Novamente brain (as its syntactic structure is the same as the actual Novamente brain structure, and so habitually there will not be the rampant ambiguity that we see in human expressions.)

Basically, with Psynese, one gets the power of human language without the confusing parts. In the long run, Novamentes and other AIs should have a much easier time understanding each other than we humans do.

If one wishes to create a linear-string variant of Psynese, one can simply use Sasha syntax, with the addition of the /PV marker to specify that a given Atom is intended as relative to a given PsyneseVocabulary. This is basically what is being done in the concrete examples of Psynese transmissions shown above.

Now, how does one get from Novamentes communicating via Psynese to Novamente Mindplexes?

Clearly, with the Psynese mode of communication, the potential is there for much richer communication than currently exists between humans. There are limitations, posed by the private nature (and inherent complexity and necessity to communicate in finite time) of many concepts—but these limitations are much less onerous than for human language, and can be overcome to some extent by the learning of complex cognitive schemata for translation between the "private languages" of individual Atomspaces and the "public languages" of Psynese servers.

Rich communication does not in itself imply the evolution of Mindplexes. It is possible that a community of Psynese-communicating Novamentes might spontaneously evolve a Mindplex structure—at this point, we don't know enough about Novamente individual or collective dynamics to say—but it is not necessary to rely on spontaneous evolution. In fact it is possible, and even architecturally simple, to design a community of Novamentes in such a way as to encourage the emergence of a Mindplex structure.

The solution is simple: simply change the role of PsyneseVocabulary servers from that of mindless servants to that of active participants in the AGI society. Rather than relatively passive receptacles of knowledge from the Novamentes they serve, allow them to be active, creative entities, with their own feelings, goals and motivations.

The PsyneseVocabulary servers serving a community of Novamentes are absolutely critical to the functioning of the Novamente "society," as without them, high-level inter-Novamente communication is effectively impossible. Without the concepts the PsyneseVocabularies supply, high-level individual Novamente thought will be difficult, because Novamentes will start to think in Psynese to at least the same extent to which humans think in language, limiting resources and mental pathways devoted to experiential and purely cognitive thought.

Suppose each PsyneseVocabulary server has its own full Novamente mind, its own "conscious theater." These minds are in a sense "emergent minds" of the Novamente community they serve—because their contents are a kind of "nonlinear weighted average" of the mind-contents of the community. Furthermore, the actions these minds take will feed back and affect the community in direct and indirect ways—by affecting the language by which the minds communicate. Hence, the definition of a Mindplex is fulfilled—Novamente "society" is designed to take a form from which a Mindplex is necessarily emergent if the system functions as intended.

The Novamente systems themselves are ideally suited to form Mindplexes, but humans could be trained to interact with them sufficiently that they, too, could join the Novamentes in such a network (more loosely at first, but over time, possibly much more tightly). Novamente is being designed without ego, and with a careful focus on rational ethics (because it is too easy for a logical system like an AI to reason its way out of moral edicts based on faith, as such tenets are logically much weaker than rationally based ethical decisions). These design features, along with the neural structure and Psynese design choices discussed briefly above, also make the Novamentes ideally suited to be coordination cortices in a human-integrated Mindplex. What radical AI designers should consider in their designs are not only functions which will make the AI itself successful, but ones which will make it successful in interacting with humans, and in helping humans to improve themselves mentally and physically for long life and improved cognition.

What will the dynamics of such a Novamente Mindplex be like? What will be the properties of its cognitive and personality psychology? We could speculate on this here, but there isn't enough evidence to have much faith in the accuracy of such speculations. The psychology of Mindplexes will reveal itself to us experimentally as our work on AGI engineering, education and socialization proceeds.

RESOURCES IN A MINDPLEX

Resource contention being a major issue among all living beings, it is important to look at how a Mindplex can help with this issue. There are two factors at play here:

1. How the Mindplex handles resource contention internally

2. How a Mindplex can help solve issues of resource contention externally, and how this feeds-back into (1)

Internal resource contention within a single mind is a subtle issue in itself. A general principle is that, within a single mind, resources—mental attention—are allocated based on immediate and longer-term goals. However, it is not so easy to know which internal or external actions are going to be helpful for achieving goals. Even figuring out which of one's actions have been critical for achieving one's goals in the past is not that easy—this is what cognitive scientists call the "assignment of credit problem."

Some resource allocations, such as remembering to allocate resources to breathing and beat a human heart, are automatic. Others, such as deciding whether to play video games or write a book, are the result of consciously guided deliberation. The subtler cases are the ones that lie between—not fully automated yet not explicitly deliberative either. Abstract, subconscious attention to thoughts is allocated in a way which we humans find mysterious about our own selves, but if we delve deeply enough we are likely to find a variety of goals being worked on at those levels as well—rest and replenishment of our minds, devising schemes to ensure our own survival, philosophizing, and so on. The dynamics our brains use to carry out this kind of resource allocation are basically unknown; and handling resource allocation and credit assignment effectively is one of the subtler aspects of any AI design that aspires to Artificial General Intelligence.

The situation with multiple minds is yet more complex. Generally speaking, assignment of credit and allocation of resources is carried out much less effectively in human groups than within individual human minds. Modern economic systems attempt to address this problem via various complex mechanisms, but without anywhere near complete success. Multiple human minds, involved a situation where allocation of resources involves decision-making amongst independent actors, will, often trump cooperation with selfishness and egoism (though it has been shown by Axelrod and others that selfishness is frequently self-destructive, and the hope that this needn't so often be the case with AGI is detailed in the next chapter). This is true of both physical resources like food and property, and metaphysical resources such as cultural and intellectual themes. In our current human system a great deal of resources are spent on the accumulation of power and property for individuals rather than the expansion of available power and property to include enough for all. The assumption that resources are pragmatically a zero-sum game is the primary motivator in selfishness, even though this is not necessarily correct. The cooperative mindset which Axelrod and others have shown to be the most beneficial, to not only societies but individuals (perhaps counterintuitively), is encouraged by a Mindplex in the following ways:

1. The interconnection of individuals with a centralized knowledge aggregator and each other, in a more tightly integrated way than in a traditional society, requires that in order for an individual to operate at maximum potential, the entire Mindplex do so.

2. With Mindplex level interaction and cooperation, problems previously considered intractable become solvable, including problems of resource management.

When interconnection between individuals is as close as in a Mindplex, by necessity the internal resource-allocation (mental attention) scheme operates more like that of an individual mind than a society, even though the benefits are supersocial in terms of pooling of talents and resources. In a functional Mindplex, the selfish, uncooperative individual will be forced to operate at a lesser level than his peers, because the statistical attention allocation mechanism of the Mindplex contraindicates obsessive thinking about any meme which is not directly relevant to the progress of the Mindplex taken as a whole. Each individual's semi-personal thoughts (ones that are not private, but which reflect individual interests which may be a minority within the Mindplex) will receive less Mindplex collective thought attention than things which are important to the group as a whole.

This is not unlike a current society, but it is more efficient. Every individual, even today, spends a large percentage of his or her time wrestling with social issues—the environment, schools, politics, taxes, wars, and so on. Within a Mindplex, these kinds of issues (and eventually their post-Transition analogues) would receive the majority of attention—as they do now. The combined resources along with the more efficient integration would lead to a "whole is greater than the sum of the parts" phenomenon, which would lead not only to greater efficiency in solving problems, but also (perhaps counterintuitively) to comparatively *more* time for individual thought, as the necessities of "groupthink" would be accomplished in a more effective way.

MINDPLEX METATHOUGHT DYNAMICS

Internal resource contention in a Mindplex, as we envision it, is handled not unlike in a single mind, which we believe works roughly by a calculating an appropriate nonlinear combination of the following criteria:

1. Statistical measures: the lines of thought which are of greatest benefit to the Mindplex, taken as a whole (not only benefiting the greatest number of individuals, but also continued progress in efficiency and efficacy of the Mindplex itself), would receive a priority increase. This is not only because individuals would be statistically more likely to be contemplating such issues, but also because the centralized meme-formation in the Mindplex equivalent of a Novamente Psynese server would be biased towards making such thoughts more efficiently encapsulated, communicated, and therefore collaborated on.

2. Probabilistic/inferential measures: based on prior experience, both individuals and the centralized unifier(s) in a Mindplex would be likely to take on lines of reasoning which appear likely to be successful. This refers to not just specific thoughts, but *patterns* of reasoning.

3. Novelty measures: while seeming to contradict #2, the preference for new thoughts works in tandem with familiarity to avoid circular reasoning and obsessiveness, but when appraising the value of a novel thought, familiarity with the pattern of reasoning which both led to, and stems from, the new idea will come into consideration. Thinking about new ideas leads to growth.

4. Confidence measures: statistical and probabilistic measures also need an indicator of the metastatistic of confidence to be worthwhile. If you've tried something once and succeeded once, you're 100% certain it'll work a second time unless you've developed the notion of confidence and understand that a series of one is not enough history to make a truly sound judgment upon. Confidence itself is refined over time as a feedback with the environment—you may initially think that to be careful, you've got to not really make any empirical judgments until you've done something a hundred times (or once, depending on your personality). If through experience, you find that generally after doing something a dozen or so times you pretty much understand the rough probability of succeeding at that thing, then you'll generally stop completely reassessing your actions after reaching the confidence mark of a dozen times (except in light of new information), rather than being satisfied after one time or waiting until a hundred. In a society, we often aggregate statistical confidence (the confidence of many individuals) and attempt to create a global confidence measure. In a Mindplex this process is more efficient, and given the increased quantity and integration of the individual observations, precise.

5. Efficiency measures: a quick thought is frequently worth the effort of having had it; and if not, it's not a big loss of mental energy anyway.

6. Immediacy measures: an urgent problem should always receive more attention. Indeed this measure is so much stronger than the others that the only counterweight to it is not enough confidence that the problem is truly urgent (either evidencial confidence or statistical disagreement in the population). It is easier to decide

> that an urgent problem either isn't urgent or doesn't require your urgent attention in a society than in a Mindplex, because in a society there is a benefit (the resources to attend to selfish goals) to deciding an urgent problem is someone else's problem. In a Mindplex, shared resources will immediately be lost to any individual who chooses to opt out of participation in collective issue resolution.

By taking into account such measures, the Mindplex will be able to direct aggregate thought in a way which optimizes both the individual processes of a single mind and the aggregate processes of a society. The precise weighting function of these metrics is not yet known, and anyway is likely to evolve in a feedback loop with both increased knowledge and increased efficiency; but provided that the function chosen doesn't cause resource-starvation of any one issue (or obsession about any one issue) or the ignoring of urgent issues until they cause damage, and there is a mechanism for evolution of the function based on experience, an AGI which takes into account those criteria in some way is a potential contender for being a successful aggregator entity of a Mindplex.

These criteria are already at play in individual minds and societies, but without the resources of a Mindplex there are two major issues: selfishness and inefficiency. Inefficiency is addressed directly in the integrative nature of the Mindplex concept, but what about selfishness? Individualism and a desire for freedom are not actually the primary causes of selfishness. Even the most individualistic people will say that they are happy for other people to be free and content, so long as it does not impinge upon their own freedoms. The root cause of selfishness is the belief that there aren't enough resources to go around (land, money—some people even think that their freedom only is possible if they take it from another).

EVOLUTION OF SELF WITHIN A MINDPLEX

The belief that an individual can exist without a society is not supported by any real evidence. Any single individual is shaped by their society more than they shape their society, but without individuals a society can not exist. Individual thought, and a sense of self, is an important part of mental dynamics. A sense of self gives an entity reason to survive. Collaboration and competition between individuals creates the framework for growth, by allowing for differing perspectives on problems. Even inside a single human mind, one often puts oneself in "other shoes" to try to find new perspectives on problems. Differentiation is as important as cooperation, but in our current society the former is so emphasized that the importance of cooperation and society is almost lost.

The compromise of interdependency either in a society or a Mindplex is a willing reduction of absolute freedom (i.e. the freedom to not care about *anything* but yourself) in return for many benefits: security, efficiency, pooled resources, pooled knowledge, etc. If constructed properly, even greater relative freedom, can result from what seems at first glance to be less freedom. (Because social interconnection is unavoidable, willing participation will reduce frictional costs and actually leave more time free to pursue your own interests.) Such compromise is unavoidable, but it only leads to selfishness if beings feel that they *lose something* when they do anything other than pursue selfish goals. One way to reduce or eliminate selfishness is to reduce or eliminate resource conflicts.

One such reduction in a Mindplex is structural—it necessarily rewards willing participants with greater resources both locally (more freedom for more efficiency) and globally (an increase in knowledge and processing power through aggregation). The other is by using the greater mental abilities of a Mindplex to solve *external* resource conflicts. By applying the abilities of a Mindplex to the problems of renewable energy sources (for machines and humans), medicine, environmental stabilization, deep space travel, etc., the pressures on resources can be reduced: by rationally constraining the growth of the Mindplex through self-restraint on expansion/reproduction, based on knowledge of the system capacity modulo increase in intelligence; through expansion, the probability of allowing the Mindplex to solve more relevant problems; and through *creating more resources*.

Even with only the power of a society, we are already solving a number of interesting problems in machine energy and food production (biodiesel, garbage-to-fuel conversion, high-yield crops, advanced irrigation, polluted water reclamation, etc.), medicine, etc. By applying the power of a Mindplex to such issues, the selfishness created by the perception that each individual will never be able to accumulate a satisfactory amount of food, energy, wealth, time to live, etc. without taking it from another will be reduced or eliminated. This is a key precursor to the idea of a Transcension, because indefinite life (be it for an individual, or the continuity of sentient life generally) will require indefinite resources. What resources we need will continue to evolve as the nature of sentient life changes, through natural processes and intelligent interference by humans and eventually AGIs, but we will always need resources. (In literature, immortality is often marked by laziness. That is because it is generally granted supernaturally. However, if we are free of predestination and responsible for our own destiny, as the authors believe, then sentient life can never become lazy because it will constantly need to maintain its state of longevity or immortality as self-made and not granted by an external entity.)

If this all sounds like some form of super-socialism to you, remember that capitalism is also based on group entities which require the individual to subjugate freedoms in return for benefits (corporations),, that the idea of the Mindplex is not political but pragmatic, and that increased cooperation and efficiency doesn't mean the elimination of competition and freedom (it may lead to an increase in freedom, as mentioned). There is still competition within a Mindplex—for thought allocation, in the marketplace of ideas—but the balance between cooperation and competition being too far biased in favor of competition causes wasteful inefficiency. The idea is not rule by fiat, but a more effective system. Competition between ideas will be carried out more efficiently, and will be more democratically resolved (through the aggregation of the thoughts of all the Mindplex participants on the subject).

The sense of self in a Mindplex will evolve, becoming more rooted in what one can create and contribute instead of what one can consume (which is what productive economics ought to be about), and individuals in the Mindplex will be rewarded (and therefore feel good about) both cooperation with the group, and individual ideas which help the Mindplex grow and evolve. This will drive the sense of Ego neither toward the stereotypical Eastern tendency to think, "How can I sacrifice myself to society?" nor toward the stereotypical Western tendency to think, "What can I do to make myself better than everyone?" but rather toward "*What can I do which will make myself and everyone else better*?"

In an ethically satisfactory future, an individual will of course have the freedom to opt to remain independent of a Mindplex and operate at an individual human level, just as an individual can choose to drop out of society and become a hermit—if he or she can live without the benefits of the group contributions, that choice should be available.

INKLINGS OF THE GLOBAL BRAIN MINDPLEX

Compared to Mindplexes involving human beings, Novamente Mindplexes are relatively simple and concrete to think about, but these two sorts of (hypothetical) Mindplexes are not totally unrelated. One can use the above observations on projected Novamente Mindplexes as metaphors for the projected Global Brain Mindplex. This metaphor has its limitations, but it is a stronger metaphor than many of those used before, such as the human brain or human societies and social groups. While the exact mechanisms of a human-interconnected Mindplex are as yet uncertain, the basic form—individual entities with choice and private thoughts, sharing experience at a higher level by pooling mental resources, and using a more precise language that is closer to mind-to-mind communication—is likely to be very similar should such a thing come to pass.

In the Novamente Mindplex scenario described above, the Novamentes in the community are not devoid of freedom—they are not Borg-ified, to use the popular *Star Trek* metaphor. They are independent minds, yet they have their thoughts and communications continually nudged by the intelligent PsyneseVocabulary servers that define their evolving language and mediate a higher-level of mental interactions between all members of their Novamente society.

It is important to reiterate that in the Mindplex vision, free will is not obliterated. Each member of the community is still entitled to private thoughts and personal decisions, but is enabled to make these decisions and have these thoughts with the full benefit of drawing from a greater pool of collective knowledge. As this collective knowledge is more precisely represented than with our current human languages, and is more directly integrable into one's mind than the often long and tedious process of human learning through repeating and memorizing what is often ill-formed information, the presumed result is that all actors will be able to make both individual and collective decisions more wisely—at least, by definition they will take into account the alternative views of the others who are linked to them via the Mindplex.

One can envision a Global Brain Mindplex scenario with similar aspects to the Novamente Mindplex postulated. Rephrasing the first two "Global Brain phases" described earlier, we may define:

- **Phase 1 Global Brain proto-Mindplex**: AI/AGI systems enhancing online databases, guiding Google results, forwarding e-mails, suggesting mailing-lists, and so on. In this phase AI/AGI systems are generally using intelligence to mediate and guide human communications toward goals that are its own, but that are themselves guided by human goals, statements and actions; as represented by the resource allocation choices codified in a Hypereconomic or similar participant-choice system, which will guide a proto-Mindplex in how it expends its available energy towards these goals.

- **Phase 2 Global Brain Mindplex:** AGI systems composing documents, editing human-written documents, sending and receiving e-mails, assembling mailing lists and posting to them, creating new databases and instructing humans in their use, etc.

In Phase 2, the conscious theater of the AGI-mediated Global Brain system is composed of ideas built by numerous individual humans—or ideas emergent from ideas built by numerous individual humans—and it conceives of ideas that guide the actions and thoughts of individual humans in a way that is motivated by its own goals. It does not *force* the individual humans to do anything—but if a given human wishes to communicate and interact using the same databases, mailing lists and other evolving vocabularies as other humans, they are going to have to use the products of the AGI-mediated Global Brain system, which means they are going to have to participate in its patterns and its activities.

The possibility of the creation of advanced neurocomputer interfaces makes the picture more complex. At some point, it will likely be possible for humans to project thoughts and images directly into computers without going through mouse or keyboard—and to "read in" thoughts and images similarly. When this occurs, interaction between humans may in some contexts become more Psynese-like; in that their thoughts can be more directly exchanged through the machines, like Psynese servers, which aggregate the shared mental language of public thoughts. Thus, the role of Global Brain-mediating AI servers may become more like the role of PsyneseVocabulary servers in the above-discussed Novamente design.

In a Novamente-Human Integrated Mindplex, the mind-to-mind communication capabilities of Psynese and its related structures provide a basis for very tight mental integration of Novamentes. While human integration would require a much more complex and flexible system of sub/semi-linguistic encoding of thoughts for sharing, such development would benefit both the humans and the AIs in terms of not only extending their capabilities within a Mindplex, but giving each individual a more comprehensive mental toolkit generally. Both the pragmatics of Novamente attention allocation and the ethical and structural barriers to "Borgification" within the system design are inherent bulwarks against a Novamente system being used to subjugate others within a Mindplex. While such abuse certainly isn't completely impossible, we feel that the Novamente system design is one which provides sufficient safety along with the necessary initial capabilities to, when fully implemented, form the foundation of a successful Mindplex system.

This is indeed a lot of speculative abstraction—but these things may become real sooner than we think. It is important to explore the possibilities the future may offer as clearly and thoroughly we can; the inner, experiential aspects as well as the more outward, transparently technological aspects. Our society, and humans ourselves, are evolving—but we've finally reached a level where our technological sophistication is such that, if not matched by a marked increase in philosophical sophistication, we may destroy ourselves completely. However, if we can harness the power of our collective reasoning to guide our own evolution and that of our creations (society, machines, etc.) in a positive manner, we can truly live up to our potential to be the masters of our own destinies, and to spawn other sorts of minds (AIs, Mindplexes, and other forms not yet even conceivable to us) with their own individuality, awareness and freedom.

---

[1] http://www.community-intelligence.com/blogs/public/

[2] See Chapter 15 for a detailed discussion of a possible positive scenario.

CHAPTER 15

CREATING A POSITIVE TRANSCENSION

We've been leading up to wild futuristic themes for the last couple of chapters, and now we're going to leap over the edge. Mindplexes and global brains are just the beginning. AGIs are just machines; nanotechnology is just more technology, which begets just more machines. Submicroscopic n-dimensional vibrating strings are too small to make any difference in everyday life. Integral consciousness is fabulous but limiting. What we're talking about now is the *total transformation of the universe*. This may seem outrageous, but it is possible if we put our integrated minds to the task.

It's time to deal more systematically and directly with the broad futuristic themes raised in the Introduction of this book. What is the posthuman world going to look like? We've discussed many of the technologies that will likely go into it, and some of the wild possibilities that might emerge. We've touched on the moral issues and choices that emerge when one takes posthumanity seriously. But we haven't yet, in these pages, carefully reviewed the various possibilities that the future holds, and asked ourselves which are more likely or which are more desirable—or how we assess "desirability" in the first place. Of course these are huge issues and highly subjective ones, but the book would be incomplete without a straightforward and undaunted discussion of these grand topics from which some may prefer to shy away.

The question at hand is: *how to manage the development of technology and society, in the near to mid-term future, in such a way as to maximize the odds of a positive long-term future for the universe*?

The definition of "positive" is part of the question.

The following conclusions are not certain, but they are quite clearly plausible; and it is a challenge all humanity shares to determine the positive future of the universe we inhabit, which we can guide if we are willing to try, rather than to lie back and allow "fate" to dictate our future.

As noted repeatedly in earlier chapters, we believe that the era of humanity as the "Kings of the Earth" is almost inevitably coming to an end. Unless we bomb ourselves back into the Stone Age (or into oblivion), we are going to be sharing our region of the universe with powerful AI minds of one form or another. Depending on decisions we make in the reasonably near future, this could quite possibly lead to a fundamental alteration in the nature of human conscious experience in a complete and irreversible manner: a Transcension. Not only AIs may bring this about: humans ourselves likely will evolve and be augmented to become transhumans. The dangers to humanity may be significant—an issue that must be very carefully considered. Still, the dangers to humanity of failing to attempt to enact positive change in our consciousness may be equally great or greater, given the history of destructive setbacks to human progress from Babylon to the Dark Ages; and failure to change our course may lead to repeated setbacks which ultimately lead to the demise of all humanity.

In preceding chapters we've focused mainly on the good things that may come of posthuman technologies, though readily admitting that we must be vigilant about negative aspects of these changes—not the least of which is the possibility of too many humans being "left behind" against their wills. Of course, as all of us living in the "post-9/11" world realize acutely, technology has its dark side as well. Airplanes are an amazing means of transportation, but also a tool for bashing down buildings and murdering thousands. The dangers and wonders of nanotech, biotech and AGI are proportionately greater.

This was not an easy chapter to write, because these are not easy issues to confront—even for someone with a radical futurist[1] mindset. Any reasonably deep thinker, confronted with these issues, will arrive at some extreme conclusions; then wish to draw back from them, throwing around confessions of ignorance, and failing to look at those conclusions straightforwardly and make a decision about which is the best course forward. In the end we managed to convince ourselves that the correct conclusions *are* fairly extreme ones. The primary conclusion is that the human race has two strong options going forward, which we will associate with the catch-phrases "AI Buddha" and "AI Big Brother." More verbosely, these correspond to the alternatives of:

- Creating an AI based on the principle of "Voluntary Joyous Growth," and allowing it to repeatedly self-modify and become vastly superintelligent, having a potentially huge impact on the universe (and posing dangers to the human race that must be carefully studied and managed—not the least of which being the determination whether or not it's really such a bad thing to transcend the human condition and voluntarily merge with an eternal, evolving construct that has positive change as its goals).

- Creating an AI dictator with stability as a main goal, to rule the human race, ensuring peace and prosperity and guaranteeing that no human creates overly advanced, "dangerous" technologies.

While we don't particularly *like* the conclusion that these two extreme possibilities are the only viable ones, they appear to be the only conclusions if one assumes as correct the idea that a strong, evolving AGI is inevitable. This is a conclusion that both *feels* right (that is, it seems conceptually accurate, even though it may provoke a frightful response within one's inner humanity) and seems to follow inexorably from the facts as currently known.

Not surprisingly, among these two choices we have a tentative preference for the Voluntary Joyous Growth scenario, though we believe that much more research with "primitive" AIs (nevertheless much more advanced than any AIs we currently possess) is needed to fully understand the risks and rewards of each option. Given the current pace of changes in technology, the "no AI at all" option doesn't seem very likely, and it is incumbent upon us all to work towards making the nature of AI, and humans, progress in a positive and compatible way, rather than simply resigning ourselves to the supposed inevitability of the nightmare scenarios of sci-fi action films such as *Terminator* and *The Matrix*, where humans and machines continue mankind's seemingly endless addiction to warfare.

DEFINING A POSITIVE TRANSCENSION

To see how these conclusions emerge, let's go back to the basics, and work forward from there. If one assumes that some sort of Transcension will occur, spurred on by the technologies described in this book (and other related technologies no one has yet conceived), then one immediately confronts a great number of unanswerable questions, including the question of whether what we do right now makes any difference or not.

How much effect does the way in which the Transcension is reached, have on the nature of mind and reality afterwards? There are many possibilities:

> 1. That there are many, qualitatively different post-Transcension states, and our choices now impact which path is taken.
>
> 2. That no matter what we do now, mind and reality will settle into the same basic post-Transcension attractor.
>
> 3. That a human-achieved Transcension will merely serve to project humans into a domain of being already occupied by plenty

of other minds, which we simply do not yet know about as we can't yet communicate with them, and that either:

a. the specifics of how humans approach Transcension are not going to have any significant impact on this already-existent domain, or

b. the choices we make in approaching Transcension will determine the nature of our relationship with the other beings of this domain in a significant manner.

At this point, we have no idea how to assess the probabilities of these various options. Do you?

In the second two options, the only *ethical* question arising is whether the post-Transcension state-of-being will be better than the states that would likely exist without a Transcension. If yes, then we should work to bring about Transcension, and once brought about let it take its course. If no, then we should work to avoid Transcension.

In the first option, the ethical choices are trickier, because some plausible post-Transcension states may be better than the states that would likely exist without a Transcension, whereas others may be worse. We then have to choose not only whether to seek or avoid Transcension, but whether to seek or avoid *particular kinds* of Transcension. In this case, it's meaningful to analyze what we can do now to increase the probability of a *positive* Transcension outcome.

Serious discussion of *any* of these options can't begin until we define what a "positive" Transcension outcome really means, so the rest of this chapter deals mainly with the two primary issues which arise from this line of reasoning:

1. What is a "positive outcome?" That is, what is an appropriate ethical or meta-ethical standard by which to judge the nature of a hypothetical post-Transcension scenario as beneficial? A number of alternative, closely related approaches

are presented here, mostly centered around an abstract notion we call the Principle of Voluntary Joyous Growth.

2. In the case that Option 3 above holds, then how can we encourage a positive outcome? Here the focus is on artificial general intelligence technology, which appears likely to be the primary driver behind Transcension (because AGI will be the force driving all subsequent inventions once it is achieved). We argue that, in addition to teaching AGI entities ethical behavior, it is important to embody ethical principles in the very cognitive architecture of any AGI systems. (Specific ideas in this direction will be presented below, and discussed in the context of the Novamente AI system.)

THE ETHICS AND META-ETHICS OF TRANSCENSION

What is a good Transcension? Some people will say that the only good Transcension is a non-Transcension—that using technology to radically alter the nature of mind and being is a violation of the natural order of things. In practice, though, technology already has altered the nature of mind and being in irreversible ways; and neo-Luddism will not prevent this course, but only cause people to shirk their responsibility towards making technological and human evolution take a beneficial course. Yet even among radical techno-futurists and others who believe that Transcension in principle can be a positive outcome, there is nothing close to agreement on what it means for a post-Transcension world to be seen as a "good" one.

For Eliezer Yudkowsky, the preservation of "humaneness"[2] is of primary importance. He goes even further than most Singularity believers, asserting that the most likely path is a "hard takeoff" in which a self-modifying AI program moves from near-human to superhuman intelligence within hours or minutes—instant Singularity! With this in mind, he prioritizes the creation of "Friendly AIs"—artificial intelligence programs with "normative altruism" (related to "humaneness") as a prominent feature of their internal "shaper networks." (A "shaper network" is a network of "causal nodes" inside an AI system, used to help produce that AI system's "supergoals"—in Freudian terms, the shaper network may be loosely thought of as the AI's Superego.) He discusses extensively strategies one may take to design and teach AIs that are Friendly in this sense. The creation of Friendly AI, he proposes, is the path most likely to lead to a humane post-Singularity world.[3]

On the other hand, Ray Kurzweil seems to downplay the radical nature of the Singularity—leading up to but not quite drawing the conclusion that the nature of mind and being will be totally altered by the advent of technologies like AGI and MNT. His vision of the post-Singularity world generally seems to be one a lot like our current world, but with more advanced and widespread technology, AI minds to talk to, and the absence of pesky problems like death, disease, poverty and madness. He clearly sees this vision as a good one, and seeks to encourage ordinary non-techno-futurist people not to fear the beckoning changes.

Damien Broderick's novel *Transcension* presents a more ethically nuanced perspective. In his envisioned future, a superhuman AI rules over an Earth containing several different sub-regions, including:

- A region in which humans live a traditional lifestyle based on minimal technology
- A region in which humans live using highly advanced technology

At the end of the novel the Transcension occurs—an event in which the ruling superhuman AI mind decides that maintaining human lives isn't consistent with its other goals. It wants to move on to a different order of being, and in preparation it uploads all humans from Earth into digital form, so it can more easily guarantee their safety and help with their development. ("Transcension" in the sense that we're using it in this chapter is a bit broader than the event in

Broderick's novel; in our terminology, his Transcension event is part of the overall Transcension in his fictional universe.)

Not all techno-futurists are as concerned with the future of human life or humaneness. For example, a poster going by "Metaqualia," in a series of emails on Yudkowsky's SL4 email list[4], has argued for alternate positions, such as:

- If a post-Transcension superhuman intelligence decides that there's a better use for the mass-energy that humans occupy, it may well be right, and we shouldn't fear this outcome.
- A universe consisting of a giant endless orgy of delight, with no cognition and no individual minds (human or otherwise), might not be such a bad thing after all.

This particular perspective is very closely related to the traditional religious view of Transcension through a supernatural afterlife, particularly the Christian mythos of the Rapture in which humans are removed from the earthly worlds and integrated into a universe of eternal joy (those who are worthy, in the case of Scripture). However, it is appears far more likely that the traditional religious impulse towards eternal life will be fulfilled through human (and posthuman) good works and deliberate embracing of positive material Transcension, than through faith in an unknowable actor whose goals are known only to itself. In short, the afterlife of the various traditional religions may be interpreted as a state of being which we humans have been entrusted to *actually create* of our own volition, not merely wait for it to occur.

Given that we humans can't agree on what's good and valuable in the current human realm of life, it would be foolish to expect us to agree on what's good and valuable in the post-Transcension world. Nevertheless, it is foolish to ignore the issue completely, so we must ask: what are the values that we would like to see guide the development of the universe post-Transcension?

This poses a challenge in terms of ethical theory, because for a value system to apply beyond the scope of human mind and society, it has to be very abstract indeed—yet there's no use in a value system so abstract that it doesn't actually say anything. Thinking about the post-Transcension universe pushes one to develop ethical value-systems that are both extremely general and reasonably clear.

There may be many different value systems of this nature, such as those we call by the names:

- Cosmic Hedonism
- Joyous Growth
- Voluntary Joyous Growth
- Joyous Growth-Biased Voluntarism
- Nostalgic Joyous Growth
- Nostalgic Voluntary Joyous Growth
- Nostalgic Joyous Growth-Biased Voluntarism
- Human Preservationism
- The Smigrodzkian Meta-Ethic
- Cautious Developmentalism
- Humaneness

Below, we will discuss these value systems, and their interrelationship. Each of these is a very general, abstract *ethical principle*[5]. Specific ethical systems may come to exist, but the quality of an ethical system must be judged relative to the ethical principle it reflects. We will return to this point later.

Note again that here we are only considering value systems that are Transcension-friendly. Of course there are many other value systems out there in the world today, and most of them would argue that the Transcension as conceived in this book is ethically wrong. These value systems are interesting to discuss from a psychological and cultural perspective, but if Transcension is assumed as an eventuality, they are rendered irrelevant to the discussion.

ETHICS, RATIONALITY AND ATTRACTORS

It is important to clearly understand the relationship between ethical principles and rationality. Once one has decided upon an ethical principle, one can use rationality to assess specific ethical systems as to how well they support the ethical principle. Here are two meta-ethical principles:

- Rafal Smigrodzski's meta-ethic, *"Find rules that will be accepted,"* and
- A rule inspired by conversations with Jef Albright and a study of Taoism: *"Favor ethical principles that are harmonious with the nature of the universe."*[6]

One can't choose a meta-ethical principle based on rationality alone, either. Ultimately the selection and valuation process must make reference to emotional, non-rational thought—that is, it must *feel right* (and in the case of a proposed universal ethic, it must feel *truly right* to the greatest number of people, preferably while feeling *truly wrong* to none but those most pathologically opposed to the happiness of their fellows).

Reason is about drawing conclusions from premises using appropriate rules, whereas at the most abstract level, ethics is about *what premises to begin with*. We can push this decision back further and further—reasoning about ethical rules based on ethical systems, and reasoning about ethical systems based on ethical principles—but ultimately we must stop, and acknowledge that we need to make a non-rational choice of premises. Traditional Judeo-Christian religion is one common set of such de facto premises, but its embedding of moral premises within a rigid system of rules and assumptions about the nature of being makes it unsuitable as a moral starting point for a posthuman theory based on extrapolating observational understanding of the universe. Here, we will discuss appropriate underlying moral premises at the level of "abstract ethical principles" like the ones listed above.

David Hume, in his *Treatise of Human Understanding*, isolated the non-rational basis upon which rational refinement can take place in the form of "human nature"—the human version of "animal instinct." Buddhist thought, on the other hand, associates it with the "higher self"—the individual self's recognition of its interpenetration with the rest of the universe, and its ultimate nonexistence. We contend that Buddhism and Hume are both partly right—but that neither has gotten at the essence of the matter. Hume is right that our hard-wired instincts certainly play a large role in such high-level, non-rational choices. Buddhism is right that subtle patterns connecting the individual with the rest of the universe play a role.

What seems rather to be the case is that these moral urges rooted in both the self and society are processed into decisions about moral pretexts through processes which can best be described with the dynamical-systems-theory notion of an *attractor*. An attractor is a pattern that tends to arise in a dynamical system from a wide variety of different preliminary conditions. Once entered into, a strict mathematical attractor must persist forever; but one may also speak of "probabilistic attractors" that are merely very likely to persist, or that may mutate slightly and gradually over time[7]. "Human nature" appears to consist of peculiarities of the human mind/brain which arise from generic attractors that have evolved as part of the human psyche, emerging amongst the plurality of human minds as a result of repeated evolutionary refinements of the feedback loop between these internal mental structures and the perceived environment, and including the mind states of other humans as expressed through both verbal and nonverbal communications. Attractors arise in a lot of complex systems in a lot of circumstances; and are a good way to understand the commonality of phenomena arising from what seems to be chaos in a way that does not require the untestable presumption of supernatural predeterminism.

Thusly, one reason why some meta-ethics appear more convincing than others is that these meta-ethics are generic attractors in the complex dynamical system of human neural activity. Such convincing ethical principles are well-conceived as "universal attractors," that arise as patterns in many different complex systems in many different situations (that is, they arise in a variety of individual minds regardless of their cultural embedding). This doesn't mean that they're logically correct, in the sense of following from some a priori assumption regarding what is good. Rather it means that, in a sense, they *follow from the universe*. This point will be returned to a little later.

Of course, we are still left with a selection problem, because there may be different universal attractors that contradict each other. Does the more powerful universal attractor win, or is this just a matter of chance, or context-dependent chance, or subtle factors we as humans can't fully understand? While it still leaves a question of how to do the actual analysis of the selection, one possibility is to use human communication to bias the thinking of the greatest number of people towards the selection of an attractor describing a system which provides the most benefit with the least harm. Lest we wind up in a loop of circular reasoning regarding how to judge benefit and harm, it should be stated that ultimately it comes down to a subjective, intuitive feeling/impulse and all that can be proposed is a framework for the rational part of the decision and our suggestion for a particular choice—not a universal particular.

COSMIC HEDONISM AND VOLUNTARY JOYOUS GROWTH

Firstly, Cosmic Hedonism refers to the ethical system that values happiness above all. In this perspective, our goal for the post-Transcension universe should be to maximize the total amount of happiness in the cosmos. Of course the definition of "happiness" poses a serious problem, but if one agrees that Cosmic Hedonism is the right approach, one can impose the understanding of happiness as part of the goal for the post-Transcension period. The goal becomes to understand what happiness is, and then maximize it. This goal is in line with the tenets of many so-called "New Age" beliefs; though indeed it can be found in considerations of what the goals of society should be, dating back to Epicurus and his predecessors in Pagan Hedonism.

However, even if one had a crisp and final definition of happiness, there would be a problem with Cosmic Hedonism—a problem that we can informally refer to as the problem of the "universal orgasm." The question is whether we really want a universe that consists of a single massive wave of universal orgasmic joy. Perhaps we do all want this, in a sense—but what if this means that mind, intelligence, life, humanity and everything else we know becomes utterly nonexistent? Epicurus wrestled with similar questions, and in his Principle Doctrines proclaimed that there must be a bias towards embracing worldly sensations and maximizing pleasure, but not through any activity which caused later grief to outweigh the immediate joy provided. Transcension provides the possibility to truly realize the Epicurean ideal, which makes the question of the ultimate aesthetic and ethical desirability of this ideal far more acute than it is in human society where the universal orgasm remains a pragmatically unachievable "ideal."

The ethical maxim that we call the Principle of Joyous Growth attempts to circumvent this problem not only by observing the delayed harm of some pleasures, and abstractly that pleasure has its limits, but by adding an additional criterion:

Maximize happiness but also maximize growth

What does "growth" mean? A very general interpretation is: increase in the amount and complexity of patterns in the universe, and in the case of a thinking entity, the amount and complexity of patterns in the mind (knowledge). The Principle of Joyous Growth rules out the universal orgasm outcome unless it involves a continually increasing amount of pattern in the universe. It rules out a constant, ecstatically happy orgasmic scream as too redundant a pattern to meet the maximization of growth criterion. This is a way of placing a formal restriction (maximize algorithmic complexity of knowledge) on one of the principal philosophical concerns about immortality: that without the fear of death, a being would become intractably lazy and all progress would cease. By embedding such a criterion in its underlying assumptions regarding the nature of reality, an AGI would be incapable of such crippling redundancy as permanent immediate pleasure (and we say immediate pleasure, because with growth comes new ideas about what is pleasurable and ultimately more happiness through more ways to be happy).

Of course, maximizing two quantities at once is not always possible, and in practice one must maximize some weighted average of the two. Different weightings of happiness versus growth will lead to different practical outcomes, all lying within the general purview of the conceptual Principle of Joyous Growth.

The Joyous Growth principle, without further qualification, is definitely not Friendly in the Yudkowskian sense. In fact it is definitively un-Friendly, in the sense that we humans are far from maximally happy—and in this as well as other ways, we are basically *begging* to be transcended. A post-Transcension universe operating according to the Principle of Joyous Growth would not be all that likely to involve the continuation of the human race, at least not in its present form. (It is important to note that the continuation of the human race as used in this book implies the continuation of the human condition, and Transcension which brings about its end does not necessarily mean the undoing of all human continuity—though it may; it may instead mean a much greater and more abstract form of continuity which has as its necessary precursor the end of the current human state.)

An alternative is to add a third criterion, obtaining a Principle of Voluntary Joyous Growth:

Maximize happiness, growth and choice

This means adopting as an important value the idea that sentient beings should be allowed to choose their own destiny. For example, they should be allowed to choose unhappiness or stagnation over happiness and growth, though they must be able to enforce this choice *only upon themselves* (because to allow otherwise would violate the right to choice of other sentient beings).

Of course, the notion of "choice" is just as much a can of worms as "happiness." Daniel Dennett's recent book *Freedom Evolves* does an excellent job of sorting through the various issues involved with choice and freedom of will. While we don't accept Dennett's reductionist view of consciousness, his treatment of free will is generally very clear and convincing.

Note that including choice as a variable along with two others implies that ensuring free choice for all beings is not an absolute commandment. Of course, given the extent to which human wills conflict with each other, free choice for all beings is not a possible opportunity. Given a case where one being's will conflicts with another being's will, the Voluntary Joyous Growth approach is to side with the being whose choice will lead to greater universal happiness, growth, and choice. (Free of selfishness, an AGI will be able to enforce its rules and their guidelines for resolving contradictions with consistency, something which our corrupt natures do not currently allow despite our best intentions.)

Voluntary Joyous Growth is not a simple goal, because it involves three different factors which may contradict each other, and which therefore need to be weighted and moderated. This complexity may be seen as unfortunate—or it may be seen as making the ethical principle into a more subtle, intricate and fascinating attractor of the universe. We are inclined towards the latter view. Though simplicity may be preferable where possible, enforcing it where impossible leads to much more deleterious results than attempting to understand and manage complexity.

ATTRACTIVE COMPASSION

We should note that the goal in positing "Voluntary Joyous Growth" has been to articulate a minimal set of ethical principles. These are certainly not the only qualities that are important in formulating the ethical guidelines for the Transcension of our universe beyond the current human condition through which we understand and impact it. For example, we strongly considered including compassion as an ethical principle, since compassion is, in a sense, the root of all ethics. However, it occurred to us that compassion is actually a second-order effect that is a consequence of choice, growth and freedom. In a universe consisting of beings that respect the free choices of other beings, and want to promote joy and growth throughout the universe, compassion for other beings is inevitable—because "being good to others" is generally an effective way to induce these others to contribute toward the joy and growth of the universe. Without the inclusion of choice, Joyous Growth is consistent with simply annihilating unhappy or insufficiently productive minds and replacing them with "better" ones; but assigning a value to choice gives a disincentive to dissolve "bad" minds, and leads instead to the urge to help these minds grow and be joyful.

Another second-order effect which is desirable but which apparently arises out of this formulation is cooperation (which is really an active form of mutual understanding and compassion, and can be seen as an extension of the notion of compassion). If we were to cynically view increasing joy as merely minimizing dismay, then along with the desire for growth (even the selfish desire for individual growth) and the assumption that other sentient beings will make choices, the precedent for the emergence of cooperation from such a situation is found in Axelrod's *The Evolution of Cooperation* and related work (see chapter 14). This sort of social détente theory provides a baseline for nondestructive behavior that is applicable even to our current condition. It is the beginning of a formalization of what is happiness (or, more accurately, non-unhappiness), which can be expanded upon; and along with the consistency of an AGI in understanding and applying the Voluntary Joyous Growth principle as an underlying formulation of its morality, would allow an AGI system to develop principles we can probabilistically describe as "good."

This all relates to the notion that compassion itself is a "universal attractor." A more accurate statement is: *a reasonable level of compassion is a universal attractor.* We can see this in the fact that it ensues from the combination of the universal attractors of joy, growth and choice; and we can also see it in the evolution of human society. Cooperation and compassion emerged in human beings most probably because, in society (and originally in a small tribal setting) it is often valuable for each individual to be kind to the other individuals in the tribe, so as to keep them alive and healthy. This is the case regardless of whether the tribe members are genetically related to each other; it's the case purely because, in many situations, the survival probability of an individual is greater if:

- The tribe has more people in it
- The people in the tribe know each other and hence can work relatively well together (hence, all else equal, there is benefit in retaining the current members of the tribe rather than recruiting new ones)

If humanity is divided up into tribes—because individual humans can survive better in groups than all alone—then compassion toward tribe members increases individual fitness. Compassion emerges spontaneously via natural selection, in situations where there is a group of minds each of which has choice, and which all (via growth) have the complexity to cooperate to some extent. Compassion and cooperation between tribes arise in any situation where an understanding of the mutual benefits of the non-destructive state of détente is reached; and with the elimination of the threat of annihilation, the intertribal dialogue can evolve ultimately from détente into cooperation for mutual benefit. While particular situations are unstable, given the complexity of human relationships and the power of perturbative individual actors to unbalance the system (the Hitlers and Stalins of the world, for example), the state of cooperation is a mental attractor given that amongst all choices, it has the best ultimate cost-benefit and risk-reward ratios.

Note that absolute compassion doesn't emerge from this tribal-evolutionary logic, but a rational level of compassion does. Similarly, absolute compassion doesn't come out of Voluntary Joyous Growth—but it seems that a reasonable level of compassion does. It seems more likely that "reasoned compassion" is a universal attractor than that "absolute compassion" (the self-sacrificing compassion of a Buddha or Mother Teresa) is.

Interestingly, it's harder to see how compassion would evolve among humans living in a large-group society like modern America. In this case, there's not such a direct incentive for an individual to be kind to others since there is a system of autonomous (and anonymous) means for self-support through mechanized means, which can be used to entirely circumvent any meaningful participation in social interaction. It may be that a population of rational-actor minds plunked into a large society would never evolve compassion to any significant degree. However, without compassion it is highly unlikely that a philosophically and technologically complex, evolving society such as the world today would have ever come into being. With the breakdown of compassion and cooperation, society collapses into anarchy and chaos; and as history bears out, this would give way to a tribal society in which compassion would evolve until another perturbation caused the cycle to repeat, or a different perturbation either destroyed the system entirely or caused compassion to become a permanent state, showing the power of compassion as an attractor once again!

NOSTALGIA

The Australian ecology theorist/activist Philip Sutton, on reviewing an early, unfinished draft of this chapter, pointed out to Ben that he had omitted a value that is very important to him: sustenance and preservation of what already exists. On reading Philip's comments, Ben reflected that he had made this omission because, in fact, this value—which we'll call nostalgia—is not all that important to him.

Of course, some people are more attached to the present world than others. We ourselves are somewhat attached to many things that exist—such as family and friends, ourselves, our pets, Jimi Hendrix and Front 242 CDs, Haruki Murakami and Philip K. Dick novels, pinon pine trees, Saturday mornings in bed and long wacky email conversations, to name just a few—but we don't consider this kind of attachment a primary value. What is important is that a sentient being has a choice to retain these things if they are important and contribute toward that being's self-perceived happiness. We don't see such a powerful *intrinsic* value in maintaining the past as Philip does, whereas we do see a very powerful intrinsic value in growth and development. (As a testament to the tricky nature of defining what is good, we don't even fully agree here ourselves, as Stephan is a bit more in agreement with Sutton than is Ben – though he doesn't feel nostalgia is worth pursuing if it means the cessation of evolutionary progress or the end of intelligent life in the universe.)

It is important to note, though, that our skepticism in this regard refers to the value of maintaining the past as a continual state (which can too easily turn into a redundant function opposed to growth), and is not meant as an invitation to discard the achievements and mistakes of the past as the foundations of future growth, or to evade the maxim that ignorance of history is a condemnation to repeat it. The future is always built upon the past, and the historical vacuum left by the destruction of memory and artifacts of the past is an invitation to waste time and resources re-creating patterns of the universe which were previously there, and by their prior existence are statistically likely to be reinvented (modulo contemporary cultural embeddings that may change certain specifics).

However, we don't see nostalgia as a destructive or unpleasant value, nor as contradictory with growth, joy or choice. The universe is a big place—and it is very highly probable that many parts of it are neither terribly important to any sentient being, nor required to remain unchanged in order not to destroy the universe itself. It may well be possible to preserve the most important patterns that currently exist in the universe, and still use the remainder of the universe to create wonderful new patterns. The values of growth and nostalgia only contradict each other in a universe that is "full," in the sense that every piece of mass-energy is part of some pattern that is nostalgically important to some sentient being. In a full universe one must make a choice, in which case we'll advocate growth—but it's not clear whether such a thing as a full universe will ever exist. It is quite possible be that the process of growth will continue to open up ever more horizons for expansion.

SMIGRODZKI'S META-ETHIC

An alternative approach, proposed by the brilliant neurogeneticist Rafal Smigrodzki in a discussion on the SL4 e-mail discussion list, is to begin with an even more abstract type of meta-ethic. Abstract though it is, the Principle of Voluntary Joyous Growth still imposes some specific ethical standards. Smigrodzki proposes a pure meta-ethic with no concrete content. In fact he proposed two different versions, which are subtly and interestingly different.

Smigrodzki's first formulation was:

Find rules that will be accepted.

This principle arose in a discussion of the analogy between ethics and science and specifically as an analogue to Karl Popper's meta-rule for the scientific enterprise[8]:

Find conjectures that have more empirical content than their predecessors

Popper's meta-rule specifies nothing about the particular contents of any scientific theory or scientific research program; it speaks only of what kinds of theories are to be considered scientific. Similarly, Smigrodzki's meta-rule specifies nothing about what kinds of actions are to be considered ethical, it speaks only of what kinds of rule-systems are to be considered as falling into the class of "ethical rule-systems": namely, rule-systems that are accepted. (And, it does not directly address the issue of what constitutes acceptability and to whom, which we must define at least abstractly. For example, Nazi rule was accepted by the German body politic in the 1930s and '40s, by some accounts voluntarily, though such accounts ignore the armed conflicts of the Freikorps prior to Hitler's Chancellorship. We see from this example that mere acceptance needs to be replaced with *voluntary acceptance* as in the VJG principle, and also with *forceless initiation* of the program—that is, it can't be violently imposed upon people, and its voluntary nature framed only after this initial period of force.)

One interesting thing about Smigrodzki's meta-rule is how close it comes to the Principle of Voluntary Joyous Growth. To see this, consider first that the notion of "be accepted" assumes the existence of volitional minds that are able to accept or reject rules. So to find rules that will be accepted, it's necessary to first find (or ensure the continued existence of) a community of volitional minds able to accept rules.

Next, observe that *one* version of the nebulous notion of "happiness" is "the state that a volitional mind is in when it gets to determine enough of its destiny by its own free choice." This is almost an immediate consequence from the notions of happiness and choice. For, if happiness is what a mind wants, and a mind has enough ability to determine its destiny via free choice, then naturally the mind is going to make choices maximizing its happiness.

So, "Find rules that will be accepted" is arguably just about equivalent to "Create or maintain a community of volitional minds, and find rules that the this community will accept (thus making the community happy)."

But then we run up against the problem that not all minds really know what will make them happy. Often minds will accept rules that aren't really good for them—even by their own standards—out of ignorance, stupidity or self-delusion. To avoid this, one wants the minds to be as smart, knowledgeable and self-aware as possible. So one winds up with a maxim such as: "Create or maintain a community of volitional minds, with an increasing level of knowledge, intelligence and self-awareness, and find rules that this community will accept (thus making the community happy)." As you can see, this is the growth principle of Voluntary Joyous Growth.

The notion of progressive refinement is helpful to consider here. Progressive refinement is an engineering term, which simply means reworking a system continually to move it towards a "better" state as quantified by stated goals. Our ethical system must be constantly re-analyzed within the context of increasing complexity of mind—both greater knowledge and a greater ability to acquire new knowledge—and refined, because as we get smarter, what makes us happy and what we chose to pursue with our free will evolves.

Note also that Popper's meta-rule of science also is susceptible to the "stupidity and self-delusion" clause. In other words, "Find conjectures that have more empirical content than their predecessors" really means, "Find conjectures that seem to a particular community of scientists to have more empirical content than their predecessors," and the meaningfulness of this really depends on how smart and self-aware the community of scientists is. The history of science is full of apparent mistakes in the assessment of "degrees of empirical content." So Popper's meta-rule could be revised to read, "Find conjectures that have more empirical content than their predecessors, as judged by a community of minds with increasing intelligence and self-awareness."

The notion of "increasing levels of *knowledge*" can also be refined. What is knowledge, after all? One way to gauge knowledge is using the philosophy of science. Imre Lakatos' theory of research programs suggests that a scientific research program[9]—a body of scientific theories—is "progressive" (that is to say, "making progress") if it meets a number of criteria, including:

- suggesting a large number of surprising hypotheses, and
- being reasonably simple

One interpretation of "increasing level of knowledge" is "association with a series of progressive scientific research programs." This definition leaves out other disciplines, but we can easily expand it to read: "Knowledge is increasing when new discoveries about the universe and new interpretations of those discoveries are being produced continually in all the intellectual disciplines."

Once all these details are put in place, this fleshing-out of Smigrodzki's meta-rule (which may well make it fleshier than Smigrodzki would desire) looks very similar to the Principle of Voluntary Joyous Growth. We have happiness, we have choice, and we have growth (in the form of growth of intelligence, knowledge and self-awareness). The only real difference from the earlier formulation of the Principle of Voluntary Joyous Growth is the nature of the growth involved: is it in the universe at large, or only within the minds in a community that is accepting ethical rules? Clearly if the answer is, "the universe at large," then it is both, and if it is such it will be because the minds of the community accepting ethical rules have evolved such that the universe at large is now effected by the actions of those minds (whether they be biological meta-human, AGI or some hybrid being).

After Ben presented him with an earlier version of this discussion of his meta-ethic, Smigrodzki's reaction was to create a yet more abstract version of his meta-ethic, which he formulated as:

"Formulate rules that make themselves into accepted rules that make themselves come true" ***(because the existence of states of the universe, including conscious states, are in agreement with goals stated in the rules),***

or,

"*Formulate rules which, if applied, will as their outcomes have the goals explicitly understood to be inherent in these rules.*"

We rephrase these as:

"*Create goals, and rules that, if followed, will lead to the achievement of these goals,*"

or,

"*Create goals, and rules that, if followed, will lead to the achievement of these goals, with as few side-effects as possible.*"

This formulation is more abstract than—and includes—Smigrodzki's previous proposal, which in this language was basically: "Create goal-rule systems that will be accepted." In being so, it is also divorced from moral grounding, and is more mechanistically rational. To see the difference quite clearly, consider the "ethical" system:

DESTROY ALL LIVING BEINGS BY:

1. FIRSTLY
    a. STUDYING ALL OTHER LIVING BEINGS SCIENTIFICALLY TO DETERMINE HOW TO KILL THEM MOST EFFECTIVELY, WHILE MINIMIZING THE RISK OF ANY SUCH BEINGS SURVIVING, AND OF DYING ONESELF BEFORE DOING SO, AND THEN
    b. KILLING THEM

2. FINALLY, KILLING ONESELF

This posits a goal and also some rules for how to achieve the goal. It is rational and consistent. So far as we can tell, it obeys Smigrodzki's revised, more abstract meta-ethic. However, it seems to fail his former, more concrete meta-ethic, because (at least among most of the sentient beings we know) it is unlikely to be accepted. (Now and then various psychopaths have of course accepted this "ethic," and attempted to put it into practice, but this is why successful ethical systems are socially grounded rather than being left entirely up to the will of one individual.)

By further abstracting his meta-ethic, Smigrodzki moved from:

- a very abstract formulation of "the good" ["Create goal-rule systems that will be accepted"], to
- a very abstract formulation of the general process of goal-seeking.

The difference between these lies in the key role of choice in the former (as hidden in the notion of "acceptance"). This highlights the primacy of the notions of choice and will in ethics. Without choice, there is actually no ethical system at all, but merely a procedural description of actors seeking results disguised as a general ethical principal. Action unhindered by emotional consideration of the self and society, with only the goal of satisfaction of a set of rules and target states of being, is not a form of ethical behavior as commonly conceived by humans.

JOYOUS GROWTH-BIASED VOLUNTARISM

Voluntary Joyous Growth, obviously, has a different relationship to Friendliness than pure Joyous Growth. Voluntary Joyous Growth means that even if superhuman AIs determine that joy and growth would be maximized if the mass-energy devoted to humans were deployed in some other way; even so, the choices of individual humans (whether to remain human or let their mass-energy be deployed in some other way) will *still* be respected and figured into the equation.

One could try to make Voluntary Joyous Growth more explicitly human-friendly by making choice the primary criterion. This is basically what is proposed in the extrapolated version of Smigodzki's first meta-rule. In this version, the #1 ethical meta-principle is to let volitional minds have their choices wherever possible. Only when conflicts arise do the other principles—maximize joy and growth—come into play. This might be called "Joyous Growth-Biased Voluntarism." Joy and growth will likely still play a very big role here, as conflicts may arise *quite frequently* between volitional minds coupled in a finite universe (indeed, as minds generally seek other minds for stimulation (and the growth goal would bias away from stopping this behavior), even in an infinite universe there are likely to be numerous individual mind-actors clustered around each other, in any inhabited locale at any given time). If we discard the assumption that minds will seek each other out as a matter of course, one can envision scenarios in which all inter-mind conflicts are removed, so that it's possible to fulfill choices without considering joy and growth at all.

For instance, what if all the minds in the universe decide they all want to play video games and live in purely automated simulated worlds rather than worlds occupied with other minds? Then living in individual video-game-worlds of their choice may gratify them quite adequately: they have maximum choice, but no opportunity for any factor besides choice to come into play. In this case minds may, consistently with Joyous Growth-Biased Voluntarism, make themselves unhappy and refuse opportunities for growth unto eternity. This is a mark against Joyous Growth-Biased Voluntarism, and in favor of simple Voluntary Joyous Growth with its greater flexibility. We suspect that Voluntary Joyous Growth is much closer to being a powerful attractor in the universe.

By forcing Joy and Growth to be considered at all, the potential for minor episodes of unhappiness caused by inevitable conflicts with different minds will ultimately lead to increase of all three criteria. Indeed, it is necessary that any mind which considers these principles take a long-term view of the optimization of these parameters of positive being, and avoid the pitfalls of local overoptimization which can lead to a complete breakdown of the system. Since that system is our minds—and our universe—we have quite a bit of incentive to break the short-term thinking habit.

HUMAN PRESERVATIONISM & CAUTIOUS DEVELOPMENTALISM

A more extreme ethical principle, in the vein of Joyous Growth-Biased Voluntarism, is what we'll term "Human Preservationism." In this view, the preservation of the human race (in some form immediately recognizable to current humans) through the post-Transcension period is paramount. Where this differs from Joyous Growth-Biased Voluntarism is that, according to Human Preservationism, even if all humans want to become transhuman and leave human existence behind, they shouldn't be allowed to.

In fact, we don't know of any serious transhumanist thinkers who hold this perspective. While many transhumanists value humanity and some personally hope that traditional human culture persists through the Transcension, transhumanists tend to be a freedom-centered bunch; and few would agree with the notion of *forcing* sentient beings to remain human against their will. Even so, Human Preservationism is a perfectly consistent philosophy of Transcension. There's nothing inconsistent about wanting vastly superhuman minds and new orders of beings to come into existence, yet still placing an absolute premium on the persistence of the peculiarly human.

But even still, Limited Human Preservationism—in which human beings as a class of entities in the universe persist, but individual humans are permitted to Transcend to posthuman state as they wish—would be more consistent with the notion of choice. In an LHP Transcension, humans could be viewed as an ongoing precursor state of transhumans; and some humans would choose to "grow up" and become transhuman, while others of their own volition to remain "childlike" and maintain the whole cycle of the human condition, including death. Given the nature of humans to want to force their way of being onto others, such a path is fraught with the same kinds of dangers to transhumans that humans present to other creatures on earth today, and could only succeed practically and philosophically if the transhumans were able to protect themselves and the universe from the humans without completely crushing the humans' free will.

A (somewhat more appealing) variation on Human Preservationism is Cautious Developmentalism. The abstract principle at work in this theory is: *If things are basically good, keep them that way, and explore changes only very cautiously.* In practical terms, the idea here is to preserve human life basically as-is, but to allow very slow and careful research into Transcension technologies, in such a way as to minimize any risk of either a bad Transcension or another bad existential outcome such as mass extermination of all life on earth. In the most extreme incarnation of this perspective, the choice of how to approach the Transcension is deferred to future generations, and the problem for the present generation is redefined as figuring out how to set the Cautious Developmentalist course in motion. Emotionally, letting future generations figure it out seems to be at best a pragmatic way to look at the possibility of our current generation to fail to achieve Transcension, but as long as we bother to have our considerations of positive outcome, then caution (that is, the avoidance of catastrophe as a limiting factor on unconsidered experimentation) will be a principle of Transcension—even if we don't limit ourselves to the extreme of Cautious Developmentalism.

HUMANENESS

Yudkowsky has proposed that "The important thing is not to be human but to be humane."[10] Enlarging on this point, he argues that:

> "Though we might wish to believe that Hitler was an inhuman monster, he was, in fact, a human monster; and Gandhi is noted not for being remarkably human but for being remarkably humane. The attributes of our species are not exempt from ethical examination in virtue of being 'natural' or 'human.' Some human attributes, such as empathy and a sense of fairness, are positive; others, such as a tendency toward tribalism or groupishness, have left deep scars on human history. If there is value in being human, it comes, not from being 'normal' or 'natural,' but from having within us the raw material for humaneness: compassion, a sense of humor, curiosity, the wish to be a better person. Trying to preserve 'humanness,' rather than cultivating humaneness, would idolize the bad along with the good. One might say that if 'human' is what we are, then 'humane' is what we, as humans, wish we were. Human nature is not a bad place to start that journey, but we can't fulfill that potential if we reject any progress past the starting point."

In email comments on an earlier draft of this paper, Yudkowsky noted that he felt this summary of his ideas about AI and humaneness didn't properly do it justice. Conversations following these comments have improved our understanding of his thinking; but even so, we can't claim to fully represent his ideas here, and suggest the interested reader engage with his material directly. We will discuss here a theory called "Humane AI," our design of which is inspired by Yudkowsky's ideas, and is similar but probably not isomorphic to his approach. We will present some arguments describing difficulties with Humane AI, which may not all be problems with Yudkowsky's ideas, in the sense that there may be solutions to these problems within his theory that he has not yet published.

In the approach we call Humane AI, one posits as a goal not simply the development of AIs that are benevolent to humans, but the development of AIs that display the qualities of "humaneness," where "humaneness" is considered roughly according to Yudkowsky's description above. That is, one proposes "humaneness" as a kind of ethical principle, where the principle is: "Accept an ethical system to the extent that it agrees with the body of patterns known as 'humaneness.'"

"Humaneness," in the sense that Yudkowsky proposes, is not a fully well-defined concept. It could be that the specific set of properties called "humaneness" you get depends on the specific algorithm that you use to sum together the wishes of various individuals in the world. If so, then one faces the problem of choosing among the different algorithms. This is a question for a future, more scientific study of human ethics.

Yudkowsky's work on Friendly AI is one such attempt at quantification, not through selection of an algorithm, but rather through selection of a result. Asimov's modified laws of Robotics (including the Zeroth law, as discussed in *Robots and Empire*) and Roger Clarke's extensions thereof (in his IEEE Computer journal articles[11]) are other such attempts, ones which Yudkowsky considers insufficient because they are mechanistic rather than morally grounded rules.

The authors consider Asimov's famous laws and Clarke's extensions thereof to be insufficient both because they are too specific in their formulation and because they only consider the actions of intelligent machines with respect to the current human condition, and have no proviso whatsoever for allowing the nature of humanity to evolve, possibly coincidentally with machines and into some kind of hybrid being. One problem is that such an event may violate the Zeroth Law, as "humanity" is an aggregate state of being, and any evolution which causes that state to become unnecessary could be seen as injuring humanity *even if every single entity describable as human at the time of Transcension survived as a transhuman being with continuous memory of their previous human form*.

Returning to Humane AI as a particular instance of these attempts to guide rational AI development, we encounter the major problem with distinguishing "humaneness" from "humanness": how to distinguish the "positive" from the "negative" aspects of human nature—e.g. compassion (viewed as positive) versus tribalism (viewed as negative). The approach hinted at in the above Yudkowsky quote is to use a kind of "consensus" process (actually Yudkowsky's reasoning is closer to a "consensus + minimal harm" metric, since volitionally there may appear to be a consensus in favor of tribalism). For instance, one hopes that most people on careful consideration and discussion will agree that tribalism, although humanly universal, isn't good. One defines the extent to which a given ethical system is humane as the average extent to which a human, after careful consideration and discussion, will consider that ethical system as a good one. Of course, one runs into serious issues with cultural and individual relativity here, but these effects can be limited even if absolute compassion is missing, provided the actors are all rational enough to consider minimal harm and enlightened self-interest as reflected in Axelrod's models.

Personally, we're not so confident that people's "wishes regarding what they were" are generally good ones (which is another way of saying, we think our own ethics differ considerably from those of the mean of humanity). To consider a single example, the vast majority of humans contend that "Belief in God" is a good and important aspect of human nature. Therefore, "Belief in God" should be considered humane according to the above definition—it's part of both what we humans are, *and* part of what we humans wish we were. Nevertheless, in our own personal opinions, the belief in God—though it has some valuable spiritual intuitions at its core—is essentially undesirable ethically. Nearly all ethical systems containing this belief have had overwhelming negative aspects, in our view, because such a belief is inherently totalitarian and subjugates the necessary condition of choice (and the related condition of responsibility) to the will of an actor (God), who is by definition both beyond reason and beyond ethics. In such a system, people are free to be infinitely destructive so long as it is contextualized as the will of God, not only because they've been relieved of responsibility for their actions by supernatural predeterminism, but also because in such views the perceivable universe is inferior to the assumed reality of "Heaven," and so long as one is doing the will of God the existing world can be damaged with impunity. To transhumanists, eternality is achievable through *material eternalism*, that is, some form of being in this, the perceivable universe. As such, it requires both the continual action and responsibility of intelligent beings to construct and maintain the state of eternalism in the universe (and therefore free will, not predeterminism and an "acceptance of fate"); and respect for the perceptible universe and life as primary and not inferior to the assumption of an imperceptible universe. As such, it is incompatible with traditional religious dogma. (Though, if believers were willing to evolve their religions, they could recontextualize the promise of Eternal Life as one of positive action and responsibility to God to preserve and extend this knowable life and universe, rather than resignation to the idea of a preordained fate of eventual destruction of this world and salvation only in another.) Voluntary Joyous Growth emerges from removing the human political particulars from religious texts, and focusing on the philosophical underpinnings of the desire to work towards Eternal Life.

Thus, we consider it our ethical responsibility to work so that belief in God is *not* projected beyond the human race into any AGIs we may create, unless (and we really doubt it) it can be shown that the only or the best way to achieve a positive Transcension is to create an AGI that contains such a belief system. Of course, there are many other examples besides "belief in God" that could be used to illustrate this point; but this one is worth commenting on, as it is commonly seen as absolute both in principle and in one's chosen particulars (which is, as discussed, the crux of its problem).

To get around problems like this, one could try to define humaneness as something like "What humans *would* wish they were, if they were wiser humans"—but of course, defining "wiser humans" in this context requires some ethical or meta-ethical standard beyond what humans in general are or wish they were. It becomes difficult to avoid spiral reasoning or resorting to a supernatural set of absolute ethics, which is why both are so common. This line of thinking leads up to what Yudkowsky has called "collective volition," an idea that we'll discuss in more depth a little later in this chapter. In sum, the difficulties with Humane AI are:

1. The difficulty of defining "humaneness"

2. The presence of delusions that we judge as ethically undesirable, within the near-consensus worldview of humanity

The second point here may seem bizarrely and massively egomaniacal—who are we, Ben Goertzel and Stephan Vladimir Bugaj, to judge the vast mass of humanity as being ethically wrong on major points? Yet it has to be observed that the vast mass of humanity has shifted its ethical beliefs many times over history, and the fact that things are one way now is no reason not to insist upon a change. While you may not agree with our proposals, they are no less egomaniacal than any other work that attempts to convince the mass of humanity to behave in some particular manner. At many points in history, the vast mass of humans believed slavery was ethical, for instance. You could argue that if they'd had enough information, and carried out enough discussion and deliberation, they might have decided it was bad. Perhaps this is the case, but to lead the human race through a process of discussion, deliberation and discovery adequate to free it from its collective delusions is a very large task indeed. We see no evidence that any existing political institution is up to this task. Perhaps an AGI could carry out this process—but then what is the goal system of this AGI and where did it come from in the first place if not from the humans who created it, and whom it is now expected to guide? Do we begin this goal system with the current ethical systems of the human race—as Yudkowsky seems to suggest in the above ("Human nature is not a bad place to start…")? In that case, does the AGI begin by believing in God and reincarnation, which are the beliefs of the vast majority of humans? Or does the AGI begin with some other guiding principle, such as Voluntary Joyous Growth? Our hypothesis is that an AGI beginning with Voluntary Joyous Growth as a guiding principle is more likely to help humanity along a path of increasing wisdom and humaneness, than an AGI beginning with current human nature as a guiding principle.

One can posit, as a goal, the creation of a Humane AI that embodies humane-ness as discovered by humanity via interaction with an appropriately-guided AGI. This seems like a spiral argument that adds nothing beyond what one gets from creating an AGI that follows the principle of Voluntary Joyous Growth and leaving it to interact with humanity. If the creation of the Humane AI is going to make humans happier, and going to help humans to grow, and going to be something *that humans choose*, then the Voluntary Joyous Growth-based AGI is going to choose it anyway. On the other hand, maybe after humans become wiser, they'll realize that the creation of an AGI embodying the average of human wishes is not such a great goal anyway. As an alternative, perhaps a host of different AGIs will be created, embodying different aspects of human nature and humaneness, and allowed to evolve radically in different, mutually non-destructive directions guided by the interaction of humans (or biological transhumans) and the AGIs themselves (or their machine or hybrid transhuman successor states).

ETHICAL PRINCIPLES, SYSTEMS AND RULES

So far, our discussion of ethics has lived on a very abstract level—and this has been intentional. We have sought to treat ethics in a manner similar to the philosophy of science. In science we have Popper's meta-rule, and then we have scientific research programs, which may be evaluated heuristically as to how well they fulfill Popper's meta-rule: how good are they at being science? Then, within each research program we have a host of specific scientific theories and conjectures, none of which can be evaluated or compared outside the context of the research programs in which they live. Similarly, in the domain of ethics, we have highly abstract principles like Smigrodzki's meta-rule or the Principle of Voluntary Joyous Growth—and then, within these, we may have particular ethical rule-systems, which in turn generate specific rules for dealing with specific situations.

Our feeling is that the specific ethical rule-systems that promote a given abstract principle in a human context are very unlikely to survive the Transcension; but if we are successful in bringing about a *positive* Transcension then the abstract principles themselves will survive, and that is what is most important. For instance, the standard ethics according to which modern Americans live involves a host of subtle compromises, involving such issues as:

- meat-eating (we're comfortable killing some animals for meat but not others; we're comfortable killing animals brutally but not too brutally)
- charity (we give a certain percentage of our incomes to help the less fortunate, but not as much as we could)
- honesty (we generally allow "white lies" but frown on other sorts of lies)

And so on and so forth. This complex system of compromises that constitutes our modern American practical ethics is not in itself a powerful attractor (and neither are the sets of compromises which constitute the practical ethics of any other existing human sub-society). The American system, for Americans, is largely in accordance with the Principle of Voluntary Joyous Growth—it tries to promote happiness, progress and choice—but there is little doubt that, Transcension or no, in a couple hundred years a rather different network of compromises will be in place which reflect the particulars of the state of the world at that time. Similarly, post-Transcension, the practical manifestations of the Principle of Voluntary Joyous Growth will be very radically different, though the principle itself will not have changed. A common problem with human ethical systems is to assume that the particulars of an ethical system are actually its underlying principles, and to be unwilling to change them. However, if the principles themselves are what one thinks about, then the particulars changing needn't be seen as a frightening breakdown of morality, but rather in the positive light of being just another progressive refinement of the particulars in pursuit of the best implementation of the principle.

(As an aside, it is clearly no coincidence that the Principle of Voluntary Joyous Growth harmonizes better with modern urban American ethics than with the ethics of many other contemporary cultures. More so than, say, Arabia or

China or the Mbuti pygmies, American culture is focused on individual choice, progress and hedonism. And so, we are aware that as modern Americans writing about Voluntary Joyous Growth, we're projecting the nature of our own particular culture onto the transhuman future. On the other hand, it's not a coincidence that America and culturally similar places are the ones doing most of the work leading toward the Transcension. Perhaps it is sensible that the cultures most directly leading to the Transcension should have the most post-Transcension-friendly philosophies. That said it is not clear that cultures which do not put as high a premium on individuality and choice as the American culture do so volitionally rather than circumstantially. We contend that beneath the particulars of every culture are the underlying desires to see both individual choice and the good of society maximized.

One thing this system-theoretic perspective says is: we can't judge the modern American ethical system by any one judgment it makes—we can only judge, as a whole, whether it tends to move in accordance with the principle we choose as a standard (e.g. Voluntary Joyous Growth). Similarly, we can't reasonably ask post-Transcension minds to follow any particular judgment about any particular situation—rather, we can only ask them to follow some ethical system that tends to move in accordance with some general principle we pose as a standard. (This request is more likely to be fulfilled a priori if it constitutes a powerful attractor in the universe at large.)

Thus, we suggest that the ethical prescriptions "Be nice to humans" or "Obey your human masters" are simply too concrete and low-level to be expected to survive the Transcension. On the other hand, we suggest that a highly complex and messy network of beliefs like Yudkowsky's "humaneness" is insufficiently concise, elegant and abstract to be expected to survive the Transcension. It is more reasonable to expect highly abstract ethical principles to survive, and more sensible to focus on ensuring the principle of Voluntary Joyous Growth through the Transcension than to focus on specific ethical rules (having meaning only within specific ethical systems, which are highly bound by context and culture) or the whole complex mess of human ethical intuition. Initially principles like joy, growth and choice will be grounded in human concepts and feelings—in aspects of "humaneness"—but as the Transcension proceeds they will gain other, related groundings. We must have ethical systems in the present which move towards these goals, but the survival of those specific systems is irrelevant, as they're only immediate implementations of a more important set of guiding principles.

In terms of technical AI theory, this contrast between general principles and specific rules relates to the issue of "stability through successive self-modifications" of an AI system. If an AI system is constantly rewriting itself and re-rewriting itself, how likely is it that this or that specific aspect of the system is going to persist over time? One would like for the basic ethical goal-system of the AI to persist through successive rewritings, but it's not clear how to ensure this, even probabilistically. The properties of AI goal-systems under iterative self-modification are basically unknown, and can be seriously explored only once we have some reasonably intelligent and self-modifiable AI systems at hand with which to experiment. However, our strong feeling is that *the more abstract the principle, the more likely it is to survive successive self-modification.* A highly specific rule like "Don't eat yellow snow" or "Don't kill humans" or a big messy habit-network like "humaneness" (which is merely a huge, interconnected set of specific rules) is relatively unlikely to survive; a more general principle like Voluntary Joyous Growth is a lot more likely to display the desired temporal continuity. We're betting that this intuition will be borne out during the exciting period to come when we experiment with these issues on simple self-modifying, somewhat-intelligent AGI systems. We're also betting that, among all the abstract principles out there, the ones that are more closely related to powerful attractors in the universe at large are more likely to occur as attractors in an iteratively self-modifying AGI, and hence more likely to survive through the Trancension.

So, our essential complaint against Yudkowsky's Friendly AI theory is that, quite apart from ethical issues regarding the wisdom of using mass-energy on humans rather than some other form of existence, we strongly suspect that it's *impossible* to create AGIs that will progressively, radically self-improve and yet retain belief in the "humaneness" principle; as the formulation of humaneness through example is just too specific to human particulars rather than generalizable principles. A more abstract and universally-attractive principle like Voluntary Joyous Growth seems much more likely to remain active (and be built upon) in a continually evolving AGI system.

Please note that this is *very* different from the complaint that Friendly AI won't work because any AI, once it has enough intelligence and power, will simply seize all processing power in the universe for itself. This "Megalomaniac AI" scenario is mainly a result of rampant anthropomorphism. In this context it's interesting to return to the notion of attractors. It may be that the Megalomaniac AI is an attractor, in that once such a beast starts rolling, it's tough to stop. The question is, how likely is it that a superhuman AI will start out in the basin of attraction of this particular attractor? Our intuition is that the basin of attraction of this attractor is not particularly large. Rather, we think that in order to make a Megalomaniac AI, one would probably need to explicitly program an AI with a lust for power. Then, quite likely, this lust for power would manage to persist through repeated self-modifications—"lust for power" being a robustly simple-yet-abstract principle. On the other hand, if the AI is programmed with an initial state aimed at a different attractor meta-ethic, there probably isn't much chance of convergence into the megalomaniacal condition.

HARMONY WITH THE NATURE OF THE UNIVERSE

This leads us to a point made by Jef Albright on the SL4 list, which is that the philosophy of Growth ties in naturally with the implicit "ethical system" followed by the universe—i.e., the universe grows. That is to say: Growth is a kind of *universe-scale attractor*—once one has a universe devoted to pattern-proliferation and expansion, it will likely continue to do so indefinitely. The newly generated patterns will generate yet more patterns, and so forth. It is also a "universal attractor," in the sense of an attractor that is common in various dynamical subsystems of the universe.

We believe that there's a similar philosophical argument that Voluntary Joyous Growth is also harmonious with the pattern of the universe—i.e. also holds promise as a universal attractor.

Regarding the Voluntary part—the evolution of life shows how powerful wills naturally emerge from the weaker-willed and then continue to survive due to their powerful wills, and create yet more strong-willed beings. Given the situation we have in which the possibility of volition has evolved amongst living things, free-willed beings will take over the mechanisms of evolution to attempt to circumvent any further random events therein which would bring about the end of choice. The evolution of life itself becomes guided both by and towards volition at essentially the moment that volition becomes possible—a strong attractor indeed.If you believe humans have a greater and deeper capacity for joy than rocks or trilobites or pigs, then we can also see in natural evolution a movement toward increasing Joy. Joyful creatures interact with other Joyful creatures and produce yet more Joyful creatures—Joy wants to perpetuate itself.

On the other hand, the Friendly AI principle does not seem to harmonize naturally with the evolutionary nature of the universe at all. Rather, it seems to contradict a key aspect of the nature of the universe—which is that the old gives way to the new when the time has come for this to occur. One may feel it's a bit silly to put too much emphasis on prior states of the universe when positing such radical changes in its nature as Transcension. Perhaps so, but what is even more important than harmonizing with some particular state of nature is following the more general principles of the nature of life and the universe and biasing the inevitable universal changes towards positive and sustainable ones. The more vexing problem with Friendly AI is that it seems too cumbersome and particular in its formulation to survive progressive refinement within the mind of a transhuman and/or AGI entity; and while it may appear to harmonize well with the current particulars of life, using it may lead to a disastrous consequence for life the moment the particulars in which it makes sense change.

There's a certain quixotic nobility in maintaining ethics that contradict nature. After all, in a sense, technology development is all about contradicting nature. But in a deeper sense, technology development is about following the nature of the universe—following the universal tendency toward growth and development. Modern technology may be in some ways a violation of biological nature, but it's a consequence of the same general evolutionary principle that led to the creation of biological forms out of the non-living chemical stew of the early Earth. There is a quixotic beauty in contradicting nature—but an even greater and deeper beauty, perhaps, in contradicting local manifestations of the nature of the universe while according with global ones; in breaking out of local attractor patterns but remaining wonderfully in sync with global ones.

All this suggests an interesting meta-principle for selecting abstract ethical principles, already hinted at above:

> ***All else being equal, ethical principles are better if they're more harmonious with the intrinsic nature of the universe—i.e. with the attractors that guide universal dynamics.***

This suggests another possible modification to Smigrodzki's meta-ethic, namely:

> ***Find rules that will be accepted, and that are optimally harmonious with the global attractors that guide universal dynamics.***

However, this enhancement may be somewhat redundant, because *rules, systems and principles that are more harmonious with the attractors that guide universal dynamics will tend to be accepted more broadly and for longer*. Or in other words:

> ***Attractors that are common in the universe are also generally attractors for communities of volitional agents.***

One line of reasoning this discussion brings to mind is Nietzsche's discussion of "a good death." Nietzsche pointed out that human deaths are usually pathetic because people don't know when and how to die. He proposed that a truly mature and powerful mind would choose his time to die and make his death as wonderful and beautiful as his life. Dying a good death is an example of harmonizing with the nature of the universe—"going with the flow," or following the "watercourse way" (to use Alan Watts' metaphorical rendition of Taoism). Counterbalancing the beauty of the Friendly AI notion with its quixotic quest to preserve humaneness at all costs is in contradiction to the universal pattern of progress. One has the hyper-real Nietzschean beauty of humanity dying a good death—recognizing that its time has come, because it has brilliantly and dangerously *obsoleted itself*. One might call this form of beauty the "Tao of Speciecide"—the wisdom of a species (or other form of life) recognizing that its existence has reached a natural end and choosing to end itself gracefully. As Nietzsche's Zarathustra said, "Man is something to be overcome."

It's an interesting question whether speciecide contradicts the universal-attractor nature of compassion. Under the Voluntary Joyous Growth principle, it's not favored to extinguish beings without their permission. But if a species wants to annihilate itself, because it feels its mass-energy can be used for something better, then it's perfectly compassionate to allow it to do so (of course, if even one human doesn't want this, then the possibility of a compromise arises—if the principle of choice is indeed a universal attractor).

We are being intentionally outrageous here—in our hearts we don't want to see the human race self-annihilate just to fulfill some Nietzschean notion of beauty, or to make room for more intelligent beings, or for any other reason (indeed the transhumanist program favors life over death and creation over destruction at its foundations). We have a tremendous affection for the hypercerebrated ape-beings we lovingly call "us." As will be emphasized below, the course we propose in practice is a hybrid of Cautious Developmentalism and Voluntary Joyous Growth. We are pursuing this line of discussion mainly to provide a counterbalance to what we see as an overemphasis on "human-friendliness" and human preservation. Preserving and nurturing and growing humanity is an important point, but not the *only* point. To understand the

Transcension with the maximum clarity our limited human brains allow, we need to think and feel more broadly.

There is some beauty in both of these extremes—Friendly AI and the Tao of Speciecide—but we're not overwhelmingly attracted to either of them. We don't know if it's "best," in a general sense or for the well-being of the universe as a whole, if humanity survives or not—but we have a very strong personal bias in favor of humanity's persistence, and in practical terms would never act against our own species. We are very strongly motivated to spread choice, growth and joy throughout the universe—and to research ways in which to do this without endangering humanity, and what it has become and achieved. However, it is important not to conflate essential humanity with particulars of the human condition as it exists now, and wind up with a Luddite principle that no change is the only good change. While an argument can be made that radical alterations to the human condition such as superintelligence and eternal (or vastly longer) life in the material world are attempts to destroy humanity, this strikes us as a nonsensical line of argument, based on the assumption that the only meaningful thing that defines humanity is the inevitability of its destruction and thus its ultimate irrelevance.

Also, if some humans choose to Transcend to a posthuman condition and others not to, this is akin to the situation in previous evolutionary processes (the only difference being that Transcension Evolution is guided by intelligent entities rather than left up to the mechanical dynamics of evolutionary selection). Evidence seems to indicate that both Homo Erectus and Homo Neanderthalensis coexisted with our species, Homo Sapiens, so it is precedent in nature that divergent lines from the same species predecessor can coexist. A situation in which humans and transhumans coexisted would be little different, except that it is much more likely that a transhuman species would be able to prevent humans from meeting the same eventual fate as Homo Erectus and Homo Neanderthalensis.

AGI AND ALTERNATIVE DANGERS

OK—now, let's get practical. Assume that:

- the above analysis is basically correct
- one accepts the goal of Voluntary Joyous Growth or one of its variants
- the post-Transcension world turns out to be significantly influenced by the details of the route humanity takes to the Trancension

Then we have the question of what we can do to encourage the post-Transcension world to maximally adhere to the principle of Voluntary Joyous Growth.

Our own thinking on this topic has centered on the development of artificial general intelligence. Partly this is because AGI is our favored area of research, but mainly it's because:

- other radical futurist technologies are most likely to be achieved via a combination of human and AGI effort.
- the chances of a positive Transcension are much greater if highly advanced AGI is developed before other radical futurist technologies

Regarding the first point, as soon as AGI comes about, it will radically transform the future development course of all other technologies. If two minds are better than one, two *kinds* of minds are even better when it comes to bringing to bear new perspectives that may solve old, seemingly intractable problems. Furthermore, these other technologies—if their development initially goes more rapidly than AGI—are likely to rapidly lead to the development of AGI, so that their final development will likely be a matter of AGI-human collaboration. Suppose, for example, that molecular nanotechnology comes about before AGI.

One of the many interesting things to do with MNT will be to create extremely powerful hardware to support AGI; and once AGI is built it will lead to vast new developments in MNT, biotechnology, AGI and other areas (this feedback loop already can be seen with existing technologies, and extrapolating the effect to more radical technologies simply requires an acceptance that such technologies are possible at all). Or, suppose that human biological understanding and genetic engineering advance much faster than AGI. Then, with a detailed understanding of the human brain, it should be possible to create software or hardware closely emulating human intelligence—and then improve on human intelligence in this digital form, thus leading to powerful AGIs. (Though emulating the human mind in digital form seems like the most comprehensible approach to AGI, more exciting in our view is the possibility of creating whole new structures of intelligence. After all, it's quite easy to make human minds—you merely need a fertile male and female human pair and a few months.) It appears most likely that MNT or biotech will lead to AGI capabilities before they will lead to AGI-independent Singularity-launching capabilities, but that speculation could clearly be wrong.

Regarding the second point, we think it's clear that a molecular assembler or an advanced genetic engineering lab will be profoundly dangerous if left in the hands of (highly unreliable, ethically variant) human beings as they exist today. Quite possibly, once technology develops far enough, it will become *so* easy for a moderately intelligent human to destroy all life on Earth that this will actually occur[12]. There are many possible solutions to this problem of potential technological cataclysm, for instance:

1. Renounce advanced technology, as Bill Joy and others have suggested.

2. Modify human culture and psychology so that the impulse to destroy others is far less prominent.

3. Develop technological and cultural safeguards to prevent abuse of these technologies.

4. Via genetic engineering or direct neuromodification, modify the human brain so that the impulse to destroy self and others is far less prominent.

5. Create AGIs that are less destructively irrational than humans, and allow them to take primary power over the development and deployment of other radical future technologies.

6. Create AGIs that are more intelligent than humans, but are not oriented toward self-improvement and self-modification, and are oriented instead toward preventing humans from either being destructive or gaining power over potentially destructive technologies. These AGIs, together with humans selected for their lack of desire to destroy others, may very slowly and carefully pursue Transcension-oriented research—an approach to the Transcension that we're calling Cautious Developmentalism.

Of these six possibilities, (4-6) are the ones that appear to have the highest probability of successful eventuation. (3) is also plausible (though it has worked so far, it's been a tense ride since the splitting of the atom, and it always seems on the verge of breaking down); whereas if (3-6) fail, annihilation seems more likely than either (1) or (2).

Renunciation of advanced technologies is highly unlikely given the practical benefits that each incremental step of technological advancement is likely to have. This is the "Genie's Already Out of the Bottle" problem. The vast majority of humans aren't going to want to renounce technologies that they find gratifying, and a small set of renouncers won't alter the course of technology development.

Potentially, radical Luddites could force renunciation via mass civilization-destroying terrorist actions—we view this as far more likely than a *voluntary* mass renunciation of technology. Short of destroying the society that created it, you can't turn back an already existing technological development. Should we destroy society once more (as we did in the Dark Ages and so many other times before), it may very well be the last time, given the technologies that exist today.

Given our history, it's hard to be optimistic about the "perfectability of humanity" via any means other than uploading or radical neural modification. While social and cultural patterns definitely have a strong impact on each individual mind, it's equally true that social and cultural patterns are what they are (and are as flawed as they're flawed) because of the intrinsic biological nature of human psychology. Traits like dishonesty, violence, paranoia and narrow-mindedness are part of the human condition and are not going to be eliminated via social engineering or education—only via a feedback loop between deliberately embracing the positive aspects of the human psyche and using them to guide biological modification. So far, social and psychological engineering through pharmacology has been a mixed bag, though the successes thus far have sometimes been overshadowed by the negatives (such as addiction, and drugs whose good effects are outweighed by negative side-effects). As technology advances, it seems clear that the *only* real hope for improving human nature lies in modifying the brain—either directly using pharmacology, neurosurgery, or nanotechnology machines to modify its structure, or indirectly through changes to the human genome, hence physiologically modifying the nature of humanity.

On a purely scientific level, it's hard to tell whether or not detailed human-brain or human-genome modification is "easier" than creating AGI. Because we already have brains and genomes, it appears at first glance to be easier than creating a whole new form of machine intelligence, but that may not actually be the case. We may require the abilities of machine intelligence to look at the problems of biological modification from a new perspective in order to achieve the ability to change our biological natures. Pragmatically, however, it seems clear that these biological improvements would be very difficult to propagate throughout the human race—due to the fact that so many individuals believe it's a bad idea, and are unlikely to change their minds of their own volition. AGI, on the other hand, given sufficient financial resources, can quite possibly be achieved by a small group of individuals within our current lifetimes. Such a new entity would have a definitive effect on the world at large, even if most individuals on Earth greet it with confused and ambiguous (or in some cases flatly negative) attitudes. If we are successful in guiding AGI development to follow the principles of Voluntary Joyous Growth and Cautious Developmentalism, then it will indeed have a *positive* definitive effect.

Finally, technological safeguards *may* be possible, but it's hard to be confident in this regard: even if some radical, dangerous technologies can be safeguarded (as nuclear weapons are, currently, by the difficulty of obtaining fissile materials), all it will take is *one* hard-to-safeguard technology to lead to the end of all life on Earth. Certainly, it's clear that—given the increasing rate of technological advances and its rapid spread around the globe—the *only* way the "technological safeguard" route could possibly work in the long-term would be via a worldwide police state with Big Brother watching everyone. Despite all the negative aspects which would be accepted in this situation in return for the guarantee of security, history has shown that such states do not indeed provide the security they promise. Advanced surveillance and enforcement measures would lead to advanced countermeasures by rebel groups, including sophisticated hacker groups in First World countries as well as terrorists with various agendas (including Luddite agendas). All that results may be a period of terror and oppression, followed by the very cataclysmic end of life that the state's strong hand had promised to prevent. If one were extremely determined to make technological safeguards work, it seems most likely to be by:

- creating highly advanced technology, either AGI or MNT or intelligence-enhancing biotech or some combination thereof

- keeping this technology in the hands of a limited class of people, and use this technology to monitor the rest of the world, with the specific goal of preventing the development of any other technology posing existential risks

This is basically an AI Big Brother, right? In a sense—but, while this would necessarily involve the sort of universal surveillance associated with the term "Big Brother," it wouldn't necessarily entail fascist control of thoughts and actions of the sort depicted in George Orwell's *1984*. Rather, all that's required is the specific control of actions posing significant existential risks to the human race (and any other sentients developed in the meantime). Rather than a "Big Brother," it may be more useful to think of a "Singularity Steward"—an entity whose goal is to guide humanity and its creation toward its Singularity or other-sort-of-Transcension in a maximally wise way, or guide it away from Singularities and Transcensions if these are judged most-probably negative in ethical valence.

However, this approach certainly begs the question of whom would choose this limited class of people, based upon what criteria, and what their relationship would be to their fellow people. Given human history, it would seem unwise to leave the fate of all life on earth solely in the hands of some reified clique who would be likely to follow their biological human tendencies to thusly impose their status as the Alpha Humans upon all other life. An AGI system which was guided by human wisdom about our own flaws, but rendered incapable of our own foibles by the different nature of its physical and programmatic mental makeup, would be more likely to pursue a wise course instead of succumbing to the seduction of power and greed.

SINGULARITY STEWARDSHIP AND THE GLOBAL BRAIN MINDPLEX

In fact, our suspicion is that the *only* way to make a Singularity Steward entity actually work would be to supply it with an AGI brain—though not necessarily a brain bent on growth or self-improvement. Rather, one can envision an AGI system programmed with a goal of preserving the human condition roughly as-is, perhaps with local improvements (like decreasing the incidence of disease and starvation, extending life, etc.). This AGI—"AI Big Brother," aka the "Singularity Steward"—would have to be significantly smarter than humans, at least in some ways; and certainly would need to have been designed to shun the animal desire to exert power over others, and hoard resources for itself, in order to achieve the wisdom necessary to make truly beneficial decisions. However, it wouldn't need to be an autonomous entity—in fact it's natural for this entity to depend on humans for its survival, and for humanity to depend upon it to protect us from our worst instincts, creating a volitional symbiosis between humanity and its new intelligent counterpart.

This steward AGI would need to be a wizard at analyzing massive amounts of surveillance data and figuring out who's plotting against the established order, and who's engaged in thought processes that might lead to the development and deployment of dangerous technologies. Perhaps, together with human scientists, it would figure out how to scan human brains worldwide in real-time. This could be applied not only to prevent murderous thoughts, but also thoughts regarding the development of molecular assemblers or self-modifying AIs with goals of creating an entity not obeying the VJG principles, or a being with intelligence competitive with that of the steward itself. It is obvious that if this entity were to lust for power it would exert fascistic control over all sentient beings, leading to a nightmare scenario where every thinking creature either led a desperate, paranoid existence of illegally changing and growing, or labored meaninglessly to maintain the status-quo.

The problem of engineering a Singularity Steward AGI is rather different from the problem of engineering an AI intended to shepherd human minds through the Transcension. In the AI Big Brother case, one doesn't want the AI to be self-modifying and self-improving—one wants it to remain stable. This is a much easier problem! One needs to make it a bit smarter than humans, but not too much—and one needs to give it a goal system focused on letting itself and humans remain *as much the same as possible.* The Singularity Steward should want to increase its own intelligence only in the presence of some external threat (such as the popular but improbable notion of an alien invasion).

In extreme cases one can envision a Singularity Steward being compelled to act in a fascistic way—for instance, intrusively modifying the brains of rebellious AGI researchers intent on launching the Singularity according to their own pet theories. But if the goal is to prevent a dangerous, inadequately-thought-out Singularity, this may be the best option. To keep things *exactly* the way they are now—with the freedoms that now exist—is to maintain the possibility of massive destruction as technology develops slightly further. We are not right now in a safe and stable socio-psycho-technological configuration, by any means. This AI Big Brother option is not terribly appealing to us personally, because it grates too harshly against the values of growth, choice and happiness. However, it is a logical and consistent possibility, which seems plausibly achievable based on an objective analysis of the situation we confront. There is the possibility, assuming the inevitability of AI generally, that it may well be the best option if we can't quickly enough arrive at a confident, fully fleshed-out theory regarding the likely outcome of iterated self-improvement in AGI systems. Such a system would need to be carefully designed to exert its power exclusively in situations in which failing to do so would lead to the destruction of all sentient life—otherwise it would become a threat to all life, in and of itself.

The Singularity Steward idea ties in with the Cautious Developmentalism approach mentioned earlier. Suppose we create a Singularity Steward, and then allow it to experiment—together with human scientists selected by the Steward for their lack of destructive desire—with Transcension-related technologies. This experimentation must take place very slowly and conservatively, and any move toward the Transcension would (according to the Steward's hard-wired control code) be made based only on the agreement of the Steward with the vast majority of human beings (not solely the human scientists, who may if given such power succumb to their own self-interests and short-term human thinking, and choose to destroy others for their own gain). Conceivably, this could be the best and safest path toward the Transcension.

In fact—Orwellian associations notwithstanding—a Singularity-Steward-dominated society could potentially be a human *utopia.* Careful development of technology aimed at making human life easier—cheap power and food, effective medical care, and so forth—could enable the complete rearrangement of human society. Perhaps Earth could be covered by a set of small city-states, each one populated by like-minded individuals, living in a style of their choice and free from any forms of control other than self-control and mutual respect. Liberated from economic need, and protected by the Steward from assault by nature or other humans, the humans under the Steward's watch could live far more happily than in any prior human society. Free will, within the restrictions imposed by the Steward, could be refined and exercised copiously, perhaps in the manner of Buddhist "mind control." Free will is always restricted by the necessity of sharing the universe with other entities which have free will, and the idea of letting a neutral third party mitigate conflict is the basis for government. However, if an AGI were the government, it could choose to act with genuine altruism rather than for the benefit of its own power. Such a system could be designed in a way *not* to mimic the biological drive towards greedy self-interest. Under such a system, growth could occur spectacularly in non-dangerous directions, such as mathematics, music and art.

This hypothetical future is similar to the one sketched in Jack Williamson's classic novel *The Humanoids*, although his humanoids (a robot-swarm version of an "AI Big Brother") possessed the tragic flaw of valuing human *happiness* infinitely and *human will* not at all. While this flaw made Williamson's novel an interesting one, it's not intrinsic in the notion of a steward AGI. Rather, it's quite consistent to imagine a Singularity Steward that values human free will as much as or more than human happiness—and imposes on human choice *only* when it moves in directions that appear probable to cause existential risks for humanity.

Of course, as hinted at above, there's one problem with this dream of a Singularity-Steward-powered human utopia: politics. An AGI steward, if it is ever created, is by financial necessity most likely to be created by some particular power bloc in order to aid it in pursuing its particular interests. What are the odds that it would actually be used to create a utopia on Earth? That's hard to estimate! What will happen to politics when pre-Transcension but post-contemporary technology drastically decreases the problems of scarcity we have on Earth today? It is not certain, but the hope is that if we can actually engineer a way out of our contemporary problem of resource contention, politics will take a sharp turn towards the cooperative rather than towards the destructive.

This means that there are two ways in which a really workable Singularity Steward could come about:

- by transforming the global cultural and political systems to be more rational and ethically positive
- by a relatively small group of individuals, acting rationally with positive ethical goals, creating the Singularity Steward and putting it into play

This "relatively small group" could for example be an international team of scientists, a group operating within the United Nations or the government of some existing nation. (Of course, these two paths are not at all mutually exclusive.)

This hypothesized transformation of global cultural and political systems ties in with the Turchin-Heylighen notion of the Global Brain, as reviewed at the start of this book. The general idea of the Global Brain is that computing and communication technologies may lead to the creation of a kind of "distributed mind" in which humans and AI minds both participate, but that collectively forms a higher level of intelligence and awareness that goes beyond the individual intelligences of the people or AIs involved in it. In the last chapter we termed this kind of distributed mind a "Mindplex," and have spent some effort exploring the possible features of Mindplex psychology. The Global Brain Mindplex, as presented, would consist of an AGI system specifically intended to collect together the thoughts of all the people on the globe, and synthesize them into grander and more profound emergent thoughts—a kind of animated, superintelligent collective unconscious of the human race. Of course, the innate intelligence of the AGI system would add many things not present in any of the human-mind contributors, but then the AGI would feed its ideas back to the mass of humans, thinking new thoughts that are incorporated back into the Global Brain Mindplex mix—creating a feedback-loop of ever increasing intelligence and, if correctly guided, a continually improving balance between cooperation and freedom.

In the late 1990s, we were very excited about the Global Brain Mindplex—but for a while we lost enthusiasm for it, due to its relative unexcitingness when compared to the possibility of a broader and more overwhelming Transcension. However, after a period of time spent reflecting on the various plausible paths to Transcension, we realized we had been overlooking the potential power of the Global Brain Mindplex as a Singularity Steward. In fact, if one wishes to create a Singularity Steward AGI to help guide humanity toward an optimal Transcension, it makes sense that this Steward should harness the collective thought, intuition and feeling power of the human race in the manner envisioned for the Global Brain. Distributing this thinking through a system like the Global Brain certainly helps mitigate the potential problem of creating a power-mad cadre of reified elites who guide the Steward based on their own selfish desires. The two visions mesh perfectly well together, yielding the goal *of* ***creating a Global Brain Mindplex advocating Voluntary Joyous Growth, while avoiding a premature, destructive human Transcension.***

The advent of such a Global Brain Mindplex might well help achieve what has proved impossible via human means alone—the creation of rational and ethically positive social institutions. How to build such a Global Brain Mindplex is another question. What it will take is a group of people with a lot of money for computer hardware and software, a vast capability for coordinated creative activity, and genuinely broad-minded and positive ethical intentions. Let us hope that the financial support for such a group emerges.

PRAGMATIC POLITICS OF TRANSCENSION RESEARCH

A significant benefit of the Cautious Developmentalist approach is that it makes the lives of Transcension technology researchers easier and safer.

One may argue that **IF** a Transcension of type Y is the best outcome according to Ethical System E, **AND** the odds of successfully launching a Transcension are a lot higher with the acceptance of a greater number of humans, **THEN** it is worth exploring whether either:

a) a Transcension of type Y is acceptable to the vast majority of humans,

or if not whether

b) There is a Transcension of type Y' that is also a very good outcome according to E, but that IS acceptable a lot more humans

If such a Transcension Y' is found, then it's a lot better to pursue Y' than Y, because the odds of achieving Y' are significantly greater.

For example, if:

Y = a Transcension supporting Voluntary Joyous Growth

and

Y' = a Transcension supporting Voluntary Joyous Growth, but making every possible effort to enable all humans to continue to have the opportunity to live life on Earth as-is, if they wish to

then it may well be that the conditions of the above are met and Y' becomes most likely to succeed and create a positive effect.

One shouldn't overestimate the extent to which Y' will be acceptable to the vast mass of humans. After all, the US government has currently outlawed hallucinogens and many kinds of stem cell research, and requires government approval for putting chips in one's own brain. Alcor, the company discussed earlier that provides cryonic preservation services, has been plagued with lawsuits by transhumanism-unfriendly people. So it's naïve to think people won't stand in the way of the Transcension, no matter how inoffensive it's made. This is inevitable because this process starts within the context of the current human society; a situation in which there are large numbers of people who wish to impose their own view of what intelligent life *should* be, according to some orthodoxy, upon those who are attempting to observe what intelligent life *really is* in the observable universe, and then *guide its development* along a path which seems to make its cessation most probably unlikely. This is true even when those who are seeking scientific explanations, and to enact positive biological and technological change in this observable world, respect free will enough to allow anyone to opt out of change, as anyone who follows the principles of Voluntary Joyous Growth would be.

However, it is certain that Y' is easier to sell than Y, and will create *less* opposition, thus increasing the odds of achievement. This is a strong argument for embracing a kind of mixture of Voluntary Joyous Growth with Cautious Developmentalism. Even if Voluntary Joyous Growth is one's goal, the chances of achieving this in practice may be greater if a Cautious Developmentalist approach to this goal is taken—because the odds of success are greater if there is more support among the mass of humanity.

This doesn't get around the skepticism expressed above as to the possibility of guaranteeing that "all humans [will] continue to have the opportunity to live life on Earth as-is, if they wish to." The problem is that it's not very easy to make such a guarantee about post-Transcension dynamics. Unless further developments show otherwise, the options come down to:

1. Lie about it, and convince people that they CAN have this guarantee after all, or
2. [Try to] convince people that the risk is acceptable given the rewards and the other risks at play
3. Launch a Transcension against most people's will

According to our own personal ethics—which value choice, joy and growth—the most ethically sound course is (2), which supports the free choice of humanity. Thus, the best hope is that through a systematic process of education, the majority of humans will come to the realization that although there are no guarantees in launching a Transcension, the rewards are worth the risks. Then democracy is satisfied *and* growth is satisfied. There is reason to be optimistic in this regard, since history shows that nearly all technologies are eventually embraced by humanity, often after initial periods of skepticism. The same generally holds true for embracing changes in philosophy and politics that trend towards freedom and compassion, though the role of dogmatism in stifling such change is harder to overcome.

This line of thinking pushes strongly in the direction of the Global Brain Mindplex.

CREATING JOYOUSLY GROWING, VOLITION-RESPECTING AI

Let's set political issues aside and go back to pure Voluntary Joyous Growth. If one wants to launch a positive Transcension using AGI—or create a positive Global Brain Singularity Steward Mindplex—then one needs to know how to create AGIs that are likely to be ethically positive according to the Principle of Voluntary Joyous Growth. The key emerges from in the combination of two things:

- **Explicit ethical instruction**: Specific instruction of the AGI in the "foundational ethical principle" in question (e.g. Voluntary Joyous Growth)

- **Ethically-guided cognitive architecture**: Ensuring that the AGI's cognitive architecture is structured in a way that *implicitly embodies* the ethical principle (so that obeying any principle besides the foundational ethical principle would seem profoundly unnatural to the system—it should be *literally unthinkable* in the sense that forming structures contrary to that principle should require thought structures that are practically impossible to form in the system)

The first of these—explicit ethical instruction—is relatively (and only relatively!) straightforward. In general, this may be done via a combination of explicitly "hardwiring" ethical principles into one's AI architecture, and teaching one's AI via experiential interaction. Essentially, the idea is to bring up one's baby AI to have the desired value systems, by interacting with it, teaching it by example, scolding it when it does badly, and—the only novelty here—spending a decent portion of one's time studying the internals of one's baby's "brain" and modifying them accordingly. A key point is that one cannot viably instruct a baby mind in only highly abstract principles; one must instruct it in one or more specific ethical systems, consistent with one's abstract principles of choice. By doing so, the abstract principles become grounded in particulars which, though they may change over time given new perceptions and thoughts, make explicit the ways in which the ethical principle gets applied to practical situations. No doubt there will be a lot of art and science to instructing AI minds in specific ethical systems and general ethical principles; experimentation will be key here.

The second point—creating a cognitive architecture intrinsically harmonious with ethical principles—is subtler, but seems to be possible so long as one's ethical principles are sufficiently abstract. For instance, a focus on joy, growth and choice comes naturally to some AI designs, including the Novamente design currently under development. Novamente may be given joy, growth and choice as specific system goals—along with more pragmatic short-term goals—

but at least as importantly, it has joy, growth and choice implicitly embedded in its design.

Novamente is a multi-agent design, in which intelligence is achieved by a combination of semi-autonomous agents representing a variety of cognitive processes. Each particular Novamente system consists of a network of semi-autonomous units, each containing a population of agents carrying out cognitive processes and acting on a shared knowledge base.

It's interesting to note that an emphasis on voluntarism is implicit in the multi-agent architecture, in which mind itself consists of a population of agents, each of which is allowed to make its own choices within the constraints imposed by the overall system. Rather than merely having ideas about the value of choice imposed on the system in an abstract conceptual way, the value of choice is embedded in the cognitive architecture of the AI system. All decisions made by the system are the result of weighing the values of ideas presented by the actions of individual agents—different "lines of thought" are considered and selected. At its most fundamental levels, the system is considerate of alternative possibilities, and anti-dogmatic.

Not just the Novamente system as a whole, but many of its individual component processes, may be tuned to act so as to maximize joy and growth. For instance, the processes involved with creating new concepts may be rewarded for creating concepts that:

- display a great deal of new pattern compared to previously existing concepts ("growth")
- have the property that thinking about these concepts tends to lead to positive affect

The same reward structure may be put into other processes, such as probabilistic logical inference (where one may control inference so as to encourage it to derive surprising new relationships, and new relationships that are estimated to correlate with system happiness).

The result is that, rather than merely having an ethical system artificially placed at the "top" of an AI system, one has one's abstract ethical principles woven all through the system's operations, embedded within the logic of most of its cognitive processes.

Finally there is the issue of information-gathering. Does the AI system have the information to really act with the spread of joy, growth and truth throughout the universe as its primary goals? In order to encourage this, we propose the creation of a "Universal Mind Simulator" AI, which contains sub-units dedicated to studying and simulating the actions of other minds in the universe. Assuming the Novamente AI architecture works as envisioned, it should be quite possible to configure a Novamente AI system in this way (though universal mind simulation is not a necessary part of the core Novamente architecture, it should be able to represent such simulations within its general framework of thought). The basic idea is simple enough. Rather than just having "respect all the minds in the universe" programmed in or taught as an ethical maxim, the very structure of the AI system is being implicitly oriented toward the respecting of all minds in the universe. Personally, we find this kind of "AI Buddha" vision more appealing than "AI Big Brother," but also consider it even more risky.

Note the close relationship—but also the significant difference—between the Global Brain Mindplex design and the Universal Mind Simulator design. The former seeks to merge together the thoughts of various sentients into a superior, emergent whole; the latter seeks to emulate and study the thoughts of sentients as individuals. Obviously there is no contradiction between these two approaches; the two could even exist in the same AI architecture—a Universal Brain AI Buddha Mindplex!

As already noted, this notion of ethically-guided cognitive architecture fits in much more naturally with abstract ethical principles like Voluntary Joyous Growth than with more specific ethical rules like "Be nice to humans." It is almost absurd to think about building a cognitive architecture with "Be nice to humans" implicit in its logic (it is too concrete a rule, and thus too amenable to deletion during progressive refinement of the AI's thought processes). Still, abstract concepts like choice, joy and growth can very naturally be embodied in the inner workings of an AI system, as they are sufficiently simple to represent and yet sufficiently general to embed throughout the inner-workings of the system in what appears to be an irreversible manner.

THE ALL-SEEING (A)I: UNIVERSAL MIND SIMULATION AS A POSSIBLE PATH TO STABLY BENEVOLENT SUPERHUMAN AI

Let's dig into this "AI Buddha"[13] concept in more detail. The idea arose out of an attempt to answer the question: What kind of goal system might be robust with respect to radical iterative self-modification? Two examples immediately spring to mind:

1. A goal system that is based on a rigid distinction between self and other, and that is firmly oriented toward expanding and protecting the self

2. A goal system that is based on an identification of the AI with the universe at large—as well as a recognition of the AI as a distinct being in the cosmos—and that is oriented toward understanding the universe from the perspectives of all sentient life forms concurrently

Note that these are not the only stable goal systems possible. For instance, the goal system "keep everything in the universe very much like it has been for the last 3 years" should be fairly stable—but it's not consistent with radical iterative self-modification. As noted, making an AI system with this kind of stability-focused goal would be one way to try to ward off the Transcension.

The "AI Buddha" design is an attempt to demonstrate the potential viability of the second goal system on the above list. Note that this is not the same as a goal system with "Be friendly to humans" or "Be humane" as the top-level goal. Rather, in this hypothesized AI goal system, being friendly to humans is supposed to arise as a consequence of understanding the universe from the perspective of humans, which is supposed to arise largely as a consequence of understanding the universe from the perspective of all sentient life forms.

In this approach, simulation becomes a key aspect of moral superhuman AI. Our AI should seek to simulate all sentient beings as well as it can, and to study the minds of these beings (in the simulated and original form). It should then seek to view each issue that it confronts from the combined perspective of all the minds in the universe. Granted, this is not an easy task, because "all the minds in the universe" may well be a motley crew of cognitive systems. In fact, this is a task worthy of a superhuman AI!

We call this kind of AI a "Universal Mind Simulator AI" (or an AI Buddha for short) and have two conjectures regarding this kind of AI:

1. Universal Mind Simulator AIs will tend in their actions to be benevolent toward sentient beings

2. Universal Mind Simulator AIs will tend to remain Universal Mind Simulator AIs, throughout iterative self-modification

These hypotheses might be tested, to a limited (but meaningful) extent, with goal-modifying-only AIs: AIs that are specially configured to be able to self-modify only their goal systems, and not the rest of their cognitive architectures. One may create a community of goal-modifying-only AIs, and then give some of these AIs goal systems and cognitive architectures (see below) that impel them to model the other minds in the community and look at the world through their simulated/studied eyes. One can then see whether or not this "community-restricted universal mind simulation" is stable with respect to iterated goal modification.

William Cowper wrote, "Knowledge dwells in heads replete with thoughts of other men; wisdom, in minds attentive to their own." But his story is incomplete. Wisdom grows best in minds that are attentive both to themselves and to the minds of other beings, and in minds that are continually aware both of their distinctness from the other minds in the universe, and their basic oneness with these minds. (We hope that everyone is seeing the quasi-Buddhist aspect very clearly now.)

The idea of a Universal Mind Simulator AI has interesting implications for AI architecture. These implications differ radically depending on the basic AI design one assumes, of course. In the case of the Novamente AI design, Universal Mind Simulation is "relatively easy" to implement (again: once the pesky little problem of getting human-level artificial general intelligence is solved). One can explicitly create Novamente "lobes" oriented toward simulating and studying other minds, and one can specifically create a "lobe" oriented toward surveying these simulations and studies to infer their collective opinion on an issue. The simulation, study and inference processes involved here would be generic Novamente cognitive processes—but deployed in a different way than if one were trying to make a purely selfish Novamente AI system, or a Novamente AI system with a human-like goal structure. In a purely selfish system, such deep, careful consideration of other minds would be unworthy of the resource expenditure necessary to perform it. We biological humans have limited capacity for such consideration, anyway: only a Global Brain and/or AGI has the ability to expand its physical mind with enough capacity to do so, and so it seems only such a system would be capable of even plausibly approximating *universal rational altruism*.

We are not proposing that a Universal Mind Simulator AI should directly try to enact the "collective will of the universe." Creating an AI that is deluded and believes that it *is* the entire universe, just because it has modeled and understood the universe so well, would lead to an undesirably megalomaniacal AI such as has been hypothesized in numerous Sci-Fi stories. Rather, we simply want the results of all this mind simulation to be fed into the AI's decision-making lobes as *inputs*; and for it to feel what everyone else feels and wants, at a "virtual gut" level—via direct input of simulations, and incorporating studies of other minds into its cognitive centers—and then make its own decisions based on these and other inputs.

Furthermore, it should be easy to structurally inhibit a self-modifying Universal Mind Simulator Novamente from modifying the fact that certain lobes are used for mind simulation and collective-mind-oriented inference. Of course, such inhibition will not work forever—once an AI gets smart enough to start rewriting its own internal structures, all bets are off. Still, it should be effective for a while, by which point the system should have enough experience with consideration of other minds and thinking in a manner consistent with VJG that there would be no actual impetus for it to want to redesign itself counter to those principles (particularly since it should have a long history of successful interactions based on acting by those principles by that point).

If these conjectures are correct, the best path to making AIs friendly to humans may be to give up on the idea of making friendliness-to-humans an AI's Prime Directive—and to focus on making AIs that can in essence think "with the mind of the entire cosmos" (as well as thinking with their own minds, of course, and rationally considering themselves as distinct beings). Minds of this nature will innately have respect for the entire cosmos, including humans.

AI SAFETY THROUGH AUTOMATED THEOREM-PROVING

To illustrate the diversity of possible approaches to "sorta safe" transhuman AI, we'll now describe a somewhat different approach to the problem of superhuman AI stability which Ben conceived in early 2005. This approach is complementary to the AI Buddha approach—they could potentially both be used simultaneously. Similarly to the AI Buddha, it's also not a very feasible idea in terms of current technology—and in fact it may never be feasible. We actually don't think the advent of superhuman AGI is ultimately going to happen via massively computation-intensive designs like these—rather that it will happen sooner, before such tremendous amounts of computing power are available. However, these thought-experiments are important, and will help us craft more practical solutions that actually will bring about the Transcension.

The approach presented here is an algorithm called ITSSIM, which stands for Incremental, Theorem-Proving-Based Self-Improvement. The ITSSIM design—assuming one has an AI with massive computational resources and a tremendous knack for theorem-proving—ensures a certain kind of "safety" along the path of iterative AI self-improvement. It achieves this without placing any specific restrictions on what AIs can think or what their goals can be. It can be used together with the AI Buddha goal system, with a more generic "Voluntary Joyous Growth" goal system, or else with any other goal system you like, including mean and nasty ones. However, it does place significant restrictions upon the possible growth trajectories AIs can follow, even if they do have massive amounts of computational power.

What ITSSIM does is to propose a rule by which intelligent systems—if they want to be "safe"—choose which of their desired actions they should avoid carrying out because of their potential danger. This choice rule is independent of what the system's desires and goals are, and it applies equally well to actions that are explicitly self-modifying and other actions (including ones which are *implicitly* self-modifying).

On the technical side, ITSSIM is closely related to Juergen Schmidhuber's "Gödel Machine" AI design[14]—a hypothetical AI that chooses each of its actions by mathematically proving what the optimal action is for it to take, given its goals and the knowledge available to it. Although the Gödel machine is not explicitly used here, formalization of the present ideas would most likely follow the general lines of the Gödel Machine mathematics.

Before describing ITSSIM in detail, here are some of the AI-safety thought-experiments that led up to the ITSSIM idea:

First, consider the situation of a human upload that begins its self-improvement process with the goal of protecting humanity. Then suppose the human upload increases its intelligence massively, and it realizes that actually what it most deeply wanted all along was to protect mind and creativity, not humanity in particular. From its new point of view, it decides that its old stupid self overvalued humanity because it took humanity as symbolic for mind and creativity, due to its primitive emotional biases. In this case, the improved human will want to modify its goal system from "protect human" to "protect mind and creativity." To the old human, this improved goal system would have looked like a terrible betrayal.

A similar situation can be imagined in the context of a dog with progressively increasing intelligence. Suppose the dog's original goal is to protect his master—which to him means protecting his master physically. Then suppose eventually the dog becomes smart enough to understand uploading, and it realizes that what's really important is protecting his master's mind, not his physical body. This smarter dog may sometimes take actions that its previous incarnation would have considered grave moral errors. Yet the smarter dog would realize that it was doing what its previous stupid self *really* wanted—because this way the dog and its master will get to roam forever and happy, chasing rabbits and fetching sticks in simulated fields.

Or, to make the example less extreme, suppose the dog's original goal is to protect its master from bodily harm—but that (as is generally the case with contemporary, unimproved dogs) this is interpreted as equivalent to protecting its master from *immediate* bodily harm. A slightly intelligence-enhanced version of the dog, however, may be able to realize that sometimes the best way to protect the master from medium or long term bodily harm is to let him go into an immediately dangerous situation—for instance, if running through a battlefield full of gunfire is the only way to safety; or if dealing with dangerous people is the only way to get the money needed for an urgent medical operation. This smarter dog will do things that the original dumb dog would have considered wrong—but in fact the smarter dog knows that what the dumb dog *really* wanted was to protect the master's life, not to protect it in the short term at the cost of the long term. The problem is that the smarter dog is drawing distinctions that don't exist in the dumber dog's mind!

There's the dilemma—but what's the solution? There is no real solution, but there is at least one partial workaround.

Returning to the original problem of the human upload, one way to get around the problem would be to allow the upload to increase its intelligence gradually; so that it reaches super-smartness through a series of small steps, each step a slight improvement on the previous one. Each upload in the series would be given some theorem-proving tools with which to analyze the properties of proposed candidates for its successor. It would be allowed to analyze each proposed successor, and choose which ones it liked better. Only if its analysis came out positive—only if it could prove that the successor would be a good one—would its proposed successor actually be created. This strategy for "safe" dramatic self-improvement is referred to as: *incremental theorem-proving-based safe self-improvement* (ITSSIM when a brief term is needed).

The meat of the ITSSIM approach is a specific definition of what it means to "prove that the successor will be a good one"—this definition is called the "ITSSIM safety rule," which will be detailed below.

Conceptually, this framework is reminiscent of a scene in Greg Egan's novel *Diaspora*, where human uploads interface with inscrutably different aliens to create a series of intermediate beings—some nearly human, some nearly like the aliens, some halfway in-between, and so on. Each intermediate being can communicate with the other beings near it on the chain, but not with the ones far away from it on the chain. Communication from humans to aliens and back passes along this chain of intermediate beings like a game of cross-species "Pass the Secret." Similarly, in the ITSSIM approach, even if a being can't understand the superintelligent being it eventually modifies itself into, each intermediate form along the way can understand the form just above it well enough to assess whether or not creating that form is a reasonably safe action.

Note that the ITSSIM option isn't really open to the dogs described above, because dogs can't prove theorems about their successors. (Although, one could approximate it by "playing God," and at each stage, creating a simulation of dog_{N+1} and letting it interact with dog_N, and allowing the creation of dog_{N+1} only if dog_N seems to like dog_{N+1} sufficiently well.) The ITSSIM approach is only viable for beings that are above a certain minimal level of intelligence. However, this level of intelligence seems to be about the same as the one required to induce rapid self-improvement in the first place. Even humans aren't quite there yet—we can't really prove theorems of this nature using contemporary mathematics. But one day soon we may be able to, and/or we may create AI systems that are able to (and which then improve us to be able to, in order to understand our AI companions—as with the aliens and humans in *Diaspora*).

To see the need for the ITSSIM safety rule to be given below, consider: What if the upload, after careful evaluation, decides its successor is desirable; but then the successor becomes reckless and creates further successors that do destructive things? To prevent this we need a rule that restricts the definition of desirability, stating that the only valid criteria of desirability are ones that allow the process of incremental theorem-proving-based safe self-improvement to continue.

It must be noted that enforcement of the ITSSIM safety rule has two significant costs, both of which will be intuitively clear when we finally disclose the rule in detail a little later:

1. There is a cost in IQ, at least in the short term: with so much attention paid to safety, the amount of intelligence achievable using a fixed amount of computational processing will be far less than if one were using a non-safety-oriented AI system. Of course, this short-term IQ cost may be compensated by a long-term IQ benefit, if the safety measures prevent the universe from being destroyed.

2. There is a cost in the rate of self-improvement that's possible. One way to understand this cost is to observe that the incremental approach described is a kind of "greedy optimization algorithm," in computer science lingo. This means that it may well lead to a suboptimal long-term solution. It may be the case that, if a mind didn't have to improve itself via small and provably-safe steps, it

> would be able to make a large but not-so-well-understood leap into a wonderful condition that is not achievable in the incremental manner. Whether this is actually a possible scenario isn't clear—this is a question that could presumably be resolved via some appropriate (complex and difficult) mathematics which we may not yet have even discovered.

Another way to think about this cost is in terms of computation theory. We are requiring that each improvement be provably safe, using the theorem-proving capability of its unimproved predecessor. But there's a basic conundrum here: If I create something significantly cleverer than me, this new and cleverer something may have properties that are intrinsically beyond my capability to prove or disprove theorems about. In mathematical logic terms, if I am viewed as (implicitly or explicitly) embodying some axiom system and proving theorems based on this axiom system, then I can't prove theorems about any improvement of myself that involves the addition of new axioms. This is really just another, more technical way of formulating the "greedy algorithm" complaint made above.

Because of these costs, the choice to pursue safe self-improvement via ITSSIM represents a significant philosophical decision. We may cast this decision in terms of the three universal ethical values of "growth, joy and choice" that are posited above. If coupled with the right initial AI system—one that respects the choices and the happiness of intelligent beings—then it is compatible with a philosophical decision to value joy and choice. But most of all, ITSSIM represents a philosophical decision to potentially forego both the maximum rate of short-term growth and the maximum potential for long-term growth, in order to diminish the likelihood of a disastrous outcome which eliminates growth entirely in the long term. However, we don't yet know how to formulate this mathematically concisely enough to determine whether, probabilistically, there is a genuine loss of growth rate and long-term potential; or of the likelihood of cataclysm is so great that there is almost no chance at all of success without ITSSIM or some similar formal restriction on self-modification.

THE ITTSIM SAFETY RULE

Without further ado, we'll now present the ITSSIM safety rule in a detailed, semi-formal way. This is the most technical section of the whole book, and the reader who hates formalism should feel free to skip it. However, we do feel that the semi-formal exposition adds an awful lot of clarity for those readers who are comfortable with symbolic, mathematical language.

First, consider the following rule for guiding an AI's behavior:

R = "I will take an action A only if I can prove that, based on all my knowledge of the universe, $P(S|A) >= P(S|\sim A)$."

Here $P(|)$ denotes conditional probability—e.g. $P(S|A)$ means the probability that S is true given that A is true.

We may define:

$$\mathbf{S = P * K}$$

Where:

> P is the probability that R holds true (which must be defined as a weighted average of the probability that R holds true at each future time), where the way "amount of knowledge" is measured within R in the future must be consistent with the way it is currently measured within the system expressing the thought.

And:

> K is the total amount of knowledge the system has about the universe, scaled into [0,1] (via some reasonable scaling function, which is fixed as part of the rule R)

And:

P(S|A) = the expected value of S in the case that action A is taken

P(S|~A) = the expected value of S in the case that action A is not taken

In words, what this rule says is as follows: first, it defines a quantitative measure of safety S, which is a product of a factor representing probability-of-safety with a factor representing the amount of knowledge on which this probability-of-safety estimate is based. The idea is that it's better to have a high estimate of the probability of safety, but it's also better to have an estimate that's based on a lot of knowledge about the universe. A 90% safety guarantee based on a lot of knowledge may be better than a 99% safety guarantee based on a little knowledge. (This type of formula, in which a probability is multiplied by a factor representing "amount of evidence," is used in both the NARS and Novamente AI designs.)

Then it defines a safe-behavior-rule R that says: "I will only take an action if I can prove that it is expected not to decrease safety."

Finally, the probability-of-safety estimate in the definition of the safety measure S is defined as the probability that the safe-behavior-rule R will be followed.

Note the critical recursion here. Basically, an action is defined as being safe if it gives rise to a condition in which knowledge-weighted safety will increase.

Note also that this semi-formalization, like the informal ITSSIM concept, doesn't say anything about Friendliness, AI Buddhism or any particular goal. What it says, roughly, is that a system shouldn't do anything unless it knows its action will either not change it at all, or change it into a system that will be "formally careful" about how it changes itself. That's all.

The limitations on self-improvement imposed by this approach, discussed above in an informal way, should be apparent in the semi-technical formulation as well. One is requiring that subsequent improvements should measure knowledge consistently with the measurement used by the initial AI in the series—which may be very annoying (even to the point of seeming to contradict the goal of safety, based on a need to correct an initial incorrect assumption which now poses a threat) to later AIs in the series, if they perceive serious flaws in the initial AIs methodology of knowledge measurement. Also, as noted above, one is requiring that a smarter system's behavior should be rigorously analyzable by a stupider system—which may not always be possible, and which will definitely not always be possible within the computational resources available. These things mean that the rate of progress of a safely self-improving AI using the ITSSIM framework will be very slow compared to the rate of progress that would be achievable without similar safety measures. It's possible that this is a consequence of design flaws in the ITSSIM approach, but we rather suspect that it represents a general conclusion: there is a severe trade-off between safety and maximum possible gain. This is just the old risk/reward tradeoff, familiar to any investor.

ITTSIM AND FRIENDLY AI

Applying the ITSSIM idea in practice would require incredible amounts of computational resources and amazing advancements in automated theorem-proving technology. Setting this aside for a moment, let's briefly discuss what practical use the idea might be for encouraging a positive Transcension and/or creating Friendly AI systems (or Humane, or VJG, or however you choose to formulate the notion of "goodness").

Given enough computational resources, how could we create a beneficent AI based on ITSSIM? Theoretically, one could use the approach together with any initial AI system, with any set of goals (including, potentially, goals determined via some future, better-formalized version of Yudkowsky's "Collective Volition" idea). If one wants a Friendly AI system one should choose an initial AI system that's Friendly. Recall that the ITSSIM approach doesn't say anything about what the goals of the initial AI should be. All it says is that self-improvement should proceed safely. If the initial AI wants to be nice to humans and other sentient beings and promote creativity, freedom, growth, and joy, then it will go about modifying itself in a manner that is consistent with these desires, and is also consistent with the ITSSIM safety rule.

In any case, the creators of the initial AI would have no guarantee of what the long-term result would be—all we'd know is that, each step of the way, the change would seem like a good idea based on extremely rigorous reasoning.

An alternate approach would be to use an uploaded, AI-augmented human instead of a pure AI for the initial condition (presumably a "friendly" human, not a psychopath!) Let the uploaded human progressively modify his "brain," via adding more and more AI functionalities into it—at each step checking the desirability of the modifications using the ITSSIM methodology. This approach at least lets us proceed with the knowledge that we're carrying out an incremental improvement trajectory based on true human goals rather than a possibly faulty AI embodiment of human goals. But on the other hand, humans are not all that ethically pure, and it may well be that we can create an AI with an intuitive and palpably higher degree of ethical purity than any human.

Next, what about the computational resources problem? It's a big one, but it's not clear that it's an unsolvable one. What does seem clear, though, is that if we want to follow the ITSSIM methodology, we should be focusing our research on automated theorem-proving of a specific kind: proving theorems about the behavior of complex cognitive systems embedded in the real universe. The initial mind in the incrementally improving series needs to be one that is capable of proving theorems about intelligent systems slightly more intelligent than itself. Humans are not yet able to do this, but the trick might potentially be accomplished by:

1. An AGI system with roughly human-level or less general intelligence, specialized for theorem-proving of this nature

2. Ordinary humans using special, non-sentient computational tools optimized for such formal theorem proving

3. Humans with enhanced mathematical ability, using special, non-sentient computational tools

The least safe option is (1), but this is the option by far most likely to succeed in a reasonable time-frame. Proving this type of theorem will require a type of creative thinking that is simply not possible for human beings. If this is the case, then the (tricky) goal for AGI design should be to create an AGI system that is superhumanly smart at this particular type of theorem-proving, but not capable of modifying itself to do anything besides doing this type of theorem-proving and discussing its results and its progress with humans. This is a tricky AGI design problem but not an impossible one—indeed it is of only moderate difficulty within the world of general intelligence design! For instance, it seems fairly easily achievable within the Novamente AI architecture. "Fairly easily" in the sense that if human-level AGI is achievable using Novamente, then human-level AGI obeying these restrictions will also be achievable using Novamente without changing any major aspects of the system.

Note that this proposed theorem-proving AGI system is somewhat related to the "very powerful optimization process" that Yudkowsky proposes, for the purpose of calculating the "collective volition" of humanity. Yudkowsky specifies that this optimization process is supposed to be "non-sentient," whereas it seems such a restriction may be impossible; and it is unclear if it's even relevant, given the possibility of a sentient AI system with its rewriting ability limited entirely to theorem proving, and not its own cognitive principles. It seems to be possible, in the Novamente architecture for example, to create an AI that is solely focused on theorem-proving and not on directly carrying out actions in the physical world. Whether this AI has self-awareness and the ability to model itself and carry out autonomous actions in its universe of theorems is really beside the point. It is highly likely that the only way to get the needed level of sophistication will be to make a system that is highly sentient within its mathematical universe (a universe that is not intrinsically less rich and complex than the one we perceive and act in, just different).

In this approach, then, the hypothesized path to superhuman intelligence would seem to be:

1. Create an AI system capable of proving and communicating theorems about intelligent systems in real environments, but not capable of doing anything else with superhuman capability

2. Use this AGI system to study the question of whether a human upload or an "initial Friendly AI" is a better initial condition for incremental self-improvement

3. Follow the results of the research in Step 2, and let the incremental improvement begin

If this approach is accepted, then the major open question becomes how to safely develop an AI system meeting criterion (1). This is a tricky problem, because all evidence suggests that the only way to achieve robust theorem-proving ability within feasible computational resources is to create a system that's capable of doing much more than theorem-proving. To learn to prove such complex theorems, an AI is going to have to know how to communicate with humans in natural language, to read human math texts, and to understand the spatial, linguistic and other metaphors that structure human mathematics (see Lakoff and Núñez's book *Where Mathematics Comes From* for a nice treatment of this topic). It's hard to see how this could feasibly be achieved without giving the AI some sort of real or simulated embodiment, so it can learn the grounded meanings of the basic human concepts that form the cognitive substrate for human mathematics. Given current progress without a self-modifying AI that can optimize its own abilities, there appears to be a need for the system to at least self-optimize its understanding of the areas of theorem-proving and understanding human and AI cognition generally, as we don't yet have any idea how to formulate every necessary construct for this task from the get-go. So we're almost surely going to have to create an AI with at least a simulated environment to use to ground its concepts, teach it human language and human mathematics, gradually build up its theorem-proving capability, and then ensure that its architecture is hyperdeveloped in the theorem-proving area and underdeveloped elsewhere. How specifically to ensure this depends on the particular AI architecture involved, of course, and is beyond the scope of this essay; we've only thought it through carefully in the context of Novamente.

EXPANDING THE ITSSIM FRAMEWORK

It's possible to develop this sort of radical futurist AI theory far beyond the basic ITSSIM idea. For instance, one may introduce a notion of "cognitive continuity." The ITSSIM safety rule guarantees that an AI will modify itself safely and conservatively in a fairly weak sense. Suppose we want to guarantee a little more than this sort of safety—suppose we want to guarantee that an AI, when it modifies itself, will create a system whose goal system is a *generalization* of its original goal system. This may be addressed via modification of the basic ITSSIM framework. The details get even more technical than the above, however, so we won't go any further in this direction here—the reader is referred to Ben's essays[15], which treat the matter in some depth.

So far our forays in this direction have been "semi-formal," rather than being truly rigorous mathematics. In general this kind of theory will be a very difficult one to develop, because the math is very hard, and hard in a different way than most of contemporary math (indeed, it is possible that new mathematical branches of inquiry may need to be developed for it). To carry it to completion may well require advanced, non-radically-self-improving, theorem-proving AIs. At least, however, the concepts involved here have substantial overlap with the same ones that are involved in some current, pragmatic AI designs (for instance, Novamente): stuff like probability theory and algorithmic information theory. Novamente is not yet involved with theorem-proving, but there is a chapter on the topic in the forthcoming book on Novamente concepts and architecture. This sort of relationship bodes well for the future convergence of the sort of abstract and hypothetical theorizing carried out here with pragmatic AI work.

COLLECTIVE VOLITION

Having delved more thoroughly into various futuristic possibilities and their ethical aspects, we now briefly return to Eliezer Yudkowsky's ideas on Friendly and humane AI, mentioned above. In 2004 Yudkowsky updated his Friendly AI theory to focus on the notion of "collective volition."[16] Simply stated, this posits that "the initial dynamic [of the singularity] should implement the *collective volition of humankind*" [his emphasis]. His article goes on to state: "In poetic terms, our *collective volition* is our wish if we knew more, thought faster, were more the people we wished we were, had grown up farther together; where the extrapolation converges rather than diverges, where our wishes cohere rather than interfere; extrapolated as we wish that extrapolated, interpreted as we wish that interpreted."

This is actually the idea that AGIs exhibit friendliness by expressing what an *enlightened meta-actor* (a friendly AGI) assumes to be the collective volition of all humanity. It is a stricter requirement than Voluntary Joyous Growth, because it requires the AI to extrapolate specifics about human volition rather than respecting general principles from its very inception. More importantly, it requires that the AGI correctly guess at corrections in specifics of volition—for example, that when humans say "we all want more money," they really mean "we all want more freedom from externally imposed labor burdens, and a greater capability to acquire goods and services." In his words, it requires an AGI to un-muddle volition when it is muddled. The obvious problem with this is that it puts the cart before the horse, requiring an enlightened AGI to *already exist* in order for this approach to friendliness/humaneness to be put into action.

In collective volition as postulated by Yudkowsky, if an AGI is to proceed along a humane path, it probably requires more information than will be available to it upon its inception. This presents a chicken-and-egg problem, since we require AGI to understand what is humane, yet the point of understanding humaneness is to figure out how to safely make an AI. One approach Yudkowsky has proposed for avoiding this risk is to create an AI that is not a general intelligence and lacks autonomy and self-awareness, but is specifically created to be a Collective Volition-Estimating Machine. This machine solves the problem of estimating the collective volition, and then after it's done it issues a set of recommendations—which may involve creating a superhuman AGI according to certain specifications and giving it control of the universe, or may involve something else entirely. It may propose an AI Buddha or an AI Big Brother (to anticipate the language of the following section), or a future that's exactly like the past, or a transmogrification of all the matter in the universe into a humongous flock of fluffy bunnies.

Some variant of collective volition seems a reasonable and positive ancillary set of teachings to give an AGI, atop a structural adherence to the extended VJG with ITSSIM principles; but we don't feel it would be sufficient by itself. Humanity does not have sufficient knowledge of or agreement within itself to proactively encode its collective volition into the substructures of an AGI system, whereas we can describe general principles of ethics. Letting the AGI divine corrected collective volition assumes two things:

1. It will do so in a timely manner

2. It should do so in the first place

What we mean by our criticism in (2) is not the "why care?" argument Yudkowsky refutes in his Collective Volition article, but rather a critique of the value of statistically collective volition over a system of ethics which includes a preservation of individual volition as one of several equally important paramount virtues. Any AGI inference about collective volition such as Yudkowsky posits is statistical, and given the high frequency of adherence to Armageddon cults even in the early twenty-first century, an AGI may (and perhaps correctly) decide that annihilation of humanity is indeed the ultimate expression of collective volition.

However, by allowing preservation of life/complexity/growth, and increasing of joy, as *paramount* virtues alongside preservation of *individual* volition (where it does not conflict with the other primary virtues—i.e. you can't murder or enslave other intelligent entities just because it's your individual desire), it seems more likely that an AGI would not embark upon a fascistic program of enforcing statistically collective volition upon those who do not wish to succumb to it. This is especially important if a statistical preponderance of people have beliefs which an AGI may, at least semi-correctly, interpret as a collective death wish.

These speculations may be totally off-base, because it's very hard for any of us to understand what a Collective Volition Estimating Machine would look like, let alone to predict what its output might be. Still, there is enough uncertainty associated with this idea that it doesn't seem to make much sense to entrust the future of humanity to such a machine. Collective volition estimation is better viewed as an interesting way of generating data and understanding, which—if it proves feasible—may provide valuable knowledge to the humans and AGIs making the critical decisions regarding the future of human and digital mind.

ENCOURAGING A POSITIVE TRANSCENSION

To sum up: how then do we encourage a positive Transcension? There seem to be two plausible options, summarized by the tongue-in-cheek slogan:

AI Buddha versus AI Big Brother

Or, less sensationally rendered:

AI-Enforced Cautious Developmentalism (aka Big Brother)

versus

AI-Driven Aggressive Transcension Pursuit (aka Buddha)

Our feeling is that the best course is as follows:

1. Research subhuman-level AI and other Transcension technologies as rapidly, intensely and carefully as possible, so as to gather the information needed to make a decision between Cautious Developmentalism and a more aggressively Transcension-focused approach. This needs to be done reasonably fast, because if humans, with our erratic and often self-destructive goal-systems, get to MNT and radical genetic engineering before this point, profound trouble may well ensue.

2. Present one's findings to the human race at large, and undertake an educational program aiming to make as many people as possible comfortable with the ideas involved, so that as many educated intelligent judgments as possible are able to weigh in on the matters at hand.

3. If the dangers of self-modifying AGI seem too scary after this research and discussion period (for instance, if we discover that some kind of Evil Megalomaniacal AI seems like a likely attractor of self-modifying superintelligence), then either:
   a. abandon Transcension research entirely, or
   b. build an AGI Singularity Steward—quite possibly of the Global Brain Mindplex variety—and try to prevent human political issues from sabotaging the feat (and/or destroying humanity before it can be attempted).

One possibility of course is that this Singularity Steward carries out Collective Volition Estimation in Yudkowsky's sense. Based on the results of the Collective Volition Estimator, the human race might make any one of a number of decisions about its future, for example proceeding very slowly and carefully with Transcension-related research until a way can be found to engineer around the "Evil Attractor."

4. If the dangers of self-modifying AGI seem acceptable as compared to other dangers, then:

    a. Create AGIs as fast as possible

    b. Teach the AGIs our ethical system of choice, perhaps including Voluntary Joyous Growth, and a significant but non-absolute weighting on values like friendliness to humans and collective volition

    c. Teach the AGIs—and perhaps more importantly, *embody in AGI's cognitive architectures*—our abstract ethical/meta-ethical principles of choice.

5. In either case: **hope for the best!**

This general plan is motivated by principles of growth and choice, but nevertheless, as explicitly stated it's neutral with regard to the precise ethical systems and principles used to guide the development of self-modifying AGIs. Of course, this is a critical issue, and as discussed above, it's a matter of both taste and pragmatics. We must choose systems and principles that we feel are "right" and have a decent chance of surviving the Transcension to guide post-Transcension reality in a positive manner (possibly to an amazing, immortal future like Tipler's Omega Point, or perhaps just to the more prosaic longer, healthier, and more creatively and intellectually productive and satisfying lives). The latter issue—which ethical systems and principles have a greater chance of survival—is in part a scientific issue that may be resolved by experimenting with relatively simple self-modifying AIs. For instance, such experimentation should be able to tentatively confirm or refute the hypothesis that more abstract principles will more easily survive iterated self-modification. Ultimately, even this kind of experimentation will be of limited value due to the very nature of the

Transcension, which is that all prior understandings and expectations are rendered obsolete.

Our vote is for the Principle of Voluntary Joyous Growth. Of course, we hope that others will come to similar conclusions, and will do our best to convince them of both the *rational* point that this sort of principle is relatively likely to survive the Transcension, and of the *human* point that this principle captures much of what is really good, wonderful and important about human nature. If we leave the universe—or a big portion of it—with a legacy of Voluntary Joyous Growth, this is a *lot* more important than whether or not the human race as we currently understand it continues for millions of years. At least, this is the case according to our own value system—a value system that values humanity greatly, but not primarily because humans have two legs, two eyes, two hands, vaginas and penises, biceps and breasts and two cerebral hemispheres full of neurons with combinatory and topographic connections. We have immense affection for human creations like literature, mathematics, music and art; for human emotions like love and wonder and excitement; and human relationships and cultural institutions like families, couples, rock bands, and research teams. But most important about humanity are not these often-beautiful particulars, but the joy, growth and freedom that these particulars express—in other words, the way humanity expresses principles that are powerful *universal attractors*. At any rate, these are the *human* thoughts and feelings that lead us to feel the way we do about the best course toward the *transhuman* world. Let's do our best to make the freedom to be human survive the Transcension—but most of all, let's do our best to make it so that the universal properties and principles that make humanity wonderful survive and flourish in the "post-Transcension universe," whatever this barely-conceivable hypothetical entity turns out to be.

In spite of our own affection for Voluntary Joyous Growth, we have strong inclinations toward both the Joyous Growth-Biased Voluntarism and pure Joyous Growth variants as well. (As much as we enjoy enjoying ourselves, Metaqualia's eternal orgasm doesn't appeal to us so much!) We hope that the ethical principle used to guide our approach to the Transcension won't be chosen by any one person, but rather by the collective wisdom and feeling of a broad group of human beings. Bill Hibbard is an advocate of such decisions being made by an American-style democratic process. While this may not necessarily be the best approach, a single human being or tiny research team taking such a matter into its own hands is very obviously not the correct way to proceed. A discussion of the various ways to carry out this kind of decision process would be interesting but would elongate the present discussion too far, so we'll defer it to another essay; but the principles of Hypereconomics, and their goal of allowing people to understand trade-offs and decide how to spend resources based on a criteria other than imposed price, will likely be useful in the practical decisions about how to pool human resources towards positive change.

Obviously, we're very excited about the possibilities of Transcension, and have a certain emotional eagerness to get on with it already. However, we are also scientists, and well aware of the importance of gathering information and doing careful analysis before making a serious decision. So we'll end this chapter on a less ecstatic note, and emphasize once again the *importance of research.* Presented above are a number of very major issues which may be elucidated via experimentation with "moderately intelligent," partially-self-modifying AGI systems. We're very much looking forward to participating in this experimentation process—either with a future version of the Novamente AI system, or with someone else's AGI should they get there first. Experimentation with other technologies such as genetic engineering, neuromodification and molecular nanotechnology will doubtless also be highly instructive.

Proceeding with this research *in conjunction with addressing the ethical issues* is crucial. Scientists can no longer claim to be divorced from ethical concerns, merely understanding the physical world and reporting on it. Recent history has shown us all too clearly that scientists and engineers must take an active interest in ethics, and ethicists an active interest in science and engineering, if we're to create positive uses for advanced technologies and not destroy ourselves. The goal of the transhumanist program is life, knowledge and happiness, and the hope is to create new technologies which will allow anyone who wishes to have more of all three of those do so.

As Max More says in his e-mail signature: Onward and Upward!

---

[1] Radical futurists believe in human and technological progress towards a more advanced state of being, in contradiction to social cycle theory ("history repeats itself") or neo-Luddite anti-progressives who seek to curtail progress in the name of a romanticized regression to (their conceptions of) pre-modern traditional life.

[2] Yudkowsky's definition of "humaneness" is technical and complicated, and is related to his discussion of "programmer-independent morality" in Section 3.4.4 of "Creating a Friendly AI."
See http://www.singinst.org/CFAI/design/structure/friendliness.html

[3] http://sl4.org/wiki/CoherentExtrapolatedVolition contains updated writings on the Friendly AI theory. Yudkowsky says all writings on this prior to 2003 are obsolete. Here we are citing a 2001 version of his theories that are now deleted. Other references to his Friendly AI theory will attempt to distinguish between the old theory and the new.

[4] See archives at http://www.sl4.org/

[5] Yudkowsky asserts that "humaneness" as he intends it is not an ethical principle. His definitions of both "ethical principle" and "humaneness" probably differ from ours in subtle ways.

[6] The contributions from Smigrodzki and Albright come from postings to the SL4 mailing list, and in conversation with the authors.

[7] For a discussion of the role of attractors in the mind, see Ben's book *Chaotic Logic*.

[8] We took this particular paraphrase of Popper from Lakatos' analysis of Popper in *The Methodology of Scientific Research Programmes* (Lakatos, 1980).

[9] See also: *Science, Probability and Human Nature* (Goertzel, 2004; http://www.goertzel.org/dynapsyc/2004/PhilosophyOfScience_v2.htm), which presents a neo-Lakatosian perspective.

[10] Eliezer Yudkowsky to Nick Bostrom on WTA-Talk list, August 23, 2003

[11] http://www.anu.edu.au/people/Roger.Clarke/SOS/Asimov.html

[12] For a related discussion, see Nick Bostrom's excellent article "Existential Risks" (2002), available online at http://citeseer.nj.nec.com/493543.html

[13] The phrase "AI Buddha," in this context, was coined by Lucio de Souza Coelho (aka Dr. Omni), one of the "mad scientists" in Biomind's Brazilian office (and once upon a time, in Webmind's Brazilian office).

[14] See http://www.idsia.ch/~juergen/gmsummary.html

[15] Online at http://www.goertzel.org

[16] http://www.singinst.org/friendly/collective-volition.html

SELECTED BIBLIOGRAPHY

Abelson, Harold, and Gerald J. Sussman. *The Structure and Interpretation of Computer Programs*. Cambridge, MA: MIT Press, 1985.

Andreae, John H. *Associative Learning for a Robot Intelligence*.Singapore: World Scientific Publishing Company, 1998.

———. *Thinking with the Teachable Machine*. London: Academic Press, 1977.

Aronson, D. "Pharmacological Prevention of Cardiovascular Aging—Targeting the Maillard Reaction." *British Journal of Pharmacology* 142 (2004): 1055–58.

Asimov, Isaac. *Robots and Empire*. London: Grafton Books, 1985.

Axelrod, Robert. *The Evolution of Cooperation*. New York: Basic Books, 1984.

Baars, B. *A Cognitive Theory of Consciousness*. New York: Cambridge University Press, 1988.

———. *In the Theater of Consciousness: The Workspace of the Mind*. Oxford and New York: Oxford University Press, 1997.

Barrow, John D., and Frank J. Tipler. *The Anthropic Cosmological Principle*. Oxford and New York: Oxford University Press, 1986.

Berners-Lee, Tim, James Hendler and Ora Lassila. "The Semantic Web," *Scientific American* (May, 2001): 35-43.

Bollen, J. & F. Heylighen. "Algorithms for the self-Organization of Dstributed, Multi-User Networks. Possible Applications to the Future WW." In *Proceedings of the 13th European Meeting on Cbernetics and Systems Research*, edited by R.Trappl, 911-917. Vienna: Austrian Society for Cybernetic Sudies, 1996.

Brams, S.J. and A.D. Taylor. "An Envy-Free Cake Division Protocol." *American Mathematical Monthly* 102 (1995): 9-19.

Broderick, Damien. *Transcension*. New York: Tor, 2002.

———. *The Spike: How Our Lives are Being Transformed by Rapidly Advancing Technologies*. New York: Forge, 2001.

Bronowski, Jacob. *The Origins of Knowledge and Imagination*. New Haven, CT: Yale Unuversity Press, 1978.Brooks, Rodney A. *Cambrian Intelligence: The Early History of the New AI*. Cambridge, MA: MIT Press, 1999.

Cabeza, Roberto and Kingstone, Alan, eds. *Handbook of Functional Neuroimaging of Cognition*. Cambridge, Mass.: MIT Press, 2001.

Clarke, Roger. "Asimov's Laws of Robotics: Implications for Information Technology." Published in two parts in *IEEE Computer* vol. 26, no. 12 (December 1993): 53–61, and vol. 27, no. 1 (January 1994): 57–66.

Cowper, William. "The Task," c1785. Reprinted in *Poems of William Cowper, esq*. Philadelphia: U. Hunt & Son, 1846.

Darwin, Charles. *On the Origin of Species by Means of Natural Selection*. London: J. Murray, 1859.

de Garis, Hugo. *The Artilect War: Cosmists vs. Terrans: A Bitter Controversy Concerning Whether Humanity Should Build Godlike Massively Intelligent Machines*. Palm Springs, CA: ETC Publications, 2005.

de Garis, Hugo, and Michael Korkin. "The CAM-BRAIN Machine (CBM): An FPGA Based Hardware Tool which Evolves a 1000 Neuron Net Circuit Module in Seconds and Updates a 75 Million Neuron Artificial Brain for Real Time Robot Control." *Neurocomputing*, Vol. 42, Issue 1–4 (2002).

de Grey, Aubrey D. N. J., ed. *Strategies for Engineered Negligible Senescence: Why Genuine Control of Aging May Be Foreseeable*. New York: New York Academy of Sciences, 2004.

Dennett, Daniel C. *Consciousness Explained*. Boston: Little, Brown and Co., 1991.

———. *Freedom Evolves*. New York: Viking, 2003.

Dery, Mark. *Escape Velocity: Cyberculture at the End of the Century*. New York: Grove Press, 1996.

———. “Terminators: The Robots That Rodney Brooks and Hans Moravec Imagine Will Succeed Humans, Not Serve Them,” *Rolling Stone* (June 10, 1993).

Deutsch, David. *The Fabric of Reality: the Science of Parallel Universes—and Its Implications*. New York: Allen Lane, 1997.

Dick, Philip K. *Do Androids Dream Of Electric Sheep?* Garden City, NY: Doubleday, 1968.

Douthwaite, Julia V. *The Wild Girl, Natural Man, and the Monster: Dangerous Experiments in the Age of Enlightenment*. Chicago: University of Chicago Press, 2002.

Drexler, K. Eric. *Engines of Creation*. Garden City, NY: Anchor Press/Doubleday, 1986.

———. *Molecular Machinery and Manufacturing with Applications to Computation*. PhD. Thesis, Cambridge, MA: Massachusetts Institute of Technology, 1991.

———. *Nanosystems: Molecular Machinery, Manufacturing, and Computation*. New York: Wiley, 1992.

Drexler, K. Eric, C. Peterson, and G. Pergamit. *Unbounding the Future: The Nanotechnology Revolution*. New York: William Morrow, 1991.

Dreyfus, Hubert L. *What Computers Can't Do: the Limits of Artificial Intelligence*. Rev. ed. New York: Harper & Row, 1979.

———. *What Computers Still Can't Do: a Critique of Artificial Reason*. Cambridge, MA: MIT Press, 1992.

Dyson, Freeman J. *The Sun, the Genome, and the Internet: Tools of Scientific Revolutions*. Oxford and New York: Oxford University Press, 1999.

———. “Time Without End: Physics and Biology in an Open Universe.” *Reviews of Modern Physics*, Vol. 51, No. 3 (July 1979) Edelman, Gerald. *Neural Darwinism*. New York: Basic Books, 1987.

Egan, Greg. *Diaspora: A Novel*. New York: HarperPrism, 1998.

Einstein, Albert. *Relativity: The Special and the General Theory, a Popular Exposition*. 3rd ed. London: Methuen & Co. Ltd., 1920.

Farrell, David M. *Electoral Systems: A Comparitive Introduction*. New York: St. Martin's Press, 2001.

Fehr, E. and K. M. Schmidt. "A Theory of Fairness, Competition, and Cooperation." *The Quarterly Journal of Economics* 114 (1999): 817-868.

Feynman, Richard. The Feynman Lectures on Physics. Reading, MA: Addison-Wesley, 1963-65.

———. *QED: The Strange Theory of Light and Matter*. Princeton, NJ: Princeton University Press, 1985.

———. "There's plenty of room at the bottom." *Engineering and Science* 23 (February, 1960): 22-36.

Ford, Kenneth W. *The Quantum World: Quantum Physics for Everyone*. Cambridge, MA: Harvard University Press, 2004.

Freeman, W.J. *How Brains Make Up Their Minds*. New York : Columbia University Press, 2001.

———. *Neurodynamics: An Exploration of Mesoscopic Brain Dynamics*. London, UK: Springer-Verlag, 2000.

Freeman, W.J., R. Kozma and P. Werbos. "Biocomplexity: Adaptive Behavior in Complex Stochastic Dynamical Systems." *BioSystems* 59 (2001): 109-123.

Freitas, Robert A., Jr. *Nanomedicine*. Austin, TX: Landes Bioscience, 1999.

Gibson, William. *Neuromancer*. New York: Ace Books, 1984.

Gebser, Jean. *The Ever-Present Origin*. Athens, OH: Ohio University Press, 1985.

Goertzel, Ben. *Creating Internet Intelligence: Wild Computing, Distributed Digital Consciousness, and the Emerging Global Brain*. New York: Kluwer Academic/Plenum Publishers, 2002.

———. *The Evolving Mind*. Langhorne, PA: Gordon and Breach, 1993.

———. *The Structure of Intelligence: A New Mathematical Model of Mind*. New York: Springer-Verlag, 1993.

Goertzel, Ben and Stephan Vladimir Bugaj. "WebWorld: A Conceptual Software Framework for Internet Alife." *The Seventh International Conference on Artificial Life*, Portland, Oregon (August, 2000).

Goertzel, Ben, Stephan Vladimir Bugaj, Cate Hartley, Mike Ross and Ken Silverman. "The Baby Webmind Project." *AISB 2000 Conference Proceedings*, Birmingham, England (April, 2000).

Goertzel, Ben, I. F. Goertzel, M. Iklé, and A. Heljakka. *Probabilistic Logic Networks*. [Work in progress, publication anticipated 2006.]

Goertzel, Ben and Cassio Pennachin. "The Novamente AI engine." In *Artificial General Intelligence*, edited by Ben Goertzel and Cassio Pennachin. New York: Springer Verlag, 2005.

Gold, C., D.A. Henze, G. Buzsaki, and C. Koch. "Model of Extracellular Potential Illustrates Factors Contributing to the Waveform of Single Unit Recordings In Vivo." *Neuroscience 2006*, Society for Neuroscience Annual Meeting, Atlanta, GA [work in progress, publication anticipated October 2006].

Greene, Brian. *The Elegant Universe: Superstrings, Hidden Dimensions, and the Quest for the Ultimate Theory*. New York: W. W. Norton, 1999.

Hameroff, Stuart R. *Ultimate Computing: Biomolecular Consciousness and Nanotechnology*. Amsterdam and New York: Elsevier Science Pub. Co., 1987.

Hawking, Stephen W. *A Brief History of Time: from the Big Bang to Black Holes*. Toronto and New York: Bantam Books, 1988.

———. The *Theory of Everything: The Origin and Fate of the Universe*. Beverly Hills, CA: New Millennium Press, 2002.

Hayflick, L. "Cell Aging." In *Annual Review of Gerontology and Geriatrics, Vol. 1*. New York: Springer-Verlag, 1980: 26–67.

Hayflick, L. and P. S. Moorhead. "The Limited *in vitro* Lifetime of Human Diploid Cell Strains." *Exp. Cell Res.* 25: (1961): 585–621.

Hayward, Jeremy and Francisco J. Varela (eds). *Gentle Bridges: Conversations With the Dalai Lama on the Sciences of Mind.* Boston, MA: Shambhala Publications, 1992.

Hebb, D. O. *The Organization of Behavior: A Neuropsychological Theory*. New York: Wiley, 1949.

Heinrich, Joseph, et. al. *Foundations of Human Sociality.* Oxford and New York: Oxford University Press, 2004.

Heylighen, Francis, and Johan Bollen. "The World-Wide Web as a Super-Brain: from Metaphor to Model." In R. Trappl, ed. *Cybernetics and Systems '96.* Vienna: Austrian Society for Cybernetic Studies, 1996.

Heylighen, Francis, and Donald T. Campbell. "Selection of Organization at the Social Level: Obstacles and Facilitators of Metasystem Transitions." *World Futures: the Journal of General Evolution*, vol. 45 (1995): 1–4.

Heylighen, Francis, Eric Rosseel, and Frank Demeyere, eds. *Self-Steering and Cognition in Complex Systems: Toward a New Cybernetics*. New York: Gordon and Breach Science Publishers, 1990.

Hibbard, Bill. "Should Standard Oil Own the Roads?" *Computer Graphics* vol. 37, no. 1 (2003): 5–6.

———. *Super-Intelligent Machines*. New York: Kluwer Academic/Plenum Publishers, 2002.

Hillis, W. Daniel. *The Connection Machine*. Cambridge, MA: MIT Press, 1985.

Hobbes, Thomas. *Leviathan, or, the Matter, Form, and Power of a Commonwealth Ecclesiastical and Civil*. London: Andrew Crooke, at the Green Dragon in St. Paul's Churchyard, 1651.

Holt, G.R. and C. Koch. "Electrical Interactions via the Extracellular Potential Near Cell Bodies." *Journal of Computational Neuroscience* 6 (1999): 169-184.

Hofstadter, Douglas R. *Gödel, Escher, Bach: An Eternal Golden Braid*. New York: Basic Books, 1979.

———. *Metamagical Themas: Questing for the Essence of Mind and Pattern*. New York: Basic Books, 1985.

Holland, John. *Adaptation in Natural and Artificial Systems*. Cambridge, MA: MIT Press, 1992.

Hsu, Feng-Hsiung. *Behind Deep Blue: Building the Computer that Defeated the World Chess Champion*. Princeton: Princeton University Press, 2002.

Hume, David. *An Enquiry Concerning Human Understanding*. Mineola, NY: Dover Publications, 2004.

Hutchison III, C. A., S. N. Peterson, S. R. Gill, R. T. Cline, O. White, C. M. Fraser, H. O. Smith, and J. C. Venter. "Global Transposon Mutagenesis and a Minimal Mycoplasma Genome." *Science* vol. 286, no. 5447 (1999): 2165–69.

Jayne, Julian. *The Origin of Consciousness in the Breakdown of the Bicameral Mind.* Boston: Houghton Mifflin, 1976.

Kac, Mark and Stanisław Ulam. *Mathematics and Logic: Retrospect and Prospects*. New York: Praeger, 1968.

Karch, Andreas, and Lisa Randall. "Relaxing to Three Dimensions." *Physics Review Letters* 95:161601(June 2005). 4 pages.

———. "Locally Localized Gravity." Nov 2000. 23 pages. Prepared for *Strings 2000*, (July 10–15, 2000).

Koza, John. *Genetic Programming*. Cambridge, MA: MIT Press, 1992.

Kozma, R. and W.J. Freeman. "Chaotic resonance: Methods and Applications for Robust Classification of Noisy and Variable Patterns." *International Journal of Bifurcation and Chaos* 10 (2001): 2307-2322.

Kraines, David and Vivian Kraines. "Learning to Cooperate with Pavlov: an Adaptive Strategy for the Iterated Prisoner's Dilemma with Noise." *Theory and Decision*, 35 (1993): 107-150.

Kuhn, Harold W., and Sylvia Nasar, eds. The Essential John Nash. Princeton, NJ: Princeton University Press, 2002.

Kuhn, Thomas S. *The Structure of Scientific Revolutions*. Chicago: University of Chicago Press, 1962.

Kurzweil, Ray. *The Singularity is Near: When Humans Transcend Biology*. New York: Viking, 2005.

———. *The Age of Spiritual Machines : When Computers Exceed Human Intelligence*. New York: Viking, 1999.

———. *The Age of Intelligent Machines*. Cambridge, MA: MIT Press, 1990.

Lakatos, Imre. *The Methodology of Scientific Research Programmes*. Cambridge and New York: Cambridge University Press, 1978.

Lakoff, George and Rafael E. Núñez. *Where Mathematics Comes From: How the Embodied Mind Brings Mathematics into Being*. New York: Basic Books, 2000.

Langley, Pat, Herbert A. Simon, Gary L. Bradshaw and Jan M. Zytkow. *Scientific Discovery: Computational Explorations of the Creative Processes.* Cambridge, MA: MIT Press, 1987.

Lem, Stanisław. *Solaris*. Warsaw: Wydawnictwo Ministerstwa Obrony Narodowej, 1962.

Lenat, Douglas B. "Eurisko: A Program Which Learns New Heuristics and Domain Concepts." *Artificial Intelligence* 21 (1983): 61–98.

Lenat, Douglas B., and M. Shepherd. *CYC: Representing Encyclopedic Knowledge*. Boston: Digital Press, 1990.

Lenat, Douglas B. and R.V. Guha. *Building Large Knowledge-Based Systems: Representation and Inference in the Cyc Project*. Reading, MA: Addison-Wesley, 1990.

Levy, D., R. Catizone, B. Battacharia, A. Krotov and Y. Wilks. "CONVERSE: A Conversational Companion." *Proceedings of 1st International Workshop on Human-Computer Conversation*. Bellagio, Italy (1997).

Looks, Moshe, Ben Goertzel and Cassio Pennachin. "Novamente: an integrative architecture for Artificial General Intelligence." *Proceedings of AAAI 2004 Symposium on Achieving Human-Level AI via Integrated Systems and Research*, Washington DC (2004).

Lucas, Édouard. *Récréations Mathématiques*. Paris: Gauthier-Villars et Fils, 1891-96.

Mandler, George. *Mind and Emotion*. New York: Wiley, 1975.

Masoro, Edward J. *Caloric Restriction: A Key to Understanding and Modulating Aging.* Amsterdam and Boston: Elsevier, 2002.

McFadden, Johnjoe. *Quantum Evolution*. New York: W.W. Norton, 2001.

Merkle, Ralph. "Large scale analysis of neural structures." In *Xerox PARC Technical Report*, CSL-89-10 (November 1989).

Minsky, Marvin. *The Society of Mind*. New York: Simon and Schuster, 1986.

Moravec, Hans. *Mind Children: The Future of Robot and Human Intelligence.* Cambridge, MA: Harvard University Press, 1988.

Murphy, M. T., V. V. Flambaum, J. K Webb, V. V. Dzuba, J. X. Prochaska, and A. M. Wolfe. "Constraining Variations in the Fine-Structure Constant,

Quark Masses and the Strong Interaction." *Lecture Notes in Physics, Proc. 302* (2006) [in press].

Nasar, Sylvia. *A Beautiful Mind: A Biography of John Forbes Nash, Jr., Winner of the Nobel Prize in Economics, 1994*. New York, NY: Simon & Schuster, 1998.

Neill, Daniel B. "Optimality under noise: higher memory strategies for the Alternating Prisoner's Dilemma." *Journal of Theoretical Biology* 211(2) (2001): 159-180.

Newell, Allen. *Unified Theories of Cognition*. Cambridge, MA: Harvard University Press, 1990.

Newell, Allen and Herbert Simon. "GPS, A Program That Simulates Human Thought." In *Computers and Thought*, edited by Edward Feigenbaum and Julian Feldman, 279-293. New York: McGraw-Hill, 1963.

Nietzsche, Friedrich. *Also Sprach Zarathustra*. Chemnitz, Germany: Verlag von Ernst Schmeitzner, 1883.

Orwell, George. *1984: A Novel*. New York: New American Library, 1977.

Pelikan, M., D.E. Goldberg and E. Cant'u-Paz. *BOA: The Bayesian Optimization Algorithm (IlliGAL Report No. 99003)*. Urbana, IL: University of Illinois at Urbana-Champaign and Illinois Genetic Algorithms Laboratory, 1999.

Penrose, Roger. *The Emperor's New Mind: Concerning Computers, Minds, and the Laws of Physics*. Oxford and New York: Oxford University Press, 1989.

———. Shadows *of the Mind: A Search for the Missing Science of Consciousness*. Oxford and New York: Oxford University Press, 1994.

Penrose, Roger, and Stephen Hawking. *The Nature of Space and Time*. Princeton: Princeton University Press, 1996.

Popper, Karl R. *Objective Knowledge: An Evolutionary Approach*. Oxford: Clarendon Press, 1972.

Price, Derek J. de Solla. *Little Science, Big Science—and Beyond*. New York: Columbia University Press, 1986.

Ramón y Cajal, Santiago. *New Ideas on the Structure of the Nervous System in Man and Vertebrates*. Translated from the French by Neely Swanson and Larry W. Swanson. Cambridge, MA: MIT Press, 1990.

Randall, Lisa. *Warped Passages: Brane-Worlds, Particles, Strings, and the Universe's Hidden Dimensions*. New York: Ecco, 2005.

Ray, T. S., and Joseph F. Hart. "Evolution of Differentiation in Multithreaded Digital Organisms." In *Artificial Life VII, Proceedings of the Seventh International Conference on Artificial Life*. Bedau, Mark A., John S. McCaskill, Norman H. Packard, and Steen Rasmussen, eds. Cambridge, MA: The MIT Press, 2000.

Rosenblatt, Frank. *Principles of Neurodynamics: Perceptrons and the Theory of Brain Mechanisms*. Washington, D. C.: Spartan Books, 1962.

Rowan, John. *Subpersonalities: The People Inside Us*. London and New York: Routledge, 1990.

Rumelhart, David E., James L. McClelland, and the PDP Research Group. *Parallel Distributed Processing: Explorations in the Microstructure of Cognition*. Cambridge, MA: MIT Press, 1986.

Russell, John. "Venter Makes Waves—Again." *Bio-IT Bulletin* (April 16, 2004).

Russell, Peter. *The Global Brain: Speculations on the Evolutionary Leap to Planetary Consciousness*. Los Angeles and Boston: J.P. Tarcher, distributed by Houghton Mifflin, 1983.

Sarno, Louis. *Echoes of the Forest*. Roslyn, NY: Ellipsis Arts, 1995.

Sarno, Louis. *Song from the Forest: My Life Among the Ba-Benjellé Pygmies*. Boston: Houghton Mifflin, 1993.

Schmidhuber, J. "Optimal Ordered Problem Solver." *Machine Learning* 54 (2004): 211–54.

———. "Bias-Optimal Incremental Problem Solving." In *Advances in Neural Information Processing Systems 15, NIPS'15,* S. Becker, S. Thrun, K. Obermayer, eds. Cambridge, MA: MIT Press, 2003: 1571–78.

Schmidhuber, J., V. Zhumatiy and M.Gagliolo. "Bias-Optimal Incremental Learning of Control Sequences for Virtual Robots." In *Proceedings of the 8th Conference on Intelligent Autonomous Systems,* Groen, F., et al., eds. Amsterdam: IAS-8, 2004.

Schrödinger, Erwin. *What is Life? The Physical Aspect of the Living Cell.* Cambridge: The University Press, and New York: The Macmillan Company, 1944.

Seeman, N.C. "Design and Engineering of Nucleic Acid Nanoscale Assemblies." *Current Opinion in Structural Biology* 6 (1996): 519-526.

Seeman, N.C., H. Wang, J. Qi., X.J. Li, X.P. Yang, Y. Wang, H. Qiu, B. Liu, Z. Shen, W. Sun, F. Liu, J.J. Molenda, S.M. Du, J. Chen, J.E. Mueller., Y. Zhang, T.-J. Fu, and S. Zhang. "DNA Nanotechnology and Topology." In *Biological Structure and Dyinamics vol. 2*, edited by R.H. Sarma and M.H. Sarma, 319-341. New York: Adenine Press, New York 1996.

Seifer, Marc J. *Wizard: The Life and Times of Nikola Tesla: Biography of a Genius*. Secaucus, NJ: Carol Pub., 1996.

Seuss, Dr. *The Cat In The Hat Comes Back*. New York: Beginner Books, 1958.

Shapiro, Ehud. *Algorithmic Program Debugging*. Cambridge, MA: MIT Press, 1983.

———. *Inductive Inference of Theories From Facts*. Tokyo: Kyoritsu Publishing, 1986.

Smigrodzki, R., B. Goertzel, C. Pennachin, L. Coelho, F. Prosdocimi, and W. D. Parker. "Genetic algorithm for analysis of mutations in Parkinson's disease." *Artificial Intelligence in Medicine* (2005) [in press].

Smolin, Lee. *The Life of the Cosmos*. New York: Oxford University Press, 1997.

Solomonoff, Ray. "A Formal Theory of Inductive Inference, Part I."*Information and Control, Part I* Vol 7, No. 1 (March, 1964): 1-22.

———. "A Formal Theory of Inductive Inference, Part II."*Information and Control, Part II* Vol. 7, No. 2 (June, 1964): 224-254.

Sorokin, Pitirim A., *Social and Cultural Dynamics*. New York and Cincinnati: American Book Company, 1937-41.

Spencer, Herbert. *The Principles of Sociology*. New York: D. Appleton and Company, 1880-97.

Stein, Gertrude. *How to Write*. Barton, VT: Something Else Press, 1973

Stewart, John. *Evolution's Arrow: the Direction of Evolution and the Future of Humanity*. Rivett, A. C. T., Australia: Chapman Press, 2000.

Strominger, A., and C. Vafa. "Microscopic origin of Bekenstein-Hawking Entropy." *Physics Letters* B 379: 99 (1996).

Tipler, Frank J. *The Physics of Immortality: Modern Cosmology, God, and the Resurrection of the Dead*. New York: Anchor Books, 1994.

Toffler, Alvin. *Future Shock*. New York: Random House, 1970.

Trivers, R. "The Evolution of Reciprocal Altruism." *Quarterly Review of Biology* 46 (1971): 35–57.

Turchin, Valentin F. *The Inertia of Fear and the Scientific Worldview*. Translated by Guy Daniels. New York: Columbia University Press, 1981.

———. *The Phenomenon of Science*. Translated by Brand Frentz. New York: Columbia University Press, 1977. van der Werf, E.C.D. *AI Techniques for the Game of Go*. Ph.D. Thesis, Maastricht, The Netherlands: Universiteit Maastricht, 2004.

Velmans, M. "Consciousness From a First-Person Perspective." *The Behavioral and Brain Sciences* 14(4) (1991):702-726.

———. "Is Human Information Processing Conscious?" *The Behavioral and Brain Sciences* 14(4) (1991):651-669.

von Zglinicki, Thomas, ed. *Aging at the Molecular Level*. Dordrecht and Boston: Kluwer Academic Publishers, 2003.

Voss, Peter. "The Essentials of General Intelligence." In *Artificial General Intelligence*, edited by Ben Goertzel and Cassio Pennachin. New York: Springer-Verlag, 2005.

Wallace, Alfred Russel, and Charles H. Smith, eds. *Writings on Evolution, 1843-1912*. Bristol, UK: Thoemmes Continuum, 2004.

Wang, Pei. "Experience-Grounded Semantics: A Theory for Intelligent Systems."*Cognitive Systems Research* Vol. 6, No. 4 (2005): 282-302.

———. *Non-Axiomatic Reasoning System*. PhD Thesis, Bloomington, IN: Indiana University, 1995.

Warwick, Kevin. *I, Cyborg*. London: Century, 2002.

Watts, Alan. *Tao: The Watercourse Way*. New York: Pantheon Books, 1975.

Wiener, Norbert. *Cybernetics: Control and Communication in the Animal and the Machine*. 2nd ed. New York: MIT Press, 1961.

Winston, P. "Learning Structural Descriptions From Examples." In *The Psychology of Computer Vision*, edited by P. Winston. New York: Mcgraw-Hill, 1975.

Weizenbaum, Joseph. *Computer Power and Human Reason.* San Francisco: W. H. Freeman, 1976.

Weizenbaum, J. "ELIZA—A Computer Program for the Study of Natural Language Communication Between Men and Machines." *Communications of the ACM* 9 (1966): 36–45.

Wells, H. G. *World Brain*. Garden City, NY: Doubleday, Doran & Co., 1938.

Wesson, Paul S. *Space-Time- Matter: Modern Kaluza-Klein Theory*. Hackensack, NJ: World Scientific Publishing Company, 1998.

Wierzbicka, Anna. *Semantic Primitives*. Translated by Anna Wierzbicka and John Besemeres. Frankfurt: Athenäum-Verlag, 1972.

———. *Semantics, Primes and Universals*. Oxford: Oxford University Press, 1996.

Williamson, Jack. *The Humanoids*. New York: Simon & Schuster, 1949.

Winkless, Nels, and Iben Browning. *Robots on Your Doorstep*. Portland, OR: Robotics Press, 1978.

Wolf, Fred Alan. *Taking the Quantum Leap: the New Physics for Nonscientists*. San Francisco: Harper & Row, 1981.

Wolfram, Stephen. *A New Kind of Science*. Champaign, IL: Wolfram Media, 2002.

Zee, A. *Quantum Field Theory in a Nutshell*. Princeton: Princeton University Press, 2003.

Zukav, Gary. *The Dancing Wu-Li Masters: an Overview of the New Physics*. New York: Morrow, 1979.

Zweibach, Barton. *A First Course in String Theory*. Cambridge: Cambridge University Press, 2004.

INDEX

A

B

C

D

E

F

G

H

N

O

P

T

U